SPRINGER
LAB MANUALS

Springer

Berlin
Heidelberg
New York
Hong Kong
London
Milan
Paris
Tokyo

Hellmut Augustin (Ed.)

Methods in Endothelial Cell Biology

With 68 Figures, 8 in Color

 Springer

Prof. Dr. Hellmut G. Augustin

Department of Vascular Biology & Angiogenesis Research
Tumor Biology Center at the University of Freiburg
Breisacher Str. 117
79106 Freiburg
Germany

e-mail: augustin@angiogenese.de

ISBN 3-540-21397-X   Springer-Verlag Berlin Heidelberg New York

Library of Congress Control Number: 2004103462

Springer-Verlag is a part of Springer Science+Business Media
springeronline.com

© Springer-Verlag Berlin Heidelberg 2004

Printed in Germany

Production and Typesetting: Friedmut Kröner, 69115 Heidelberg, Germany
Cover design: *design & production* GmbH, 69126 Heidelberg, Germany

31/3150 YK – 5 4 3 2 1 0 – Printed on acid free paper

Foreword

The field of vascular biology and its major branch, angiogenesis research, have expanded steadily since the early 1970s. Two seminal events set these fields in motion – the publication in 1971 of the hypothesis that tumors are angiogenesis-dependent and the first isolation and culture of endothelial cells 2 years later. After decades of studies by investigators in laboratories throughout the world, both fields are now solidly established. Furthermore, principles of vascular biology and angiogenesis research have been validated in the clinic. New drugs are preventing or delaying atherosclerosis and providing a novel modality for treating cancer. As a result of rapid progress in both fields, there has been an urgent need for a book that accurately describes both the standard methods and the new techniques being developed. It will be a valuable resource for any scientist working with endothelial cells.

Hellmut Augustin has now edited such a book. He has recruited authors who are themselves experts and pioneers in the many aspects of vascular biology and angiogenesis. For scientists coming into these research areas, the book offers entrée into the breadth of subjects that can be mastered, and it will also help those trying to adapt endothelial cell methods to suit their own research. Topics include among others, isolation and analysis of circulating endothelial precursor cells, stem cell-derived endothelial cells and lymphatic endothelial cells, plus methodologies that employ RNA-interference, confocal microscopy, and immortalization of endothelial cells. So, the book is a valuable contribution to these areas of research.

Methods in Endothelial Biology is also timely for another reason, the gradual realization by biomedical scientists and clinicians that disruptions of normal endothelial cell biology underlie many diseases across the whole spectrum of medical specialties. The United States National Institutes of Health recently put their imprimatur on this new paradigm by inaugurating a bold new effort, a Trans-Institute Angiogenesis Research Program. Having a book that clearly and concisely embodies the experience of scientists doing vascular biology and angiogenesis research is an important step into the future.

Boston, Massachusetts
June 2004

Judah Folkman
Children's Hospital and
Harvard Medical School

Preface

The vascular endothelium lines all blood and lymphatic vessels. The more than 10^{12} endothelial cells in the human body cover a surface area of more than 1000 m². As such, the vascular endothelium can be considered a systemically disseminated organ system that is directly or indirectly involved in most of the important human diseases. Primary vascular dysfunctions cause atherosclerotic changes in the vessel wall and coagulation disorders. Inflammatory responses are controlled by chemokines and adhesion molecules that act on the vessel wall or are expressed by endothelial cells, respectively. Angiogenesis is a prerequisite for the malignant growth of tumors and is also involved in numerous other physiological and pathological processes such as reproductive function as well as (among others) skin pathologies (e.g., psoriasis) and eye disorders (e.g., retinopathies). Lastly, vascular complications are associated with a number of chronic diseases, including diabetes and arthritis. Collectively, the functional involvement of the vessel wall in some of the most devastating human diseases opens not only an exciting field of basic biomedical research, but also offers an attractive window of opportunity for the development of novel therapeutic avenues for many diseases.

The field of vascular biology has grown rapidly within the last 20 years. I recall my first Gordon Conference in the Summer of 1990 when I was a graduate student at Cornell University. I was sent to a remote place in New Hampshire to learn about endothelial cells. We were studying the mechanisms of site-specific metastasis. Significant progress had been made in the understanding of the mechanisms of leukocyte homing and it was hypothesized that tumor cells similarly home to certain organs by using surface molecules of endothelial cells as an adhesive anchor. The mechanisms of site-specific metastasis are still largely elusive, but a lot has been learnt since then about the heterogeneity and plasticity of endothelial cells in different vascular beds. Following the (still ongoing) molecular characterization of distinct situation-specific endothelial cell phenotypes such as the inflammatory or the angiogenic phenotype of endothelial cells, the molecular identification of the constitutive organ-specific and calibre-specific vascular body map is slowly taking shape. Advanced genomic and proteomic as well as antibody and phage-display approaches are revolutionizing our molecular understanding of constitutive organotypic and calibre-specific endothelial cell differentiation. Ultimately, these experimental approaches may lead to the development of novel therapeutic strategies for the organ or site-specific delivery of biological or pharmacological therapeutic agents. Similarly, the therapeutic targeting of endothelial cells in distinct vascular beds, such as the vasculature in fat tissues, is gradually taking shape.

Methods in Endothelial Cell Biology is intended to serve as a practical hands-on guide for the continuously growing vascular biology community in general and the endothelial cell biology community in particular. It aims to serve as a reliable reference to some of the most important in vitro and in vivo endothelial cell biology techniques that have stood the test of time or which have more recently evolved with great potential to further advance the discipline. The manual is supposed to similarly provide the expert as well as the novice in the field with a toolbox of reliable protocols to isolate and culture endothelial cells, to detect endothelial cells, and to functionally manipulate endothelial cells in isolated cellular systems as well as in the whole body.

When Springer first asked me to edit this manual, I was sceptical for two reasons: First, I thought that techniques are too short-lived these days so that a manual will hardly have a reasonable half-life time. Secondly, I was concerned that the worldwide web with its numerous cell and molecular biological protocol collections would comprehensively cover the subject. It turned out that I was wrong on both accounts: There is a considerable degree of technical heterogeneity within the discipline. For example, more than a 1000 vessel density counting studies have been published in the last 15 years. Meta-analytical comparison of these studies has turned out to be a very complex challenge as a consequence of the enormous variability of experimental approaches with very little standardization. Likewise, when carefully screening the web with a number of different search engines for endothelial cell biology techniques, it was surprising to see how little reliable information is actually available and that the field is far from being adequately covered. Apparently, vascular biology is still a heterogeneous community that is influenced by many individual organ disciplines. Yet, the increasing recognition of the vessel wall as a systemically distributed organ system may gradually lead to a greater recognition of the need for a strong and coherent discipline. It would be a very satisfying accomplishment if *Methods in Endothelial Cell Biology* could serve the endothelial cell biology community to contribute to a greater degree of technical coherence in the discipline.

I would like to thank the many people who supported the writing and editing of this manual. I would like to acknowledge the many colleagues, including all of the contributors of individual chapters, for their suggestions for topics to be covered and for their comments to the chapters. Likewise, I would like to thank all members of the Vascular Biology & Angiogenesis Laboratory at the Tumor Biology Center Freiburg, most notably Dr. Ulrike Fiedler, Dr. Thomas Korff, Dr. Jens Kroll, Dr. Cynthia Obodozie, Dr. Yvonne Reiss, and Dr. Holger Weber for their contributions to this manual, which were not just restricted to the proofreading of individual chapters, but also to spending considerable time for practically testing many of the protocols for reproducibility, robustness, and reliability. Specific thanks go to Manuela Fellmann whose assistance in the editing of the manual was vital for the project. Lastly, I would like to thank Dr. Jutta Lindenborn and Friedmut Kröner from Springer for their patience and endurance in the production of the manual.

Freiburg, Germany   Hellmut Augustin
June 2004   Tumor Biology Center Freiburg

Contents

Part IV: Detection of Endothelial Cells

Appendix

Contributors

Aicher, Dr. Alexandra
Medizinische Klinik IV,
Klinikum der J.W. Goethe-Universität
Theodor-Stern Kai 7,
D-60596 Frankfurt am Main, Germany
Phone: +49-69-6301-7357,
Fax: +49-69-6301-7113
E-mail: alexandra.aicher@em.uni-
frankfurt.de

Alitalo, Prof. Dr. Kari
Molecular/Cancer Biology Laboratory
Biomedicum Helsinki and Haartman
Institute, University of Helsinki
P.O.B. 63, SF-00014 Helsinki, Finland
Phone: +358-9-1912-5512,
Fax: +358-9-1912-5510
E-mail: Kari.Alitalo@Helsinki.FI

Asahara, Prof. Dr. Takayuki
Institute of Biomedical Research
and Innovation / RIKEN Center
for Developmental Biology
2-2 Minatojima-Minamimachi, Chuo-ku,
Kobe, 650-0047, Japan
Phone: +81-78-304-5772,
Fax: +81-78-304-5263
E-mail: Asa777@aol.com

Augustin, Prof. Dr. Hellmut G.
Department of Vascular Biology
& Angiogenesis Research,
Tumor Biology Center
Breisacher Str. 117,
D-79106 Freiburg, Germany
Phone: +49-761-206-1500,
Fax: +49-761-206-1505
E-mail: augustin@angiogenese.de

Beer, Dr. Stefani
Institut für Kardiovaskuläre Physiologie,
Universität Frankfurt
Theodor-Stern Kai 7,
D-60596 Frankfurt am Main, Germany
Phone: +49-69-6301-6995,
Fax: +49-69-6301-7668

Betsholtz, Prof. Dr. Christer
Institute of Medical Biochemistry,
The Sahlgrenska Academy at Göteborg
University, P.O. Box 440
S-405 30 Gothenburg, Sweden
Phone: +46-31-773-3496,
Fax: +46-31-416-108
E-mail: christer.betsholtz@medkem.gu.se

Brandes, Dr. Ralf P.
Institut für Kardiovaskuläre Physiologie,
Universität Frankfurt
Theodor-Stern Kai 7,
D-60596 Frankfurt am Main, Germany
Phone: +49-69-6301-6995,
Fax: +49-69-6301-7668
E-mail: r.brandes@em.uni-frankfurt.de

Breier, Prof. Dr. Georg
Universitätsklinikum Carl Gustav Carus,
Institut für Pathologie, TU Dresden
Fetscherstr. 74, D-01307 Dresden, Germany
Phone: +49-0351-458-5278,
Fax: +49-0351-458-4328
E-mail: georg.breier@uniklinikum-
dresden.de

Burri, Prof. Dr. Peter H.
Institute of Anatomy, University of Bern,
Bühlstrasse 26, CH-3012, Switzerland
Phone: +41-31-631-4641,
Fax: +41-31-631-3410
E-mail: burri@ana.unibe.ch

Buschmann, Dr. Ivo
Research Group for Exp. and
Clin. Arteriogenesis, Dept. of Cardiology,
University of Freiburg
Hugstetter Str. 55,
D-79106 Freiburg, Germany,
Phone: +49-761-270-3441,
Fax:+49-761-270-3596
Email: buschmann@med1.ukl.uni-
freiburg.de

Cantara, Dr. Silvia
Dept. Molecular Biology,
University of Siena
Via Aldo Moro 2, 53100 Siena, Italy
Phone: +39-0577-234-439,
Fax: +39-0577-234-343

Chae, Dr. Sung-Suk
Center for Vascular Biology,
University of Connecticut Health Center
Farmington, CT 06030-3501, USA
Phone: +1-860-679-4128,
Fax: +1-860-679-1201

Clauss, Prof. Dr. Matthias
Indiana Center for Vascular Biology &
Medicine, Indiana School of Medicine
975 W. Walnut Street IB 433, Indianapolis,
IN 46202, USA
Phone: +1-317-278-2837,
Fax: +1-317-278-0089
E-mail: mclauss@iupui.edu

Dejana, Prof. Dr. Elisabetta
Laboratory of Vascular Biology, IFOM, Firc
Institute for Molecular Oncology
Via Adamello 16, I-20139 Milano, Italy
Phone: +39-02574303-601,
Fax: +39-02574303-244
E-mail: dejana@ifom-firc.it

Dimmeler, Prof. Dr. Stefanie
Molecular Cardiology,
University of Frankfurt
Theodor-Stern-Kai 7,
D-60596 Frankfurt/Main, Germany
Phone: +49-69-6301-7440,
Fax: +49-69-6301-7113
E-mail: dimmeler@em.uni-frankfurt.de

Djonov, Dr. Valentin
Institute of Anatomy, University of Bern,
Bühlstrasse 26, CH-3012, Switzerland
Phone: +41-31-631-4641,
Fax: +41-31-631-3410
E-mail: djonov@ana.unibe.ch

Engelhardt, Prof. Dr. Britta
Theodor-Kocher Institute,
University of Bern
Freiestr. 1, CH-3012 Bern, Switzerland
Phone: +41-31-631-4141,
Fax: +41-31-631-3799
E-mail: bengel@tki.unibe.ch

Farhadi, Dr. Mohammad
Department of Neurosurgery, Klinikum
Mannheim, University of Heidelberg
Theodor-Kutzer-Ufer 1–3,
D-68167 Mannheim, Germany
Phone: +49-621-383-2360,
Fax: +49-621-383-2004

Furneaux, Dr. Henry
Center for Vascular Biology,
University of Connecticut Health Center
Farmington, CT 06030-3501, USA
Phone: +1-860-679-4128,
Fax: +1-860-679-1201

Gerhardt, Dr. Holger
Institute of Medical Biochemistry,
The Sahlgrenska Academy at Göteborg
University, P.O. Box 440
S-405 30 Gothenburg, Sweden
Phone: +46-31-773-3463,
Fax: +46-31-416-108
E-mail: holger.gerhardt@medkem.gu.se

Goede, Dr. Valentin
Department of Hematology and Oncology,
University Hospital Cologne
Joseph-Stelzmann Str. 9
50931 Cologne, Germany
Phone: +49-221-478-5044
Fax: +49-221-478-86793
E-mail: valentin.goede@uk-koeln.de

Hagedorn, Dr. Martin
INSERM E0113 "Molecular mechanisms of
angiogenesis", University of Bordeaux I
Avenue des Facultes, 33405 Talence, France
E-mail: m.hagedorn@croissance.u-
bordeaux.fr

Haigh, Dr. Jody J.
Department of Molecular Biomedical
Research
Cardiovascular Biology Unit
Ghent University
"Fiers-Schell-Van Montagu" building
Technologiepark 927
B-9052 Ghent, Belgium
Phone: +32-9-33 13 730
Fax: +32-9-33 13 609
E-mail: Jody.Haig@dmbr.UGent.be

Hallmann, Prof. Dr. Rupert
Department of Experimental Pathology,
Lund University Jubileumsinstitute
Soelvegatan 25, S-22362 Lund, Sweden
Phone: +46-46-173419, Fax: +46-46-158202
E-mail: rupert.hallmann@pat.lu.se

Hammel, Dr. Markus
Department of Experimental Pathology,
Lund University Jubileumsinstitute
Soelvegatan 25, S-22362 Lund, Sweden
Phone: +46-46-173-419,
Fax: +46-46-158-202
E-mail: Markus.Hammel@pat.lu.se

Heidenreich, Dr. Regina
Universitäts-Hautklinik,
Eberhard-Karlos-Universität
Liebermeisterstr. 25, D- 72076 Tübingen,
Germany
Phone: +49-7071-298-4585
E-mail: regina.heidenreich@med.uni-
tuebingen.de

Hinsbergh, Prof. Dr. Victor W.M. van
Laboratory of Physiology,
Institute of Cardiovascular Research,
VU University Medical Center
Van der Boechorststraat 7,
NL-1081 BT Amsterdam, The Netherlands
Phone: +31-71-518-1818,
Fax: +31-71-518-1904
E-mail: hinsbergh@physiol.med.vu.nl

Hla, Prof. Dr. Timothy
Center for Vascular Biology, University of
Connecticut Health Center
Farmington, CT 06030-3501, USA
Phone: +1-860-679-4128,
Fax: +1-860-679-1201
E-mail: hla@nso2.uchc.edu

Hoefer, Dr. Imo
Perfusion Technologies GmbH
Stefan-Meyer-Strasse 8, 79104 Freiburg,
Germany
Phone: +49-761-203-5498 ,
Fax: +49-761-203-5499
Email : mail@perfusion-technologies.com

Issbrücker, Dr. Katja
Institute of Physiology
and Pathophysiology
University of Heidelberg
Im Neuenheimer Feld 326
69120 Heidelberg, Germany
Phone: +49-6221-54-4052
Fax: +49-6221-54-4561
E-mail: k.issbrueckeer@pio1.uni-
heidelberg.de

Jean-Guillaume, Dr. Danielle
Regeneron Pharmaceuticals Inc.
777 Old Saw Mill River Rd., Tarrytown,
NY 10591, USA
Phone: +1-914-997-9881,
Fax: +1-914-347-5045

Kiefer, Dr. Friedemann
Department of Vascular Cell Biology,
Max-Planck-Institute for Vascular Biology
Von-Esmarch-Straße 56,
D-48149 Münster Germany
Phone: +49-251-835-2221,
Fax: +49-251-835-8616
Email: fkiefer@gwdg.de

Kochanek, Prof. Dr. Stefan
Sektion Gentherapie
Universität Ulm
Helmholtz Str. 8/1
D-89081 Ulm, Germany
Phone: +49-731-500-33647
Fax: +49-731-500-33664
E-mail: stefan.kochanek@medizin.uni-
ulm.de

Korff, Prof. Dr. Thomas
Abt. Herz- und Kreislaufphysiologie,
Universität Göttingen
Humboldtallee 23,
D-37073 Göttingen, Germany
Phone: +49-551-39-5891,
Fax: +49-551-39-5883
Email: korff@veg-physiol.med.uni-
goettingen.de

Kroll, Dr. Jens
Department of Vascular Biology & Angio-
genesis Research, Tumor Biology Center
Breisacher Str. 117,
D-79106 Freiburg, Germany
Phone: +49-761-206-1501,
Fax: +49-761-206-1505
E-mail: kroll@tumorbio.uni-freiburg.de

Lampugnani, Dr. Maria Grazia
Laboratory of Vascular Biology, IFOM, Firc
Institute for Molecular Oncology
Via Adamello 16, I-20139 Milano, Italy
Phone: +39-02574303-601,
Fax: +39-02574303-244
email: lampugnani@ifom-firc.it

Laurens, Dr. Nancy
Department of Vascular and Connective
Tissue Research, Gaubius Laboratory,
TNO-PG
Zernikedreef 9, NL-2333 CK Leiden,
The Netherlands

Lindner, Prof. Dr. Volkhard
Center for Molecular Medicine, Maine
Medical Center Research Institute
81 Research Drive, Scarborough,
ME 04074, USA
Phone: +1-207-885-8143,
Fax: +1-207-885-8179
E-mail: lindnv@mmc.org

Mäkinen, Dr. Taija
Department of Molecular Neurobiology,
Max-Planck-Institute of Neurobiology
Am Klopferspitz 18A,
D-82152 Martinsried, Germany
Phone: +49-89-8578-3187,
Fax: +49-89-8578-3152
E-mail: makinen@neuro.mpg.de

Marti, Prof. Dr. Hugo H.
Institut für Physiologie und Patho-
physiologie, Universität Heidelberg
Im Neuenheimer Feld 326,
D-69 120 Heidelberg, Germany
Phone: +49-6221-544-035,
Fax: +49-6221-544-038
E-mail: hugo.marti@pio1.uni-
heidelberg.de

Morawietz, Prof. Dr. Henning
Department of Vascular Endothelium
and Microcirculation,
University of Technology Dresden
Fetscherstrasse 74,
D-01307 Dresden, Germany
Phone +49-345-557-1454,
Fax +49-345-557-1404
E-mail: henning.morawietz@mailbox.tu-
dresden.de

Morbidelli, Dr. Lucia
Dept. Molecular Biology,
University of Siena
Via Aldo Moro 2, 53100 Siena, Italy
Phone: +39-0577-234-439,
Fax: +39-0577-234-343
E-mail: morbidelli@unisi.it

Murayama, Dr. Toshinori
Translational Research Center,
Kyoto University Hospital
54 Shogoin-Kawaharacho, Kyoto 606-8507,
Japan

Nagy, Prof. Dr. Andras
Samuel Lunenfeld Research Institute,
Mount Sinai Hospital
600 University Avenue, Toronto, Ontario,
Canada
Phone: +1-416-586-8455,
Fax: +1-416-586-8588
E-mail: nagy@mshri.on.ca

Nicosia, Prof. Dr. Roberto F.
Division of Pathology and Laboratory
Medicine, Veterans Affairs Puget Sound
Health Care System
Seattle, WA, USA
Phone: +1-206-764-2284,
Fax: +1-206-764-2001
E-mail: Roberto.Nicosia@med.va.gov

Nishikawa, Prof. Dr. Shin-Ichi
Stem Cell Biology Laboratory,
Center for Developmental Biology
Riken 2-2-3, Minatojima-minamimachi,
Chuou-ku, Kobe 650-0047, Japan
Phone: +81-78-306-1893,
Fax: +81-78-306-1895
E-mail: nishikawa@cdb.riken.go.jp

Obodozie, Dr. Cynthia
Department of Vascular Biology & Angio-
genesis Research, Tumor Biology Center
Breisacher Str. 117,
D-79106 Freiburg, Germany
Phone: +49-761-206-1501,
Fax: +49-761-206-1505
E-mail: obodozie@tumorbio.uni-
freiburg.de

Paik, Dr. Ji-Hye
Center for Vascular Biology,
University of Connecticut Health Center
Farmington, CT 06030-3501, USA
Phone: +1-860-679-4128,
Fax: +1-860-679-1201
E-mail: paik@student.uchc.edu

Passaniti, Prof. Dr. Toni
Department of Pathology, Biochemistry &
Molecular Biology, University of Maryland
Greenebaum Cancer Center,
Bressler Research Building, Room 7-021
655 W. Baltimore St., Baltimore,
MD 21201, USA
Phone: +1-410-328-5470,
Fax: +1-410-328-6559
E-mail: apass001@umaryland.edu

Reiss, Dr. Yvonne
Department of Vascular Biology & Angio-
genesis Research, Tumor Biology Center
Breisacher Str. 117,
D-79106 Freiburg, Germany
Phone: +49-761-206-1510,
Fax: +49-761-206-1505
E-mail: reiss@tumorbio.uni-freiburg.de

Rudge, Dr. John S.
Regeneron Pharmaceuticals Inc.
777 Old Saw Mill River Rd., Tarrytown,
NY 10591, USA
Phone: +1-914-997-9881,
Fax: +1-914-347-5045
E-mail: John.Rudge@Regeneron.com

Schirmer, Dr. Stephan
Research Group for Exp. and
Clin. Arteriogenesis, Dept. of Cardiology,
University of Freiburg
Hugstetter Str. 55,
D-79106 Freiburg, Germany,
Phone: +49-761-270-3441,
Fax:+49-761-270-3596
E-mail: schirmer@med1.ukl.uni-
freiburg.de

Schubert, Dr. Andreas
Heart Center, University Leipzig
Russenstrasse 19,
D-04289 Leipzig, Germany
Phone: +49-341-865-1620,
Fax: +49-341-865-1452
E-mail: schua@medizin.uni-leipzig.de

Schwinn, Dr. Karsten
Department of Neurosurgery,
Klinikum Mannheim,
University of Heidelberg
Theodor-Kutzer-Ufer 1-3,
D-68167 Mannheim, Germany
Phone: +49-621-383-2360,
Fax: +49-621-383-2004

Shubert-Coleman, Dr. Jonathan
Center for Vascular Biology,
University of Connecticut Health Center
Farmington, CT 06030-3501, USA
Phone: +1-860-679-4128,
Fax: +1-860-679-1201

Tepper, Dr. Oren M.
New York University Medical Center
550 First Avenue, New York
NY 10016-6497, USA
E-mail: teppeo01@med.nyu.edu

Thurston, Dr. Gavin
Regeneron Pharmaceuticals Inc.
777 Old Saw Mill River Rd., Tarrytown,
NY 10591, USA
Phone: +1-914-997-9881,
Fax: +1-914-347-5045
E-mail: gavin.thurston@regeneron.com

Urbich, Dr. Carmen
Molecular Cardiology,
University of Frankfurt
Theodor-Stern-Kai 7,
D-60596 Frankfurt/Main, Germany
Phone: +49-69-6301-7440,
Fax: +49-69-6301-7113
E-mail: urbich@em.uni-frankfurt.de

Vajkoczy, Dr. Peter
Department of Neurosurgery, Klinikum
Mannheim, University of Heidelberg
Theodor-Kutzer-Ufer 1-3,
D-68167 Mannheim, Germany
Phone: +49-621-383-2360,
Fax: +49-621-383-2004
E-mail: peter.vajkoczy@nch.ma.uni-
heidelberg.de

Vitolo, Dr. Michele I.
Department of Biochemistry & Molecular
Biology, University of Maryland
School of Medicine
655 W. Baltimore St., Baltimore,
MD 21201, USA
E-mail: mvitolo@umaryland.edu

Volpers, Dr. Christoph
CEVEC Pharmaceuticals GmbH
Gottfried-Hagen-Str. 62
51105 Cologne, Germany
Phone: +49-221-46020800
Fax: +49-221-46020801
E-mail: volpers@cevec-
pharmaceuticals.com

Wilting, Prof. Dr. Jörg
Pediatrics I, Children's Hospital,
University of Göttingen
Robert-Koch-Strasse 40,
D-37075 Göttingen, Germany
Phone: +49-551-39-6270,
Fax: +49-551-39-6231
E-mail: joerg.wilting@med.uni-
goettingen.de

Yamashita, Dr. Yun
Department of Molecular Genetics,
Kyoto University Graduate School
of Medicine
53 Shogoin Kawahara-cho,
Sakyo, Kyoto 606-8507, Japan
Phone: +81-75-751-4163,
Fax: +81-75-751-4169
E-mail: juny@kuhp.kyoto-u.ac.jp

Zhu, Dr. Wen-Hui
Department of Pathology,
University of Washington
Seattle, WA 98195-7470 , USA

Ziche, Prof. Dr. Marina
Dept. Molecular Biology,
University of Siena
Via Aldo Moro 2, 53100 Siena, Italy
Phone: +39-0577-234-439,
Fax: +39-0577-234-343
E-mail: ziche@unisi.it

Zilles, Dr. Olaf
Department of Experimental Pathology,
Lund University Jubileumsinstitute
Soelvegatan 25, S-22362 Lund, Sweden
Phone: +46-46-173419,
Fax: +46-46-158202
E-mail: ozilles@gmx.de

Part I:
Isolation and Culture of Endothelial Cells

Isolation, Purification and Culture of Human Micro- and Macrovascular Endothelial Cells

Nancy Laurens and Victor W.M. van Hinsbergh

Introduction

Endothelial cells, lining the inside of all blood vessels, are involved in many physiological processes, such as haemostasis, vasoregulation, and angiogenesis, but also play an important role in pathophysiological processes, such as inflammation, vascular leakage and tumor development. Therefore, the study of endothelial cells both in vitro and in vivo, gives investigators the ability to understand and to get more insight into these processes. It was in the early 1970s when the first endothelial cell culture was established (Jaffe et al. 1973; Gimbrone et al. 1974). Since then, the development of molecular and cellular techniques has improved and these tools are used to study extra- and intracellular responses of endothelial cells in vitro as well as in vivo.

Human microvascular endothelial cells (hMVECs) can be isolated from different human tissues, such as foreskin (Davison et al. 1980; Voyta et al. 1984), adult dermis (Davison et al. 1983), lung (Carley et al. 1992) glomerulus (van Setten et al. 1997), endometrium (Koolwijk et al. 2001), and brain (Gerhart et al. 1988). The umbilical cord can be used to isolate endothelial cells from the vein as described by Jaffe et al. (1973) and Gimbrone et al. (1974), but the arteries in the cord are also useful for isolating endothelial cells (van Hinsbergh et al. 1990). Endothelial cells from adult human aorta and vein can be obtained by a procedure comparable to the isolation of human umbilical vein endothelial cells (van Hinsbergh et al. 1987). This chapter describes the isolation of micro- and macrovascular endothelial cells and shows that a successful isolation can lead to a considerable number of endothelial cells derived from only one foreskin or umbilical cord.

The cells thus isolated can be used to study the properties of endothelial cells in a detailed biochemical way, without the interference of tissues and other cells. This has brought us in the last decade to a greater understanding of processes like vasoregulation, angiogenesis, barrier function, and leuko-

Springer Lab Manual
H. Augustin (Ed.)
Methods in Endothelial Cell Biology
© Springer-Verlag Berlin Heidelberg 2004

cyte endothelium interaction. However, one has to realize that during isolation and culture, the differentiated properties of the endothelial cells may be lost. Co-cultures between endothelial cells, astrocytes, pericytes, or smooth muscle cells may partly overcome this loss. Although in vivo experiments are needed to test the in vitro results, the in vitro systems will always be an important tool for getting more insight into the molecular properties and metabolic regulation of endothelial cells.

Human microvascular endothelial cells

Materials

Cells

- Human foreskin microvascular endothelial cells (hMVECs)

Reagents

- Cord buffer (4 mM KCl, 140 mM NaCl, 10 mM Hepes, 11 mM D-Glucose, 100 IU/ml penicillin, and 0.1 mg/ml streptomycin (p/s), pH at 7.4 and sterilized by filtration)
- M199 medium (Biowhittaker, www.biowhittaker.be) + 100 IU/ml penicillin and 0.1 mg/ml streptomycin (M199/p/s) (Biowhittaker)
- 0.4 % collagenase type II (GibcoBRL, lot number 1120228; www.gibco-brl.com) dissolved in M199/p/s and sterilized by filtration. Before use, an incubation period of 30 min at 37 °C is recommended to destroy contaminating proteolytic activity in some purified collagenase batches that may damage the isolated cells.
- M199/p/s supplemented with 10 % Human Serum, heat-inactivated, pH 7.4 (HS$_i$, heat-inactivation means an incubation period of 30 min at 56 °C, i.e., the serum has a temperature of 56 °C at the start of the 30 min period)
- M199/p/s supplemented with 20 % HS$_i$, 10 % Newborn Calf Serum heat-inactivated (NBCS$_i$, inactivation for 30 min at 56 °C with serum that already has a temperature of 56 °C), 5 U/ml heparin, 10 ng/ml bFGF, pH 7.4 (culture medium I)
- M199/p/s supplemented with 20 % HS$_i$, 10 % NBCS$_i$, 5 U/ml heparin, 5 ng/ml bFGF, and 0.0375 mg/ml ECGF, pH 7.4 (culture medium II)
- M199/p/s supplemented with 10 % HS$_i$, 10 % NBCS$_i$, 5 U/ml heparin, 0.075 mg/ml ECGF pH 7.4 (culture medium III)
- Human fibronectin (BD Biosciences, www.bdbiosciences.com) 2.5 µg/ml dissolved in M199/p/s and sterilized by filtration
- Solution A (137 mM NaCl, 5.4 mM KCl, 4.3 mM NaHCO$_3$, 5 mM D-glucose, and 0.002 % (w/v) phenol red, pH 7.4)

- 0.05 % (w/v) Trypsin/EDTA solution (137 mM NaCl, 5.4 mM KCl, 4.3 mM NaHCO$_3$, 5 mM D-glucose, and 0.67 mM EDTA, pH 7.4)
- M199/p/s supplemented with 0.1 % (w/v) pyrogen-free human serum albumin, pH 7.4 (HSA) (Sanquin, www.sanquin.nl)
- Mouse monoclonal antibody CD31 (2 µg/ml in M199/p/s supplemented with 0.1 % (w/v) HSA filtered by 0.22 µm filter)
- Dynal goat anti-mouse Ig coated beads (1 × 10^7 beads/ml in M199/p/s supplemented with 0.1 % (w/v) HSA)
- 1 % gelatin (Merck, www.merck.com) in water (Milli-Q)

Equipment

- CO$_2$ incubator (water-conditioned)
- Laminar flow hood (preferably down-flow)
- Inverted phase contrast microscope
- Laboratory centrifuge
- Thermostatic water bath
- Clean laboratory coat
- Sterile gloves
- Sterile endotoxin-free pipettes
- Sterile scissors
- Sterile rubber plug, 6–8 cm diameter
- Sterile needles
- Sterile scalpel
- Tweezers
- 50 ml sterile centrifuge tubes (e.g., Falcon)
- Tissue culture 6-well dishes (e.g., Falcon or Costar)
- Dynal Magnetic Concentrator
- 12 ml sterile tubes

Procedure

Isolation of microvascular endothelial cells

1. Collect one foreskin in 20 ml cord buffer and preserve it at 4 °C for 1 or 2 days before isolating the endothelial cells, in order to obtain a high yield.

2. Perform the isolation in a sterile environment, e.g., a down-flow laminar flow hood. Use sterile gloves and a laboratory coat to prevent contamination of the cultures, and to protect yourself.

3. Stretch the foreskin using needles onto a sterile rubber plug.

4. Put a few drops M199/p/s on the foreskin to prevent drying.

5. Use a scalpel to scrape the upper layer of the inside of the foreskin from the outside to the middle of the piece.

6. Put the collected layer in a 50 ml tube containing 5 ml of 0.4 % collagenase solution mixed with 5 ml M199/p/s and incubate 2 hours at 37 °C under continnous shaking. Add subsequently 5 % HS$_i$ to protect the cells.

7. Pellet the cells by centrifugation for 5 min at 200 × g.

8. Resuspend the pellet in culture medium I and seed them into 2–4 fibronectin-coated 10 cm² culture dishes, depending on the size of the pellet.

9. Put the culture dishes in a (water-conditioned) incubator at 37 °C under a 5 % CO$_2$/95 % air atmosphere.

10. Add fresh culture medium I to the wells with the attached ('endothelial') cells.

11. Renew the medium after 2 days with culture medium I (1.5 ml/10 cm²). Visually inspect the cells by using an inverted phase contrast microscope.

12. After a few days, replace culture medium I by culture medium II and inspect the cell growth every two or three days.

13. When contaminating cells start to overgrow the endothelial cells, select the endothelial cells by treating them with CD31 IgG and an anti-IgG-coated dynabeads solution (see below).

Purification by α-CD31 dynabeads

1. Wash the cells with 1.0–1.5 ml Solution A per 10 cm² well and detach the cells with 0.5 ml trypsin per 10 cm² well. Add 5 volumes M199/p/s supplemented with 10 % HS$_i$ to the well.

2. Collect the medium with cells and centrifugate for 5 min at 200 × g.

3. Wash the cells by resuspending the pellet in 10 ml M199/p/s supplemented with 0.1 % (w/v) HSA by resuspending the medium.

4. Centrifugate for 5 min at 200 × g.

5. Dissolve the pellet in 5 ml M199/p/s supplemented with 0.1 % (w/v) HSA and divide the medium over at least three 12-ml tubes (depending on the amount of cells).

6. Centrifugate for 5 min at 200 × g.

7. Dissolve the pellet in 200 µl CD31 IgG solution.

8. Incubate the cells for 30 min at 4 °C under end over rotation.

9. Add 10 ml 0.1 % HSA to each tube and resuspend the medium.

10. Centrifugate for 5 min at 200 × g.

11. Dissolve the pellet in 5 ml M199/p/s supplemented with 0.1 % (w/v) HSA and resuspend the medium.

12. Add several µl IgG-coated beads solution (for example, for 20 cm² confluent cells with approximately 50 % endothelial cells, add 40–60 µl of 1 × 10⁷ beads/ml) to each tube and incubate for 30 min at 4 °C under regular but gentle shaking.

13. Coat 10 cm² wells with 0.5 ml fibronectin and incubate the wells for at least 5 min at 37 °C. Remove the medium before the cells are plated.

14. Wash the cells 3–5 times with 5 ml M199/p/s supplemented with 0.1 % (w/v) HSA per tube by bringing the tube into the magnetic concentrator for 1 min per wash. Remove the supernatant and add again 5 ml M199/p/s supplemented with 0.1 % (w/v) HSA per tube.

15. After 3–5 washing rounds of purification the CD31-positive cells are resuspended in culture medium II and plated on fibronectin- or gelatin-coated wells. After several days, the cells can be cultured with culture medium III.

Human macrovascular endothelial cells

Materials

Cells

- Human umbilical vein macrovascular endothelial cells (HUVECs)

Reagents

- Cord buffer (4 mM KCl, 140 mM NaCl, 10 mM Hepes, 11 mM D-Glucose, 100 IU/ml penicillin, and 0.10 mg/ml streptomycin (p/s), pH at 7.4 and sterilized by filtration)
- M199 medium (Biowhittaker) + 100 IU/ml penicillin and 0.1 mg/ml streptomycin, pH 7.4 (Biowhittaker) (M199/p/s)
- 0.2 % collagenase type II (GibcoBRL, lotnumber 1120228) dissolved in M199/p/s and sterilized by filtration. Before use, an incubation period of minimal 30 min at 37 °C can be helpful to destroy contaminating proteolytic activity in some purified collagenase batches that may damage the isolated cells.

- M199/p/s supplemented with 10 % Human Serum heat-inactivated (HS_i), 10 % Newborn Calf Serum heat-inactivated ($NBCS_i$) (heat inactivation means an incubation period of 30 min at 56 °C with serum that already has a temperature of 56 °C), 5 U/ml heparin, 0.075 mg/ml ECGF, pH 7.4 (culture medium III)
- 1 % Gelatin (Merck) in water (Milli-Q)
- Solution A (137 mM NaCl, 5.4 mM KCl, 4.3 mM $NaHCO_3$, 5 mM D-glucose, and 0.002 % (w/v) phenol red, pH 7.4)
- 0.05 % (w/v) Trypsin/EDTA solution (137 mM NaCl, 5.4 mM KCl, 4.3 mM $NaHCO_3$, 5 mM D-glucose, and 0.67 mM EDTA pH 7.4)
- 0.5–1 L 0.9 % NaCl in water in a sterile beaker

Equipment

- CO_2 incubator (water-conditioned)
- Laminar flow hood (preferably down-flow)
- Inverted phase contrast microscope
- Laboratory centrifuge
- Thermostatic water bath
- Clean laboratory coat
- Sterile gloves
- Sterile endotoxin-free pipettes
- Sterile kocher scissors, scalpel
- Surgical silks
- 50 ml sterile centrifuge tubes (e.g., Falcon)
- 20 ml sterile syringes
- 2 sterile cannulas
- Sterile waste beaker
- Tissue culture 6-well dishes (e.g., Falcon or Costar)

Procedure

Collection and storage of umbilical cords

1. After delivery, cut the umbilical cord from the placenta, and place it in ice-cold cord buffer.

2. Store it in a refrigerator at 4 °C until use.

Isolation and culture of endothelial cells

1. The isolation procedure must be performed in a sterile environment, e.g., a down-flow laminar flow hood. Use sterile gloves and a laboratory coat to prevent contamination of the cultures, and to protect yourself.

2. Inspect the umbilical cord for clamped or otherwise damaged areas. Discard these areas and cut a small piece off one end of the cord so that a fresh cut is obtained.

3. Insert the sterile cannula, to which a small tube (filled with cord buffer) is connected, into the vein, and fix it tightly in place using surgical silk.

4. Connect a syringe filled with 20–25 ml of cord buffer to the tube joining the cannula. Remove the blood from the vein by slowly flushing the blood vessel and draining the contents into a sterile waste beaker. Repeat if necessary to remove all blood. Ensure that no air is introduced into the lumen of the vessel.

5. Cut a small piece off the other end of the umbilical cord. Insert the second cannula into the vein and fix it tightly in place using surgical silk. Fill the vein with collagenase solution via the small tube that is joined to the cannula. Clamp the tubes on both sides with kocher scissors.

6. Incubate the filled (slightly distended) vessel for 15–20 min at 37 °C in a beaker containing 0.9 % NaCl.

7. Remove the clamps and collect the contents of the vein lumen into a 50 ml sterile plastic tube. Flush the vessel with an additional 20 ml of M199/p/s medium.

8. Centrifuge the tube for 5 min at $200 \times g$ and resuspend the pelleted cells (often a mixture of groups of endothelial cells and red blood cells) in 3–5 ml of culture medium III.

9. Seed the cells into fibronectin or gelatin-coated multiwell dishes (total area 20–30 cm²).

10. Put the dishes into a (water-conditioned) incubator at 37 °C with an atmosphere of 5 % CO_2/95 % air.

11. Wash the adherent cells one or two days after isolation and renew the culture medium III (1.5 ml/10 cm²). Inspect the cells using an inverted phase-contrast microscope.

12. Renew the culture medium and inspect the growth of the cells every 2–3 days.

Serial propagation of endothelial cells

1. When the cells have become confluent, they can be detached and seeded into new dishes.

2. Wash the cells with Solution A and detach them from the culture vessel by 0.05 % (w/v) trypsin/EDTA. Observe the cell detachment using a

phase contrast microscope. As soon as the cells start to detach, gently tap the side of the culture dishes to ensure detachment of all the cells.

3. Immediately add culture medium III to inhibit any further trypsin activity which will damage the cells.

4. Slowly resuspend the cell suspension using a sterile pipette, and transfer the cells into new fibronectin or gelatin-coated dishes (split ratio 2:1) containing an adequate amount of culture medium III.

Remarks

- hMVECs can be seeded for approximately 10 passages, while HUVECs start to increase in size, grow more slowly and lose specific functions after passage 5.
- Isolation of HUVECs yields more cells if the umbilical cord has been stored at 4 °C for 24 to 72 h.

Expected Results

The purity of hMVECs directly after isolation varies considerably. Isolation with a purity of >50 % hMVECs are preferentially used for further purification of hMVEC (about 50 % of the primary isolates). Purification by magnetic beads for at least two times is often necessary to remove all the contaminating cells. Stromal cells will overgrow the endothelial cells and therefore it is necessary to start the purification one or two days after the isolation. This procedure needs to be repeated until the culture reaches a purity of 99 %. To identify and characterize the endothelial nature of the cells obtained in culture, the cells are stained for von Willebrand Factor (vWF) and CD31. The presence of vWF (factor VIII related antigen) in specific organelles in endothelial cells is widely used as a marker to determine the endothelial nature of the cells. CD31 (PECAM-1) is a marker that is present in the cell junction of endothelial cells and can be stained with CD31 IgG.

During the culture, the quality and growth of the cultured cells is being evaluated using an inverted phase contrast microscope. Endothelial cells display a cobblestone phenotype when they reach confluence after 7–10 days of culture (Fig. 1).

The isolation of HUVECs is much easier and quicker and purification by CD31 magnetic beads will not be necessary. Just like the hMVECs, HUVECs display a cobblestone morphology at confluence. HUVECs grow rather fast and confluence will be reached after a few days.

After plating, both hMVECs and HUVECs start to grow as single cells, but they will form small clusters after a while. Upon prolonged maintenance in the confluent state, some cells acquire a 'sprouting' phenotype and start to

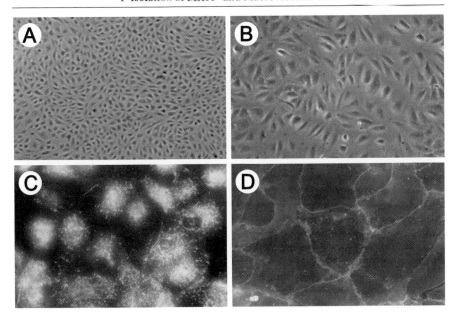

Fig. 1A-D. Phase contrast photomicrographic pictures from human microvascular endothelial cells 5× (**A**), 25× (**B**) and stained for von Willebrand Factor (**C**) and PECAM-1 (**D**)

infiltrate under the other cells. This disorganization of the monolayer increases with time; it is reduced or delayed by the presence of ECGF (or pure aFGF or bFGF) in the culture medium. When the cells are detached by trypsin/EDTA and transferred to new fibronectin or gelatin-coated dishes, the cells grow normally again.

Troubleshooting

Low yield

The effectiveness of isolating hMVECs depends on the quality and morphology of the foreskin. The foreskin may not be too old (two days at the most) and the more vascularized the foreskin is, the more cells you will isolate. In case of a low yield after purification with CD31 and magnetic beads, check the supernatant for remaining endothelial cells. When you have a low yield of HUVECs, you may incubate the umbilical cord a few minutes longer at 37 °C, however, this may also increase the number of contaminating cells. Therefore, incubation should be prolonged by only few minutes. Before you flush the vessel, push gently on the cord to detach the cells from the vein.

Purity

After isolation, the upgrowing hMVECs should be monitored at least every other day by phase contrast microscopy. Purification of hMVECs with CD31 and magnetic beads should be carried out before contaminating cells start to overgrow the endothelial cells. You do not have to wait until confluence is reached to start the purification. Another way to isolate endothelial cells from other cells is selective detachment. A short period with trypsin/EDTA should detach the endothelial cells and not the contaminating cells.

The endothelial nature of the cells can be demonstrated by immuno-staining for vWF or PECAM-1 (as indicated above; these stainings also recognize platelets) and by staining with Ulex europeus lectin-1 (recognizes also epithelial cells) or DiI-acetylated LDL (recognizes also monocytes and macrophages). Contamination by smooth muscle cells or pericytes can be demonstrated by anti smooth muscle cell actin IgG; contamination of epithelial cells by staining for cytokeratins 8 and 18.

The presence of lymphatic endothelial cells can be demonstrated by the presence of LYVE-1 or flt-4 in these cells (see Chap. 5, this Volume, Mäkinen and Alitalo).

Culture

HMVECs and HUVECs need ECGF and serum for their growth. The quality of the serum batch used is critical. Different batches of human and calf serum should be compared to identify a suitable batch of serum. After such identification, one is advised to buy a large quantity of the good batch of serum and store this at –80 °C until use. If the growth is slow (more than two weeks until confluence is reached), one may add more ECGF or serum to the medium.

▓ References

Carley WW, Niedbala MJ, gerritsen ME (1992) Isolation, cultivation, and partial characterization of microvascular endothelium derived from human lung. Am J Respir Cell Mol Biol 7:620–630

Davison PM, Bensch K, Karasek MA (1980) Isolation and growth of endothelial cells from the microvessels from the newborn human foreskin in cell culture. J Invest Dermatol 75:316–321

Davsion PM, Bensch K, karasek MA (1983) Isolation and long-term serial cultivation of endothelial cells from the microvessels of the adult human dermis. In Vitro 19:937–945

Gerhart DZ, Broderius MA, Drewes LR (1988) Cultured human and canine endothelial cells from brain microvessels. Brain Res Bull 21:785–793

Gimbrone MA Jr, Cotran RS, Folkman J (1974) Human vascular endothelial cells in culture. Growth and DNA synthesis. J Cell Biol 60:673–684

Jaffe EA, Nachman RL, Becker CG, Minick CR (1973) Culture of human endothelial cells derived from umbilical veins. Identification by morphological and immunologic criteria. J Clin Invest 52:2745–2756

Koolwijk P, Kapiteijn K, Molenaar B, van Sponsen E, van der Vecht B, Helmerhorst FM, van Hinsbergh VWM (2001) Enhanced angiogenic capacity and urokinase-type plasminogen activator expression by endothelial cells isolated from human endometrium. J Clin Endocrinol Metab 86:3359–3367

van Hinsbergh VW, Binnema D, Scheffer MA, Sprengers ED, Kooistra T, Rijken DC (1987) Production of plasminogen activators and inhibitor by serially propagated endothelial cells from adult human blood vessels. Arteriosclerosis 7:389–400

Van Hinsbergh VWM, Scheffer MA, Langeler EG (1990) In Cell culture techniques in heart and vessel research (ed HM Piper), pp 178–204. Springer-Verlag, Berlin

Van Setten PA, van Hinsbergh VWM, van der Velden TJ, van de Kar NC, Vermeer M, Mahan JD, Assmann KJ, van den Heuvel LP, Monnens LA (1997) Effects of TNF alpha on verocytotoxin cytotoxicity in purified human glomerular microvascular endothelial cells. Kidney Int 51:1245–1256

Voyta JC, Via DP, Butterfield CE, Zetter BR (1984) Identification and isolation of endothelial cells based on their increased uptake of acetylated-low density lipoprotein. J Cell Biol 99:2034–2040

Chapter 2

Isolation of Mouse Cardiac Microvascular Endothelial Cells

RALF P. BRANDES, STEFANI BEER, and ALEXANDRA AICHER

Introduction

Culturing of endothelial cells is a pre-requisite to study many aspects of endothelial cell biology. To determine the specific function of a protein, cellular responses are usually compared between control cells and cells in which the protein of interest is either inactivated, deleted, suppressed or overexpressed. In addition to pharmacological tools, such as inhibitors and inducers, this goal is achieved by molecular biological techniques including antisense oligonucleotides or plasmids, interference RNA, dominant–negative mutants or over-expressing constructs. Unfortunately, with the exception of stable transfection and cell lines, selection of any of these approaches yield only transient changes and are hampered by limitations such as low transfection efficiency or toxicity of the transfection agent.

Transgenic mice offer a unique source of cells with permanent deletion or overexpression of genes. In addition to studying the phenotype of the transgenic animal, it is frequently required to study the effect of the genetic alteration in simpler and better defined model systems, including cell culture, which are more accessible to aggressive interventions and exhibit less compensatory phenomena. Consequently, with the increasing number of transgenic mice available, culture of mouse endothelial cells is a frequent need in many laboratories.

The relative small size of the mouse grossly limits the amount of endothelial cells available from these animals. Although culture of macrovascular endothelial cells, e.g., from the aorta is feasible, using enzyme digestion or outgrowth techniques in Matrigel, both techniques are time and animal consuming and yield only few cells which are usually contaminated by fibroblasts. During the subsequent expansion, endothelial cells are easily overgrown by contaminating cells. Outgrowth techniques have also been suggested for lung endothelial cells. Although these techniques yield more cells, separation of the cells is advisable and pulmonary endothelial cells certainly differ in many aspects from other capillary endothelial cells.

Springer Lab Manual
H. Augustin (Ed.)
Methods in Endothelial Cell Biology
© Springer-Verlag Berlin Heidelberg 2004

To have a higher number of cells available which are not derived from the pulmonary circulation, we isolated microvascular endothelial cells from the mouse heart (MMCEC: Mouse microvascular cardiac endothelial cells). The technique consists of two steps. An initial isolation of the non-myocyte cell fraction, similar to the method reported by Li et al. (2001) and subsequent separation of the endothelial cells using the CD31/ECAM–1 antigen and magnetic beads.

Materials

Animals

- Male C57BL/6 mice (4–6 weeks, Charles River, www.criver.com)
- Narcotics: Isoforene (Cat. No. B506, Abbott, www.abbott.com)

Buffers/Media

- Dulbecco's Phosphate Buffered Saline (Cat. No. BE17-512F, Bio Whittaker Europe [Cambrex], www.cambrexbioproductseurope.com)
- Dulbecco's Modified Eagle's Medium (Cat. No. 51965-039, Gibco BRL, www.gibcobrl.com)
- Modified Hank's buffer (in g/L): KCl 0.4, KH_2PO_4 0.06, NaCl 8.0, $NaHCO_3$ 0.35, Na_2HPO_4 0.048, Taurin 2.5, $MgCl_2$ 0.284, BSA 1.0, pH 7.25.

Enzymes

- Trypsin EDTA (Cat. No. BE17-161E, 0.05 % trypsin, 0.1 mmol/L EDTA, Bio Whittaker Europe [Cambrex])
- Collagenase (type II, approx. 270 U/mg (Cat. No. 171-01-015, Invitrogen, www.invitrogen.com)
- Accutase solution (Cat. No. L11-007, PPA Laboratories, www.paa.at)
- DNAse (D4236, Sigma, www.sigmaaldrich.com)

Antibodies/Magnetic beads

- Fc-block (Cat. No. 553142, BD Biosciences, www.bdbiosciences.com)
- Rat anti-mouse F4/80 Antibody (Cat. No. MCAP497, Serotec, www.serotec.com
- Rat anti-mouse CD31 (PECAM-1, MEC 13.3, Cat. No. 553370, BD Pharmingen)
- Sheep anti-rat IgG Dynabeads (Cat. No. M450, Dynal Biotech, www.dynal.no)

Others

- Magnetic stand and Dynalmixer (Dynal Biotech)
- Penicillin-Streptomycin-mix (Cat. No. DE17-602E, Bio Whittaker Europe [Cambrex])
- Fetal calf serum
- Fibronectin (human, natural, Cat. No. 356008, BD Biosciences)
- Human recombinant epidermal growth factor (EGF, Cat. No. C-10015-SS, Cell Concepts, www.cellconcepts.com)
- Human recombinant basic fibroblast growth factor (bFGF, Cat. No. C-10018-BF, Cell Concepts)
- Bovine serum albumin (BSA, Cat. No. 15260-037, Gibco BRL Life; or Cat. No. #4003, CellSystems Biotechnologie, www.cellsystems.de), Chamber Slides (Cat. No. 354108, BD Biosciences)
- All other compounds were obtained from Sigma.

Procedure

Isolation of cardiac microvascular endothelial cells

1. 10 to 15 mice are anaesthetized with isoforene in groups of five animals and killed by decapitation. Skin and chest are opened and the still beating hearts without the atria are removed quickly using separate sets of sterile forceps and scissors and transferred into a petri-dish containing ice-cold phosphate buffered saline (PBS).

2. The ventricles are slit with a scalpel and joggled to remove blood. Valves and connective tissue are removed and the hearts are dipped in 2 dishes of PBS, 1 dish of ethanol (70 %, 2 s) and 3 dishes of PBS to remove potential bacterial contaminations and to de-vitalize the epicardium, endocardium and mesothelium.

3. Hearts are minced sequentially using a large scalpel and forceps in a sterile petri dish without buffer to a pulp with pieces of max. 1 mm^3 and collected in a flask containing 25 ml of modified Hank's buffer.

4. Following gentle mixing and sedimentation, the PBS is removed and replaced by 10 ml of pre-warmed (37 °C) collagenase pre-digestion buffer (collagenase type II suspended in modified Hank's buffer, final concentration 270 U/ml equal roughly up to 1 mg/ml).

5. Digestion is carried out for 5 min at 37 °C in a pre-warmed water bath with mixing every 30 s by pipetting the solution up and down several times.

6. The supernatant is removed using a pipette and discarded.

7. Pre-digestions are repeated two more times using 10 ml of collagenase buffer (identical as under point 4) and incubation times of 10 min with mixing every 3 min. The myocytes detach during this procedure and are removed with the supernatant, whereas the endothelial cells remain in the interstitial fraction which sediments rapidly.

8. This fraction is then digested using trypsin EDTA buffer (37 °C pre-warmed, 8.0 ml, 0.05 % trypsin, 0.1 mmol/L EDTA, enriched with 2.0 mg/ml glucose and 1.0 µg/ml DNAse) for 10 min at 37 °C with mixing by pipetting every 3 min.

9. The supernatant containing the detached endothelial cells was transferred into a tube (12 ml) containing 1.0 ml of fetal calf serum (FCS) to stop digestion.

10. Subsequently, this preparation is centrifuged at $25 \times g$ for 3 min to spin down remaining myocytes and debris and the supernatant containing the endothelial cells is transferred and collected into a 50 ml tube on ice.

11. The trypsin digestion of the remaining interstitial fraction (points 8–10) is repeated two more times.

12. The 50 ml tube containing the detached endothelial cells from the three digestions is centrifuged at $120 \times g$ for 7 min at RT to sediment the endothelial cells.

13. The supernatant is discarded and the pellet containing the cells is re-suspended in 10 ml of culture medium (Dulbecco's Modified Eagle's Medium, supplemented with 20 % FCS, penicillin (50 U/ml), streptomycin (50 µg/ml), EGF (0.1 ng/ml) and bFGF (1ng/ml)) and seeded onto a 10 cm tissue culture dish, pre-coated with fibronectin (25 ng/ml in PBS, 4.0 ml per 10 cm dish for 20 min)

14. After exactly 30 min the medium is carefully removed and replaced by 10 ml of fresh culture medium (as above). This step is essential as fine debris still present in the preparation will slowly sediment after this time and prevent growth of the cells.

15. Medium should be changed following washing twice with PBS the next day if a lot of debris is observed in the culture dish. Otherwise medium replacement is not necessary.

Three days after the isolation, cells should be almost completely confluent. At this day, endothelial cells are separated from fibroblasts by the aid of magnetic cell separation using anti-PECAM-antibody labelled magnetic beads.

Separation of cardiac microvascular endothelial cells

For separation of endothelial cells, the CD31/PECAM-1 surface antigen is used. However, as this antigen is also expressed on macrophages, these contaminating cells are isolated prior to the isolation of PECAM-1 positive cells. Macrophages are identified using the F4/80 antigen.

1. Cells are washed once with PBS (6.0 ml, 37 °C) and detached using 6.0 ml of accutase solution for approx. 10 min at 37 °C until the cells have rounded up. (Accutase is preferred over trypsin as the latter digests many surface markers on the cells)

2. Cells are detached by repetitive flushing of the plate using a pipette and transferred into a tube containing 6.0 ml of medium containing FCS (20 %).

3. Cells are sedimented at $200 \times g$ for 4 min at RT.

4. The supernatant is discarded and cells are resuspended in 5.0 ml PBS/BSA (0.5 %).

5. Cells are sedimented at $200 \times g$ for 4 min at RT, the supernatant is discarded and the cells resuspended in 0.5 ml PBS/BSA (0.5 %).

6. Cells are blocked by adding Fc-Block (10 μl of purified rat IgG anti-mouse CD16/CD32 [Fcγ III/II receptor], final concentration approx. 1 μg/10^6 cells/100 μl, www.bdbiosciences.com)

7. Following mixing and incubation for 15 min, contaminating macrophages and monocytes are labelled with 50 μl of rat anti mouse F4/80 antibody (IgG anti-macrophage, 10 μl of antibody per 10^6 cells in 100 μl)

8. The antibody is allowed to bind for 30 min at 4 °C and the preparation is mixed every 10 min by pipetting.

9. Subsequently, non-binding antibody is removed by two cycles of centrifugation ($200 \times g$) and re-suspension in 5.0 ml of PBS/BSA (0.5 %) and finally cells are taken up in 1.0 ml PBS/BSA (0.5 %).

10. F4/80 labelled cells are removed using magnetic beads (Dynabeads sheep anti-rat IgG). Stem suspension of magnetic beads (60.0 μl containing approx. 2.5×10^7 beads) are washed three time in 1.0 ml PBS/BSA (0.5 %) to remove the storage medium and recovered by the magnetic stand. The washed beads were resuspended in 60.0 μl of PBS/BSA (0.5 %) and 30.0 μl of the beads suspension is added to the cells.

11. Binding of the beads is allowed for 20 min at RT with mixing of the preparation in the "Dynalmixer".

12. The mix is diluted with 5.0 ml of PBS/BSA (0.5 %) and split into 6 Eppendorf tubes (1.5 ml) positioned in the magnetic stand with the magnet in place. After 1 min the non-binding fraction which is now free of F4/80 positive cells is collected into a tube, centrifuged and re-suspended in 600 µl of PBS/BSA (0.5 %).

13. Endothelial cells are labelled using the CD31/PECAM-1 antigen. For this purpose, 12 µl of anti-PECAM-1 antibody (purified rat anti-mouse CD31, approx. 1 µg/10^6 cells) is added to the cell suspension.

14. Following incubation for 30 min at 4 °C and mixing every 10 min, non-binding antibody is removed by two cycles of centrifugation, and resuspension in PBS/BSA (0.5 %).

15. Cells are taken up in 1.0 ml of PBS/BSA (0.5 %) and the remaining 30.0 µl of magnetic beads are added.

16. Following binding of the beads for 20 min at RT in the "Dynalmixer" the preparation is diluted with 5.0 ml PBS/BSA (0.5 %) and distributed to 6 Eppendorf tubes mounted in the magnetic stand with the magnet in place.

17. Following binding of the beads for 1 min, the supernatant with the CD31-negative fraction is removed.

18. The magnet is removed and the remaining cells are resuspended in the stand with 1.0 ml PBS/0.5 % BSA.

19. This washing procedure is repeated twice.

20. Finally, the endothelial cells are resuspended in 1.0 ml of cultured medium and seeded in a single fibronectin-coated well of a 12-well plate. A small aliquot of the CD31-negative fraction and also of the CD31-positive fraction is seeded onto a chamber slide for later characterization of the cells.

Characterization of cardiac microvascular endothelial cells

MMCEC are identified by their ability to take up fluorescent labeled acetylated low density lipoproteins (Dil-Ac-LDL, 10 µg/ml, 4 h). For this purpose, cells are incubated with the compound for 4 h and subsequently studied by fluorescence activated cell sorting (FACS) analysis or fluorescence microscopy. FACS analysis is also carried out using the F4/80 and anti-PECAM antibodies used for magnetic cell separation.

Expected Results

One day after isolation, the plate should be covered roughly to 10–20 % by cells and only few have divided. At day two, the plate should be 10–40 % confluent. Three to four days after isolation, cells should be 90 %–100 % confluent. In our experience and in contrast to the original report of the method (Li et al. 2001), only about 5 %–40 % of the cells are Dil-Ac-LDL positive and thus either endothelial cells or phagocytes (Fig. 1). Following magnetic cell sepa-

Brightfield   DilAcLDL

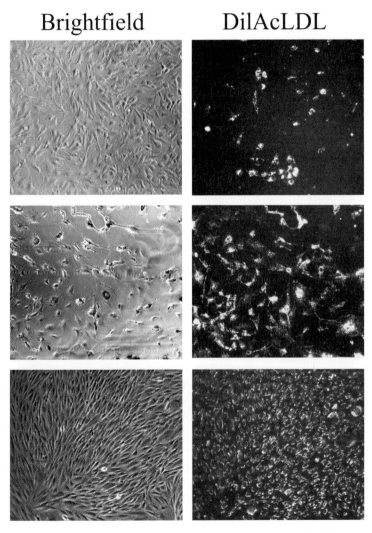

Fig. 1. Images of mouse microvascular cardiac endothelial cells. *Left column* Brightfield image (magnification × 20), *right column* fluorescence images after labeling the cells for 4 h with fluorescent Dil-Ac-LDL; *first row* 2nd day after isolation, *middle row* 2nd day after sorting, *bottom row* passage 5

ration, only few cells will be recovered in the PECAM-positive fraction. At day 2, these cells have spread and can be easily identified by the magnetic beads still attached on their surface. Approx. three days after isolation, cells start to grow but passaging is not advisable until the well is completely confluent, which may take up to 6 days. Using this technique, approx. 95 % Dil-Ac-LDL positive cells can be achieved as determined by FACS analysis at passage 5. Omission of the F4/80 isolation step results in contamination with macrophages/monocytes (12 % at passage 2 as determined by FACS).

As determined by immunohistochemistry, cardiac mouse microvascular endothelial cells express endothelial NO synthase, however at low level. Measurement of intracellular calcium by fluo-4 fluorescence in cells cultured to passage 3 revealed responsiveness to thrombin (1.0/ml), vascular endothelial growth factor (30 ng/ml), histamine (100 nmol/L), and serotonine (100 nmol/L) but not angiotensin II (^{5}val-angiotensin II, 100 nmol/L) or bradykinin (100 nmol/L). This suggests that MMCEC undergo de-differentiation during prolonged culture.

Troubleshooting

Comparison of 4 different mouse endothelial cells isolation techniques (out growth from aorta and lung, isolation of neo-natal endothelial cells and with the method reported herein) showed that mouse endothelial cells are not trivial to passage, particularly at low passages. Cells should be passaged only after reaching confluence, which may take up to 11 or 12 days. The split ratio should not be greater than 1:2 as cells stop growing if grown too sparsely. These problems could hitherto not be overcome by different coatings or media supplements.

Acknowledgments. We are indebted to Sina Bätz and Isabel Winter for excellent technical assistance. This study was supported by the Deutsche Forschungsgemeinschaft.

Reference

Li J M, Mullen AM, Shah AM (2001) Phenotypic properties and characteristics of superoxide production by mouse coronary microvascular endothelial cells. J Mol Cell Cardiol 33:1119–1131

Isolation, Culture, and Differentiation of Circulating Endothelial Precursor Cells

Carmen Urbich and Stefanie Dimmeler

Introduction

Adult and embryonic stem cells hold great promise for regenerative medicine. Since embryonic stem cells are legally protected, the use of adult stem/progenitor cells is an important therapeutic option to heal damaged tissues. Adult blood vessel formation occurs through arteriogenesis, angiogenesis, or vasculogenesis. Arteriogenesis describes the growth of collateral vessels, whereas angiogenesis refers to the growth of new capillaries by sprouting of pre-existing vessels through migration and proliferation of mature endothelial cells. Vasculogenesis includes the mobilization of bone marrow-derived endothelial stem cells which home to sites of ischemia and contribute to new blood vessel formation. The finding that vasculogenesis also contributes to postnatal neovascularization offers novel therapeutic strategies for the use of ex vivo cultured circulating endothelial progenitor cells or their precursors for cell therapy of tissue ischemia.

Endothelial precursor cells (EPCs) can be grown from purified populations of CD34+ or CD133+ hematopoietic cells, purified CD14+ monocytes, or total peripheral blood mononuclear cells (MNCs) (Asahara et al. 1997, Gehling et al. 2000, Murohara et al. 2000, Kalka et al. 2000, Dimmeler et al. 2001, Vasa et al. 2001). Cultured EPCs grown from different starting populations have been shown to express endothelial marker proteins such as von Willebrand factor, VEGF-receptor 2 (KDR), VE-cadherin, CD146 (MCAM), CD31, and eNOS (Kalka et al. 2000, Dimmeler et al. 2001). In addition, these cells have been characterized by their functional capacity to form endothelial cell colonies and improve neovascularization in animal models of hind limb ischemia and myocardial infarction (Shi et al. 1998, Kawamoto et al. 2001, Kocher et al. 2001). Likewise, a recent clinical study suggests that intracoronary infusion of blood-derived EPCs can be used to improve coronary flow and cardiac function in patients after acute myocardial infarction (Assmus et al. 2002).

To study the basic principles of EPCs, namely the intracellular signal transduction leading to the functional activity of the cells, requires the isola-

Springer Lab Manual
H. Augustin (Ed.)
Methods in Endothelial Cell Biology
© Springer-Verlag Berlin Heidelberg 2004

tion and ex vivo culture of the cells. Based on these findings, it may be possible to further optimize the functional activity of EPCs during ex vivo culture. To obtain cultured endothelial precursor cells, the following starting material can be used (Fig. 1): a) Peripheral blood mononuclear cells can be isolated from peripheral blood (freshly drawn or buffy coats). b) Purified CD34+ or CD133+ cells can be isolated from peripheral blood mononuclear cells or bone marrow mononuclear cells and used as starting populations to culture EPCs. The technique consists of a density gradient centrifugation and the subsequent ex vivo expansion. Alternatively, CD34+ or CD133+ cells can be separated with magnetic beads from mononuclear cells or bone marrow aspirates. The following chapter describes one standard protocol to isolate and ex vivo culture EPCs. However, other groups have established alternative protocols resulting in the culture of cells with similar endothelial phenotypes and functional activities (Eggermann et al. 2003).

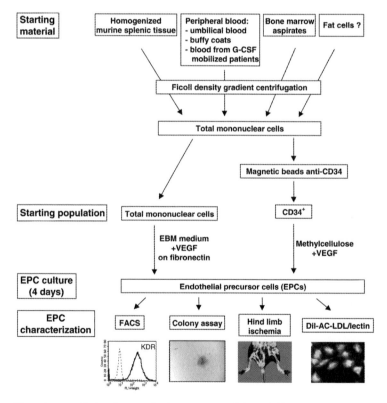

Fig. 1. Flow chart outlining the EPC isolation protocol from different starting populations

▨ Materials

Starting material

- Peripheral blood (freshly drawn or buffy coats obtained from the blood donation center), bone marrow aspirates or minced murine splenic tissue (Fig. 1).

Buffer/Media

- Dulbecco's phosphate buffered saline (Cat. No. H-15-002, PAA, www.paa.at)
- EGM bullet kit (EBM medium) (Cat. No. CC-3124, CellSystems, www.cellsystems.de)
- Fetal bovine serum (Cat. No. 10270-106, Invitrogen, www.invitrogen.com
- Methocult GF H4434 (Cat. No. 04444, CellSystems)

Antibodies/Magnetic beads

- FcR blocking reagent (Cat. No. 130-046-702, Miltenyi Biotec, www.miltenyibiotec.com)
- Mouse anti-human CD34 Microbeads (Cat. No. 130-046-702, Miltenyi Biotec, www.miltenyibiotec.com)
- Mouse anti-human CD31 (PECAM-1) (Cat. No. CBL468F, Dianova, www.dianova.com)
- Mouse anti-human CD146 (MCAM) (Cat. No. MAB16985F, Chemicon International, www.chemicon.com)
- Mouse anti-human VEGFR2/KDR (Cat. No. 101-M22, ReliaTech, www.reliatech.de)
- Mouse anti-human CD105 (Endoglin) (Cat. No. MS-1290-P0, Dianova)
- Rabbit anti-human von Willebrand factor (Cat. No. PC313, Oncogene/Merck Biosciences, www.merckbiosciences.de)
- Rabbit anti-mouse immunoglobulins-FITC (Cat. No. F0261, DakoCytomation, www.dakocytomation.com)
- Swine anti-rabbit Immunoglobulins-FITC (Cat. No. F0205, DakoCytomation)
- FITC-conjugated mouse IgG1 monoclonal isotype control (Cat. No. 556028, BD Biosciences)
- Lectin from *Ulex europaeus* (Cat. No. L-9006, Sigma)

Others

- Biocoll separating solution, density: 1.077 (Cat. No. L6115, Biochrom, www.biochrom.de)

- AutoMACS (Cat. No. 201-01, Miltenyi Biotec)
- Human fibronectin, 1 mg/ml (Cat. No. F-0895, Sigma)
- Human recombinant vascular endothelial growth factor (VEGF) (Cat. No. 100–20, Cell Concepts, www.cellconcepts.com)
- 1,1'-dioctadecyl-3,3,3',3'-tetramethylindocarbocyanine-labeled acetylated low-density lipoprotein (Dil-Ac-LDL) (Cat. No. 4003, CellSystems)
- EDTA disodium salt dihydrate (Cat. No. A1104, AppliChem, www.applichem.de)
- TÜRK'S solution (Cat. No. 1.09277.0100; Merck, www.merck.com)

Procedure

Isolation of endothelial progenitor cells from peripheral blood

Separation of mononuclear cells (MNC) with ficoll gradient centrifugation from freshly collected peripheral blood or buffy coats from the blood donation center:

1. Provide 15 ml Biocoll separation solution per 50 ml tube.

2. Dilute the peripheral blood (PB) with PBS (PB 1:1 or buffy coats 1:4).

3. Carefully and slowly overlay 25 ml of the diluted blood on 15 ml of the Biocoll separation solution.

4. Centrifugate the tube at 800 × g for 20 min at RT **without** brake. This is an important step to separate the mononuclear cells (in the interphase) from erythrocytes and granulocytes (pellet) and platelets in the upper serum phase.

5. Meanwhile: Coating of the wells with 10 µg/ml human fibronectin in PBS, incubate the wells for at least 30 min at RT.

6. Pipet the mononuclear cells of the interphase carefully in a new 50 ml tube.

7. Add PBS to 50 ml to wash the cells.

8. Centrifugate the cells at 800 × g for 10 min at RT (with brake).

9. Remove the supernatant and resuspend the cell pellet in 50 ml PBS.

10. Centrifugate the cells at 800 × g for 10 min at room temperature (with brake).

11. Remove the supernatant and resuspend the cell pellet in 10 ml PBS. Dilute an aliquot of the cells (50 µl; 1:10) with TÜRK'S solution and count cells.

12. Add PBS to the remaining cells in the 50 ml tube to wash the cells again.

13. Centrifugation: 800 × g, 10 min, RT, with brake.

14. The washing steps should be performed for at least 3 times, but should be repeated until the supernatant becomes clear (altogether 3–5 times).

15. Remove supernatant and resuspend the cell pellet in culture medium [endothelial basal medium supplemented with 20 % FBS, epidermal growth factor (10 µg/ml), bovine brain extract (3 µg/ml), gentamicin (50 µg/ml), hydrocortisone (1 µg/ml), VEGF (100 ng/ml) to a cell concentration of 8×10^6 cells/ml medium].

16. Remove the fibronectin from the dishes.

17. Pipet the cells to the fibronectin-coated wells at a density of approx. 2.1×10^6 cells/cm^2:
    – per 24-well plate: 4×10^6 cells in 500 µl medium per well
    – per 12-well plate: 8×10^6 cells in 1 ml medium per well
    – per 6-well plate: 20×10^6 cells in 2.5 ml medium per well.

18. Incubate the cells for 3 days at 37 °C and 5 % CO_2.

19. Remove the nonadherent cells three days after seeding by thoroughly washing the cells with PBS. Add fresh culture medium for 24 h before using the cells for experiments. Approximately 0.5–1 % of the initially applied mononuclear cells will become adherent EPCs.

Note: For clinical applications, the mononuclear cells have to be cultured according to Good Manufacture Practice (GMP) guidelines in X-VIVO 15™ Serum-free Medium supplemented with 1 ng/ml carrier-free human recombinant VEGF, 0.1 µmol/L atorvastatin, and 20 % human serum drawn from each individual patient.

Isolation of endothelial precursor cells from murine splenic tissue

The relative small size of mice limits the amount of endothelial precursor cells from peripheral blood. It is therefore more feasible to isolate total mononuclear cells from minced murine splenic tissue by density gradient centrifugation. For this purpose, mice have to be sacrificed to obtain specimen of spleen. Spleens are mechanically minced using syringe plungers and laid over Ficoll to isolate mononuclear cells (splenocytes) as described above. Splenocytes (8×10^6 cells/ml culture medium) are seeded on fibronectin-coated culture plates with 50 ng/ml recombinant murine VEGF. Medium should be changed after 3 days in culture.

Separation of CD34+ hematopoietic cells from peripheral blood or bone marrow MNCs

The isolation of CD34$^+$ hematopoietic progenitor cells is performed by positive selection of CD34 expressing cells. For this purpose, mononuclear cells from peripheral blood, cord blood, bone marrow aspirates, or apheresis harvest should be obtained as starting population by density gradient centrifugation with Ficoll separation solution. Of note, the rate of yield depends on the starting population used: for peripheral blood a rate of yield of 0.05–0.2 % or bone marrow MNCs a rate of yield of 0.5–3 % is to be expected.

1. Freshly aspirate the bone marrow or collect venous peripheral blood samples.

2. Isolate the total mononuclear cells by density gradient centrifugation using Ficoll separation and wash the cells three times with PBS as described above in point 1.

3. Resuspend the mononuclear cells obtained from the gradient centrifugation in a volume of 300 µl PBS/1 × 10^8 cells.

4. Block cells by adding FcR Blocking Reagent (100 µl FcR Blocking Reagent/1 × 10^8 cells) to inhibit unspecific binding.

5. Incubate the mixture for 10 min at RT.

6. Add 100 µl anti-human CD34 microbeads/1 × 10^8 mononuclear cells to magnetically label the CD34$^+$ cells and incubate for 30 min at 4 °C.

7. Remove non-binding antibody by washing the cells with PBS and a following centrifugation at 800 × g for 10 min at RT.

8. Filter the labeled cells through a 40 µm cell strainer to remove clumps.

9. Run cells over a magnetic cell separation device for positive selection of CD34$^+$ cells.

10. Isolated CD34$^+$ cells are seeded in methylcellulose (see Protocol c) to analyze the colony forming units.

Characterization of endothelial precursor cells

To characterize the endothelial phenotype of EPCs use the following methods (Fig. 1):
a) Uptake of Dil-Ac-LDL and double-staining with lectin.
b) FACS analysis of endothelial surface markers.
c) Colony forming activity.
d) Improvement of neovascularization after hind limb ischemia.

a) Uptake of Dil-Ac-LDL and lectin staining

EPCs are identified by their ability to take-up fluorescent labeled acetylated low density lipoproteins (Dil-Ac-LDL) and binding to lectin (Fig. 1):

1. Wash endothelial progenitor cells with PBS at day 4.

2. Add fresh culture medium for 1 h.

3. Add Dil-Ac-LDL (2.4 µg/ml medium) and incubate the cells for 1 h at 37 °C to take-up Dil-Ac-LDL.

4. Wash the cells with PBS.

5. Fix the cells with 4 % formaldehyde in PBS for 10 min at RT in the dark.

6. Wash the cells with PBS.

7. Add FITC-labeled *Ulex europaeus* agglutinin I (lectin: 10 µg/ml) for 1 h at RT in the dark.

8. Identify EPCs by fluorescence microscopy and count in three randomly selected high-power fields by two independent investigators.

b) FACS analysis of endothelial surface markers

Analyze the endothelial phenotype of EPCs by FACS using different endothelial markers such as VEGFR-2 (KDR), von Willebrand factor (vWF), CD105, CD146 (MCAM) and CD31:

1. After 4 days in culture, adherent EPCs are thoroughly washed with PBS to remove contaminating non-adherent cells.

2. Adherent cells are gently detached by incubating the cells with 1 mM EDTA in PBS (pH 7.4) for 15–20 min at 37 °C.

3. Detach cells by repetitive flushing of the plate.

4. Sediment cells at 800 × g for 10 min at RT.

5. Discard the supernatant, suspend cells in PBS and block by adding FcR Blocking Reagent to prevent unspecific binding of the antibodies.

6. Incubate cells for 30 min at 4 °C with the following antibodies:
   – FITC-labeled mouse anti-human CD31
   – FITC-labeled mouse anti-human CD146 (MCAM)
   – FITC-labeled mouse IgG monoclonal isotype control
   – Mouse anti-human VEGFR-2 (KDR) + rabbit anti-mouse Immunoglobulins-FITC
   – Mouse anti-human CD105 + rabbit anti-mouse Immunoglobulins-FITC
   – Rabbit anti-human vWF + swine anti-rabbit Immunoglobulins-FITC

7. Perform single- and two-color flow cytometric analyses using a FACS SCAN flow cytometer (BD) and Cell Quest software (BD). Each analysis should include at least 20,000 events.

c) Colony forming activity

The colony forming activity is a characteristic feature of stem and progenitor cells. In order to analyze the proliferative capacity of EPCs and the progenitor character, perform endothelial colony forming assays as follows (Fig. 1):

1. After 4 days in culture, wash adherent cells (EPCs) thoroughly twice with PBS to remove non-adherent cells.

2. Detach adherent cells gently by incubating the cells with 1 mM EDTA in PBS (pH 7.4) for 15–20 min at 37 °C.

3. Detach cells by repetitive flushing the plate.

4. Sediment cells at 800 × g for 10 min at RT.

5. Discard the supernatant and resuspend cells in 10 ml PBS. Count an aliquot (20 µl) of the cells. Sediment the remaining cells by centrifugation (800 × g for 10 min at RT).

6. Seed cells (1×10^5 cells in 100 µl culture medium) in 1.5 ml semi-solid methylcellulose (Methocult) supplemented with 100 ng/ml VEGF in 3.5 cm cell culture wells and incubate for 14 days at 37 °C and 5 % CO_2.

7. Study the plates by phase contrast microscopy, and count colonies after 7 and 14 days of incubation by two independent investigators.

8. To confirm the endothelial phenotype of the colonies (CFU-EC), stain the cells with Dil-Ac-LDL and lectin. For this purpose, it is necessary to extensively wash the plates with PBS to remove the methylcellulose before incubating the colonies with culture medium and Dil-Ac-LDL (see Protocol a).

d) Improvement of neovascularization after hind limb ischemia

Use the murine model of hind limb ischemia to investigate the capacity of EPCs to improve neovascularization. For this purpose, 5×10^5 EPCs per mice are intravenously injected 24 h after induction of hind limb ischemia. Of note, the cell number used for injection is critical. For example, the transplantation of 5×10^5 bone marrow mononuclear cells is as effective as 5×10^6 cells, however a lower dose of 5×10^4 bone marrow cells is ineffective, demonstrating the dose-dependency of the injected cells. To compare different cell types (e.g., cells with different sizes), it is necessary to optimize the cell number in the hind limb ischemia model. Neovascularization can be determined by laser Doppler-derived blood flow measurements or analysis of capillary density of sections of the adductor and semimembranous muscles. Endothelial cells are stained with CD146-FITC and HLA-APC to detect the transplanted human cells.

Expected Results

After 3 days in culture, adherent EPCs will remain on the cell culture dish. In our hands, about 0.5–1% of the initially seeded mononuclear cells will become attached EPCs after 3–4 days in culture. Using this technique, approx. 90% of the adherent EPCs express endothelial marker proteins as determined by FACS analysis or Dil-Ac-LDL/lectin-double-staining. As determined by RT-PCR and Western blot analysis, EPCs express the endothelial NO synthase (eNOS), however at lower levels as compared to mature endothelial cells (e.g., HUVEC). Further culture or reseeding in methylcellulose enhances the expression of endothelial marker proteins. At day 4 EPCs still express CD45 and CD14$^{(low)}$, consistent with their progenitor cell characteristics. Additional culture of EPCs up to 7 days leads to an additional increase in eNOS as well as KDR expression, indicating endothelial differentiation during prolonged culture. However, fully differentiated endothelial cells do not lead to an improvement of neovascularization after i.v. infusion in a hind limb ischemia model, suggesting that the fully differentiated endothelial phenotype is not suitable for cell therapy.

Troubleshooting

The ex vivo expansion of EPCs depends heavily on the cytokines and growth factors supplemented to the medium. In order to overcome these problems, it is important to test different charges of medium, FBS and supplements. Additionally, it is important to change the medium after 3 days in culture to remove dead cells or contaminating cells such as monocytes/macrophages or platelets, which can negatively interfere with the culture of EPCs. Platelets can be removed during isolation of peripheral blood MNCs by low speed centrifugation. However, this low speed centrifugation at 200 × g will lead to a significant loss of mononuclear cells, and should only be used with EPCs from high platelet counts (or activated platelets binding to other cells). The cell density is also a critical point, since a higher cell density will inhibit EPC expansion, whereas a lower cell density may negatively influence the survival of EPCs due to the loss of paracrine effects. Another point is the dependency of the blood donors, since medication or diseases may influence the quality and quantity of EPCs in culture. This is underlined by the findings that the number and functional activity of EPCs inversely correlates with risk factors for coronary artery disease (Vasa et al. 2001).

Acknowledgments. We are indebted to Christiane Mildner-Rihm, Andrea Knau and Melanie Näher for excellent technical assistance. This study was supported by the Deutsche Forschungsgemeinschaft.

References

Asahara T, Murohara T, Sullivan A, Silver M, van der Zee R, Li T, Witzenbichler B, Schatteman G, Isner JM (1997) Isolation of putative progenitor endothelial cells for angiogenesis. Science 275:964–967

Assmus B, Schachinger V, Teupe C, Britten M, Lehmann R, Dobert N, Grunwald F, Aicher A, Urbich C, Martin H, Hoelzer D, Dimmeler S, Zeiher AM (2002) Transplantation of progenitor cells and regeneration enhancement in acute myocardial infarction (TOP-CARE-AMI). Circulation 106:3009–3017

Dimmeler S, Aicher A, Vasa M, Mildner-Rihm C, Adler K, Tiemann M, Rutten H, Fichtlscherer S, Martin H, Zeiher AM (2001) HMG-CoA reductase inhibitors (statins) increase endothelial progenitor cells via the PI 3-kinase/Akt pathway. J Clin Invest 108:391–397

Eggermann J, Kliche S, Jarmy G, Hoffmann K, Mayr-Beyrle U, Debatin KM, Waltenberger J, Beltinger C (2003) Endothelial progenitor cell culture and differentiation in vitro: a methodological comparison using human umbilical cord blood. Cardiovasc Res 58:478–486

Gehling UM, Ergun S, Schumacher U, Wagener C, Pantel K, Otte M, Schuch G, Schafhausen P, Mende T, Kilic N, Kluge K, Schafer B, Hossfeld DK, Fiedler W (2000) In vitro differentiation of endothelial cells from AC133-positive progenitor cells. Blood 95:3106–3112

Kalka C, Masuda H, Takahashi T, Kalka-Moll WM, Silver M, Kearney M, Li T, Isner JM, Asahara T (2000) Transplantation of ex vivo expanded endothelial progenitor cells for therapeutic neovascularization. Proc Natl Acad Sci U S A 97:3422–3427

Kawamoto A, Gwon HC, Iwaguro H, Yamaguchi JI, Uchida S, Masuda H, Silver M, Ma H, Kearney M, Isner JM, Asahara T (2001) Therapeutic potential of ex vivo expanded endothelial progenitor cells for myocardial ischemia. Circulation 103:634–637

Kocher AA, Schuster MD, Szabolcs MJ, Takuma S, Burkhoff D, Wang J, Homma S, Edwards NM, Itescu S (2001) Neovascularization of ischemic myocardium by human bone marrow-derived angioblasts prevents cardiomyocyte apoptosis, reduces remodeling and improves cardiac function. Nat Med 7:430–436

Murohara T, Ikeda H, Duan J, Shintani S, Sasaki K, Eguchi H, Onitsuka I, Matsui K, Imaizumi T (2000) Transplanted cord blood-derived endothelial precursor cells augment postnatal neovascularization. J Clin Invest 105:1527–1536

Shi Q, Rafii S, Wu MH, Wijelath ES, Yu C, Ishida A, Fujita Y, Kothari S, Mohle R, Sauvage LR, Moore MA, Storb RF, Hammond WP (1998) Evidence for circulating bone marrow-derived endothelial cells. Blood 92:362–367

Vasa M, Fichtlscherer S, Adler K, Aicher A, Martin H, Zeiher AM, Dimmeler S (2001) Increase in circulating endothelial progenitor cells by statin therapy in patients with stable coronary artery disease. Circulation 103:2885–2890

Vasa M, Fichtlscherer S, Aicher A, Adler K, Urbich C, Martin H, Zeiher AM, Dimmeler S (2001) Number and migratory activity of circulating endothelial progenitor cells inversely correlate with risk factors for coronary artery disease. Circulation Research 89:E1–7

Chapter 4

Embryonic Stem Cell-Derived Endothelial Cells

Jun Yamashita and Shin-Ichi Nishikawa

▦ Introduction

Embryonic stem (ES) cells established from the inner cell mass of blastocysts possess pluripotent differentiation potential and unlimited self-renewal capacity. Thus, ES cells serve as a promising cell source for regenerative medicine. Another indispensable role of ES cells is its potency to dissect differentiation processes of mammalian cells in vitro. Until now, embryoid bodies, which form as aggregates of ES cells, are often used to induce various cell types including endothelial cells, neural cells, cardiomyocytes, and blood cells (Rathjen et al. 2001) (see Outline). Though the embryoid body method is convenient for inducing differentiation, it has several limitations in dissecting cellular and molecular mechanisms during differentiation such as: i) Difficulty to dissect the differentiation mechanisms by highlighting cells and signals of interest among the complicated cellular interactions in embryoid bodies; ii) Difficulty to directly observe differentiating cells at the cellular level by microscopy; iii) Inability to conduct single cell analysis of differentiation; iv) Difficulty in dissociating cell aggregates and obtaining single cell suspensions of induced cells. To overcome these difficulties in embryoid body cultures, we developed a novel ES cell differentiation system employing 2-dimensional culture conditions.

Vascular endothelial growth factor receptor-2 (VEGFR-2, also known as Flk-1) is designated as the earliest functional differentiation marker for blood and endothelial cells (Yamaguchi et al. 1993). VEGFR-2-deficient mice failed to develop blood and endothelial cells (Shalaby et al. 1995). As VEGFR-2 is expressed not only in progenitors for endothelial and blood cells but also in the lateral plate mesoderm in the early murine embryo, VEGFR-2 can be considered a mesoderm marker (Kataoka et al. 1997). Mural cells in blood vessels are thought to originate in the mesoderm as well as in the neural crest and from proepicardial cells. These findings imply that both vascular cell types, endothelial and mural cells, can differentiate from the VEGFR-2$^+$ cell population.

We focused on the significance of VEGFR-2 in vascular development and established a novel in vitro differentiation system for blood vessels

Springer Lab Manual
H. Augustin (Ed.)
Methods in Endothelial Cell Biology

(Yamashita et al. 2000) including both endothelial cells (Nishikawa et al. 1998) and mural cells from ES cells. We employ towards this end a differentiation method without embryoid bodies (i.e., 2-dimensional culture) and FACS (Fluorescence Activated Cell Sorting) using anti-VEGFR-2 antibody to obtain progenitor cells at an intermediate stage of differentiation into endothelial cells (Hirashima et al. 1999). We first induce VEGFR-2-positive (R2+) cells from ES cells and then purify R2+ cells by FACS. Vascular cells are then induced by several culture conditions using purified R2+ cells as starting material (Fig. 1). All cellular components of blood vessels (i.e., endothelial cells, mural cells, and blood cells) can be differentiated from the common progenitor R2+ cells, and vascular structures can be formed both in vitro and in vivo. Our differentiation system enables the observation and dissection of endothelial differentiation from a new point of view.

Outline

Brief summary of embryoid body method:
1. Harvest undifferentiated ES cells with trypsin/EDTA and suspend with differentiation medium.

2. Plate into petri dishes or glass-bottom dishes at the density of 2×10^5 cells/ml. Small embryoid bodies will be visible within 24 h.

3. Change medium every 3 or 4 days.

4. (For many applications): Transfer aggregates to culture dishes and culture them in adherent condition to induce further differentiation.

Brief summary of our novel in vitro differentiation system for blood vessels (see Fig. 1):

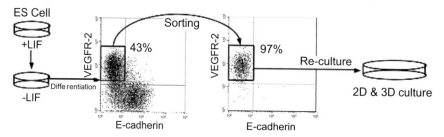

Fig. 1. In vitro blood vessel differentiation system using ES cells. Mouse ES cells remain undifferentiated in the presence of LIF (leukemia inhibitory factor). When cells are cultured in the absence of LIF, they start to differentiate spontaneously even under 2-dimensional culture condition. Approximately 4 days after induction, R2+/E-cadherin (undifferentiated ES cell marker)-negative cells appear (approximately 40 % of total cells). R2+/E-cadherin⁻ cells are purified to high purity (> 95 %) by FACS using anti-VEGFR-2 antibody. Purified R2+ cells are re-cultured under various conditions to examine the vascular cell differentiation

▓ Materials

Cells

- ES cells: Use ES cells maintained under feeder-free conditions to avoid contamination by feeder cells. Various kinds of ES cell lines can be applied to this culture system. (e.g., CCE, R1, J1, D3, E14 derivatives, CGR8, etc.).
- OP9 cells: Stroma cell line established from a newborn B6C3F1-op/op mouse calvaria (Kodama et al. 1994). The op/op mice lack M-CSF (macrophage colony stimulating factor).
- F-2 cells: Tumorigenic murine vascular endothelial cell line established from an ultraviolet-induced tumor (Toda et al. 1990).

Media

ES cell maintenance medium: Refer to maintenance protocol of each individual ES cell line.
- For E14 derivatives: GMEM (Glasgow minimum essential medium, Cat. No. 11710-035, GibcoBRL (Invitrogen), www.gibcobrl.com; www.invitrogen.com)
- 1 % FCS
- 10 % Knock-out serum replacement (Gibco, Cat. No.10828-028)
- 0.1 mM 2-mercaptoethanol (Gibco, Cat. No. 21985-023)
- 0.1 mM non-essential amino acids (Gibco, Cat. No. 11140-050)
- 1 mM sodium pyruvate (Cat. No. S8636, Sigma, www.sigmaaldrich.com)
- 50,000 units/L penicillin and 50 mg/L streptomycin

ES cell differentiation medium:
αMEM (minimum essential medium alpha medium, Gibco, Cat. No. 11900-024)
- 10 % FCS, 0.05 mM 2-mercaptoethanol
- 50,000 unit/L penicillin and 50 mg/L streptomycin

OP9 maintenance medium: αMEM
- 20 % FCS
- Penicillin-streptomycin

Note: Cell growth and differentiation are greatly affected by serum lots. Every researcher must select appropriate lots of serum for ES cell maintenance, ES cell differentiation, and OP9 maintenance according to the purpose (see Troubleshooting).

SFO3 medium: S-clone (Sanko Junyaku, Cat. No. SS1303), 0.1 % BSA, 0.05 mmol/ of 2-mercaptoethanol

Antibodies

- Anti-VEGFR-2 antibody (AVAS12α1, Cat. No. 555307, PE-conjugated Cat. No. 555308, BD Pharmingen, BD Biosciences, www.bdbiosciences.com)
- Anti-E-cadherin antibody (ECCD2) (Shirayoshi et al. 1986)
- Anti-CD31 antibody (Mec13.3, BD Pharmingen, Cat. No. 553370, 1:200)
- Anti-smooth muscle actin antibody (1A4, Sigma, Cat. No. A-2547, 1:400)
- Goat anti-rat IgG-alkaline phosphatase conjugated (Cat. No. 62-9522, 1:200, Zymed, www.zymed.com)
- Goat anti-mouse IgG-HRP conjugated (Zymed, Cat. No. 62-6520, 1:200)

Reagents

- Leukemia inhibitory factor (Gibco, Cat. No. ESG1107)
- human VEGF165 (R&D, Cat. No. 293-VE, R&D, www.rndsystems.com)
- human VEGF121 (R&D, Cat. No. 298-VS)
- human PlGF (R&D, Cat. No. 264-PG)
- human PDGF-BB (Gibco, Cat. No. 13244-033)
- Cell dissociation buffer (Gibco, Cat. No. 13150-016)
- 0.25 % Trypsin-EDTA (Gibco, Cat. No. 25200-072)
- 10× HBSS (Hanks' balanced salt solution, Gibco, Cat. No. 14185-052)
- BSA (bovine serum albumin, fraction V, Sigma, Cat. No. A-7906)
- Propidium iodide (Sigma, Cat. No. P-4170)
- normal mouse serum (Cat. No. M9-P20, OEM Concepts, www.oemconcepts.com, can be substituted by lab-made mouse serum)
- Type-I collagen gel (Cell Matrix™ typeI-A, Nitta Gelatin)

Supplies

- Collagen IV-coated culture plate or dish (Beckton Dickinson, www.bd.com)
- 10 cm-diameter dish (Beckton Dickinson, Cat. No. 35 4453)
- 6-well plate (Beckton Dickinson, Cat. No. 35 4428)
- 24-well plate (Beckton Dickinson, Cat. No. 35 4430)
- 96-well plate (Beckton Dickinson, Cat. No. 35 4429)
- Petri dish (Cat. No. 35 1007 or E. coli grade dishes)
- Cell strainer (40 μm pore, Beckton Dickinson, Cat. No. 35 2340)
- Tip-cut 1000 μl pipet tips (lab-made)

Equipment

- FACS Vantage (Beckton Dickinson)
- Clonecyte (Beckton Dickinson)

▨ Procedure

Induction of R2+ cells from ES cells

1. Dissociate and collect undifferentiated ES cells with trypsin treatment, and suspend with differentiation medium.

2. Plate ES cells onto type IV collagen-coated dish.
    – 1×10^4 cells for one well of 6-well plate. Add 3 ml of differentiation medium/well.
    – 6–10×10^4 cells for 10 cm-diameter dish. Add 18 ml of differentiation medium/dish.
    Incubate at 37 °C with 5 % CO_2 for 96–108 h.
    Note: OP9 cell feeder layer can substitute for collagen. Culture conditions are the same as above.

3. Aspirate supernatant and wash twice with Ca^{++}/Mg^{++}-free PBS.
    Dissociate cells with trypsin treatment or with cell dissociation buffer (GIBCO) followed by mechanical stress by several-times pipetting to obtain single cell suspension.

4. Filter cells with cell strainer (40 μm pore) to remove undissociated cell aggregates.
    Centrifuge at 1200 rpm for 5 min. Discard supernatant.
    Suspend in 10 ml of differentiation medium and count cell number.

5. Incubate the cells at 37 °C with 5 % CO_2 for 30 min to allow for VEGFR-2 antigen recovery.
    Note: This step can be skipped when cells were dissociated without trypsin, i.e., non-enzymatic dissociation or with weaker enzymes such as dispase or collagenase.

6. Centrifuge 1200 rpm for 5 min. Discard supernatant.
    Suspend cells with mouse serum (1×10^8 cells/ml). Place on ice for 20 min for blocking non-specific protein binding.

7. Add appropriate volume of fluorescent-labeled anti-VEGFR-2 antibody and anti-E-cadherin antibody (E-cadherin is used as a marker for undifferentiated ES cells) and mix gently by pipetting. Place on ice for 20 min.

8. Wash with HBSS/1 % BSA twice. Suspend cells with HBSS/1 % BSA containing 2.5 μg/ml of propidium iodide (approximately at the density of 10^6 cells/ml).
    Apply cells to FACS analysis and sort $R2^+$/E-cadherin⁻ population.

Please refer issues concerning set-up and details of FACS analysis to other technical guides for FACS (e.g., Chap. 15).

Induction of endothelial cells from R2+ cells

A   Induction on collagen-coated dish

1. Suspend purified $R2^+$/E-cadherin⁻ cells with differentiation medium and plate them onto type IV collagen (or 0.1 % gelatin)-coated 24-well at a cell density of $1–2 \times 10^4$ cells/cm².

2. Culture the cells in 500 µl of medium/well with 50 ng/ml of VEGF at 37 °C with 5 % CO_2. Change medium every 2 days (see Fig. 2).

Note: Endothelial cells can be observed from day 1 in culture.

B   Induction on OP9 stroma cells

1. Plate purified $R2^+$ cells on confluent OP9 feeder layer at the cell density of $0.5–1.5 \times 10^3$ cells/cm².

2. Culture the cells in 500 µl of differentiation medium/well (24-well dish). No VEGF administration is required for endothelial cell induction.

Alternative method:

1. Plate undifferentiated ES cells on OP9 feeder layer (1×10^4 cells/well (6-well plate)), and culture for 5–7 days.

Note: Endothelial cells appear 5 days after plating the ES cells on OP9.

C   Serum-free culture of R2+ cells

As serum includes various kinds of growth factors, it is difficult to dissect the signals that contribute to cell differentiation in serum-containing culture condition. Serum-free culture is required to precisely define the factors required for differentiation.

1. Plate purified $R2^+$ cells on type IV collagen-coated dish at a density of $1–2 \times 10^4$ cells/cm² with serum free medium SFO3 (Sanko Junyaku) with 0.1 % BSA and 0.05 mM 2-Mercaptoethanol.

2. Add 50 ng/ml of VEGF and/or 10 ng/ml of PDGF-BB for endothelial cell or mural cell induction, respectively.

3. Change medium at least every 2 days.

Note: Endothelial cells predominate in VEGF-stimulated conditions, but some mural cells co-exist. Under PDGF-BB-administration conditions, no endothelial cells are observed.

Even with repetitive administration of VEGF, endothelial cell viability is reduced after 3 days of culture, especially at lower cell density. Administration of VEGF-E instead of VEGF is effective for prolonged endothelial cell survival and growth even in serum-free conditions (Hirashima et al. 2003).

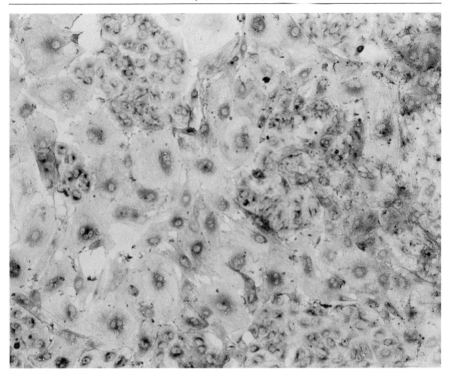

Fig. 2. Vascular cell differentiation from R2+ cells. CD31+ endothelial cell sheets (*dark*, small) and surrounding α-SMA+mural cells (*light*, large) appeare when R2+ cells are cultured with 50 ng/ml of VEGF and 10 % serum for 3 days (cells are double immunostained with the mixture of Mec13.3 and 1A4 followed by the secondary antibodies anti-rat IgG alkaline phosphatase-conjugated and anti-mouse IgG HRP-conjugated). Note that almost all cells become either endothelial cells or mural cells under these culture conditions

D Single cell culture of R2+ cells

To determine the differentiation potential of stem/progenitor cells, it is necessary to analyze cell differentiation at the single cell level. R2+ cells can survive and grow on type IV collagen-coated dishes even from a single cell.

1. Prepare type IV collagen-coated 96-well dish containing 100 µl of differentiation medium with 50 ng/ml VEGF.

2. Sort single R2+ cells into every well using the Clonecyte™ system (Beckton Dickinson).

3. Culture for 3–4 days and double stain with endothelial cell markers (e.g., CD31) and mural cell markers (e.g., smooth muscle actin; 1A4).

In vitro vascular formation in 3-dimensional culture of R2+ cells

Induction of cell differentiation is merely the first step of organ development or regeneration. Formation of complex structures as a functional organ is a much more complicated process. We succeeded in reproducing early vascular organization processes in vitro using 3-D culture of R2$^+$ cells. R2$^+$ cells differentiate into both endothelial and mural cells and form 3-dimentional vascular structures in vitro. To obtain highly organized structures such as blood vessels, it is necessary to start the culture from R2+ cell aggregates to increase cell-cell interactions during vascular structure formation.

1. Sort and purify R2$^+$ cells.

2. Culture purified R2$^+$ cells in non-adherent conditions. Plate R2$^+$ cells onto a petri dish or glass bottom dish at a density of 4×10^5 cells/ml in differentiation medium with 50 ng/ml of VEGF for 12–18 h. R2$^+$ cells will adhere each other and form cell aggregates.

3. Collect and transfer cell aggregates to 15 ml Falcon tube with tip-cut 1000 µl tip to avoid destruction of cell aggregates.
    Allow the aggregation to sink to the bottom for 15–30 min.

4. Aspirate supernatant and resuspend aggregates at $2–5 \times 10^5$ cells/ml with 2× differentiation medium (differentiation medium made at twice the concentration) and pipett gently using tip-cut 1000 µl tips. Place tube on ice.

5. Add isovolume of type I collagen gel (Cell Matrix™ typeI-A, Nitta Gelatin, 3 mg/ml) to the aggregate solution and mix gently but thoroughly until aggregate solution and gel are homogenized by tip-cut 1000 µl tip. (final collagen concentration is 1.5 mg/ml).

6. Apply 300 µl of gel solution to the well (24-well).

7. Incubate at 37 °C with 5 % CO_2, 15–30 min to allow gel polymerization.

8. Add 700 µl of differentiation medium with VEGF (final concentration 50 ng/ml).

9. Change medium every 2 days.

Note: Cells start to migrate out after 2–3 days, and vascular structures are formed after approx. 5–7 days of culture.

Expected Results

Vascular cell differentiation from R2+ cells

When R2+ cells are cultured in differentiation medium with VEGF on type IV collagen-coated dishes, both endothelial and mural cells appear. Fig. 2 shows double immunostaining of induced vascular cells with CD31 (dark, small) and smooth muscle actin (α-SMA). CD31+ endothelial cell sheets and surrounding α-SMA+ mural cells (light, large) are observed. Note that almost all cells are either endothelial cells or mural cells under these culture conditions.

Serum-free culture of R2+ cells leads predominantly to endothelial cell differentiation upon VEGF induction, whereas PDGF-BB stimulation induces mural cell differentiation. Simultaneous administration of VEGF and PDGF-BB induces both cell types (Fig. 3).

Single cell culture of R2+ cells demonstrates that three kinds of colonies, pure endothelial cells, pure mural cells, and mixture of both, appear from single R2+ cells (Fig. 4).

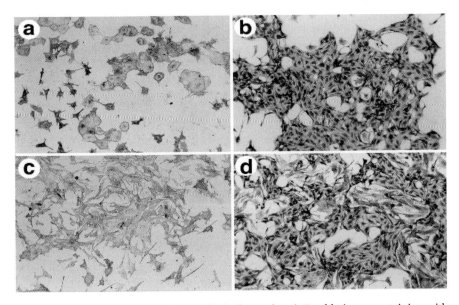

Fig. 3a-d. Serum-free culture of R2+ cells (3 days culture). Double immunostaining with CD31 and α-SMA. **a** Vehicle: Only few α-SMA+ cells survive when no growth factors are added; **b** VEGF (50 ng/ml): CD31+ endothelial cells predominate; **c** PDGF-BB (10 ng/ml): α-SMA+ cells proliferate and are spindle-shaped, resembling cultured vascular smooth muscle cells. No endothelial cells appeared under these conditions; **d** Simultaneous administration of VEGF and PDGF-BB: The frequency of both vascular cell types increases

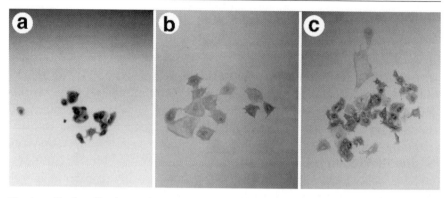

Fig. 4a-c. Single cell culture of R2⁺ cells (4 day culture, CD31/α-SMA staining). Three types of colonies are observed when single R2⁺ cells are cultured with VEGF and serum on type IV collagen-coated dishes. **a** endothelial cells only, **b** mural cells only, **c** mixture of both

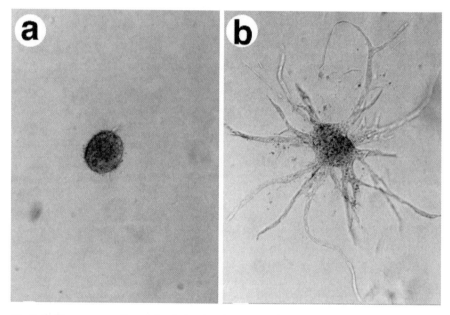

Fig. 5a,b. Time course of vascular formation from R2⁺ cell aggregate in 3-dimensional type I collagen culture. **a** R2⁺ cell aggregate, Day 0; **b** Sprouting of vascular cells after 5 days of culture

In vitro vascular formation in 3-D culture

When an aggregate of R2+ cells is cultured in a 3-dimensional type I collagen gel, cells start to migrate out from the aggregate within 2–3 days and vascular-like structures are formed after approximately 5 days (Fig. 5). Double immunostaining demonstrates that the structure consists of CD31+ endothelial cell tubes with attached α-SMA+ mural cells similar to the structure of a capillary vessel in vivo (data not shown).

Troubleshooting

Low percentage of R2+ cells upon induction of ES cell differentiation

Various parameters are involved in the efficacy of R2+ cell induction. Every researcher should carefully optimize induction conditions.

▸ Cell clones: Differentiation properties of ES cells are very different not only among cell lines but also among cell clones even from the same cell line. Good clones of following cell lines should give rise to R2+ cells with the indicated efficiency; CCE: 30–40+%; R1, J1, D3, CGR8: 15–30+%; E14 derivatives: 10–30+%.

▸ Serum lots: Induction potential up to 40 % of R2+ cells is largely affected by the serum (from several percentage to 40 %). We usually test approximately 20–30 lots of serum and choose good batches by cell growth and percentage of R2+ cells (i.e., yield of R2+ cell collection).

▸ Culture period: The percentage of R2+ cells rapidly increases after 96 h of differentiation induction. Thus, the culture period requires little more than 96 h. As differentiated endothelial cells (i.e., VE-cadherin+ or CD31+ cells) are observed among R2+ cells after 108 h, the differentiation time should not exceed 108 h.

▸ Cell number: Growth rate of ES cells is different among the clones, and cell density during differentiation affects differentiation efficacy. We usually test the cell number at the starting point ranging from 1–3 × 10⁴ cells/well (6-well dish) and optimize the initial cell density.

▸ Medium: Freshness of the differentiation medium is important for differentiation (especially 2-mercaptoethanol). Differentiation efficacy is reduced when old medium (more than 1 month old) is used. A large amount of medium (e.g., 18 ml per 10 cm-diameter dish) seems to be beneficial for differentiation.

▸ Extracellular matrices (ECM) & feeder cells: Type IV collagen has the most efficient R2+ cell induction capacity among different ECM that we test (gelatin, type I or IV collagen and fibronectin). OP9 feeder cells should be effective to induce R2+ cells from ES cell clones that do not give rise to R2+ cells on ECM.

▶ FACS (staining failure): Every researcher must optimize the staining procedure to obtain good results. In particular, cell dissociation method and amount of the antibody greatly affect the quality of staining. Mouse endothelial cell lines, such as F2 cells (or other endothelial cell lines), should be used as positive controls for VEGFR-2 staining.

Poor endothelial cell differentiation of R2+ cells

VEGF is essential for the differentiation of endothelial cells in this culture system. Responsiveness of cells to VEGF varies among different ES clones. Usually, 30–50 ng/ml VEGF is sufficient for induction. 100 ng/ml has maximum effect. Lower concentrations may cause less endothelial cell development. VEGF165 (human) is more potent than VEGF121 (human) to induce endothelial cell differentiation. PlGF has no effect. Different extracellular matrices (e.g., type I collagen, type IV collagen, gelatin, fibronectin, etc.) affect the shape of endothelial cell sheets but have almost no effect on induction efficacy.

Poor vascular structure formation in 3-D culture

▶ Presence of VEGF is essential to form vascular structures in 3D cultures. VEGF needs administration to be supplied from the beginning of the floating culture of R2+ cells.

▶ Lot and concentration of the collagen gel seem to affect vascular structure formation. Higher concentrations seem to be better for tube formation, but cause higher background in immunostaining. Selection of a good batch of collagen is also required to obtain reproducible results.

References

Hirashima M, Ogawa M, Nishikawa S, Matsumura K, Kawasaki K, Shibuya M, Nishikawa SI (2003) A chemically defined culture of VEGFR2+ cells derived from embryonic stem cells reveals the role of VEGFR1 in turning the threshold for VEGF in developing endothelial cells. Blood 101:2261–2267

Hirashima M, Kataoka H, Nishikawa S, Matsuyoshi N, Nishikawa S (1999) Maturation of embryonic stem cells into endothelial cells in an in vitro model of vasculogenesis Blood 93:1253–1263

Kataoka H, Takakura N, Nishikawa S, Tsuchida K, Kodama H, Kunisada T, Risau W, Kita T, Nishikawa SI (1997) Expressions of PDGF-receptor alpha, c-kit, and FLK1 genes clustering in mouse chromosome 5 define distinct subsets of nascent mesodermal cells. Dev Growth Differ 39: 729–740

Kodama H, Nose M, Niida S, Nishikawa S, Nishikawa S (1994) Involvement of the c-kit receptor in the adhesion of hematopoietic stem cells to stroma cells. Exp Hematol 22:979–984

Nishikawa SI, Nishikawa S, Hirashima M, Matsuyoshi N, Kodama H (1998) Progressive lineage analysis by cell sorting and culture identifies FLK1+VE-cadherin- cells at a diverging point of endothelial cells and hemopoietic lineages. Development 125:1747–1757

Rathjen J, Rathjen PD (2001) Mouse ES cells: experimental exploitation of pluripotent differentiation potential. Curr Opin Genet Dev 11:587–594

Shalaby F, Rossant J, Yamaguchi TP, Gertsenstein M, Wu XF, Breitman ML, Schuh AC (1995) Failure of blood-island formation and vasculogenesis in Flk1-deficient mice. Nature 376:62–66

Shirayoshi Y, Nose A, Iwasaki K, Takeichi M (1986) N-linked oligosaccharides are not involved in the function of a cell-cell binding glycoprotein E-cadherin. Cell struct funct 11:245–25

Toda K, Tsujioka K, Maruguchi Y, Ishii K, Miyachi Y, Kuribayashi K, Imamura S (1990) Establishment and characterization of a tumorigenic murine vascular endothelial cell line (F-2). Cancer Res 50:5526–5530

Yamashita J, Itoh H, Hirashima M, Ogawa M, Nishikawa S, Yurugi T, Naito M, Nakao K, Nishikawa SI (2000) Flk1-positive cells derived from embryonic stem cells serve as vascular progenitors. Nature 408:92–96

Yamaguchi TP, Dumont DJ, Conlon RA, Breitman ML, Rossant J (1993) Flk1, a flt1-related receptor tyrosine kinase is an early marker for endothelial precursors. Development 118:489–498

Chapter 5

Detection, Isolation and Culture of Lymphatic Endothelial Cells

Taija Mäkinen and Kari Alitalo

Introduction

Lymphatic vessels are essential for the maintenance of normal tissue fluid balance and immune surveillance, but they also provide a pathway for metastasis in many types of cancers (reviewed in Alitalo and Carmeliet 2002). In spite of the importance of lymphatic vessels in medicine, the cell biology of this part of the vascular system has received little attention until recently. Only few lymphatic endothelial cell lines have been available for molecular biological studies, and these were mainly derived from lymphatic tumors. However, the identification of lymphatic specific markers during the past few years and the isolation and maintenance of primary cultures of lymphatic endothelial cells have enabled studies of the molecular properties of these cells.

At the molecular level, several lymphatic specific markers have been identified, including the VEGFR-3 receptor tyrosine kinase, the Prox-1 transcription factor, the hyaluronan receptor LYVE-1, and the membrane mucoprotein podoplanin (Kaipainen et al. 1995, Banerji et al. 1999, Breiteneder-Geleff et al. 1999, Wigle and Oliver 1999). These genes are also found to be selectively expressed in cultured lymphatic (LEC) but not in blood vascular endothelial cells (BEC), and antibodies against podoplanin, VEGFR-3, or LYVE-1 have been used both for the detection of lymphatic endothelium in tissues and in cell culture as well as for the isolation of the LECs (Kriehuber et al. 2001, Mäkinen et al. 2001, Petrova et al. 2002, Hirakawa et al. 2003).

Different isolation procedures described in the literature include fluorescence-activated cell sorting (FACS) and magnetic separation employing antibodies against lymphatic endothelial cell specific proteins. Kriehuber et al. (2001) used multicolor flow cytometry and antibodies against the endothelial cell-specific molecules CD31 or CD34 and against the lymphatic endothelial specific podoplanin antigen for the separation of the BECs and LECs from freshly prepared dermal cell suspensions. We describe here the immunomagnetic separation of BECs and LECs from microvascular

Springer Lab Manual
H. Augustin (Ed.)
Methods in Endothelial Cell Biology
© Springer-Verlag Berlin Heidelberg 2004

endothelial cells. Immunomagnetic isolation of LECs has been successfully carried out using antibodies against VEGFR-3 or podoplanin (Mäkinen et al. 2001), or against LYVE-1 (Hirakawa et al. 2003).

Materials

Cells

- Microvascular endothelial cells (human cells available, e.g., from Promocell, www.promocell.com, or Clonetics [Cambrex], www.cambrex.com)

Reagents

- Trypsin-EDTA (TE; 0.05 % trypsin, 0.53 mM EDTA)
- Endothelial cell culture medium (e.g., from PromoCell, Cat. No. C-22220 and C-39220; Endothelial Cell Basal Medium supplemented with 5 % Fetal Calf Serum, Endothelial Cell Growth Supplement/Heparin (0.4 %), EGF (10 ng/ml), Amphotericin B (50 ng/ml), Gentamicin (50 µg/ml) and Hydrocortisone [1 µg/ml]).
- Suitable antibodies against lymphatic endothelial cell-specific surface antigen. The antibodies that we have used for LEC purification (podoplanin, VEGFR-3) are described in more detail in Mäkinen et al. (2001).
- Magnetic microbeads conjugated to appropriate secondary antibodies (e.g., from Miltenyi Biotech, www.miltenyibiotec.com)
- Fibronectin (e.g., from Sigma, www.sigmaaldrich.com)
- Phosphate-buffered saline (PBS)
- Bovine serum albumin (BSA)
- EDTA
- Optional: recombinant VEGF-C protein (Cat. No. RDI-R20-014, Research Diagnostics Inc., www.researchd.com) or VEGF-D (R&D Systems, www.rndsystems.com), required for optimal growth of the LECs. VEGF-C (C156S; the VEGFR-3 specific mutant form of VEGF-C) is also commercially available from R&D Systems.

Equipment

- Magnetic cell separator [e.g., MiniMACS and MidiMACS separators from Miltenyi Biotech, similar magnetic cell separation systems are available from Dynal Biotech, BD Biosciences, Qiagen (www.qiagen.com), and StemCell Technologies (www.stemcell.com)]
- Magnetic columns (e.g., LD and MS columns from Miltenyi Biotech)
- Pre-separation filters, mesh size 30 µm (e.g., from Miltenyi Biotech)

▓ Procedure

The method given here is based on the magnetic cell separation equipment, reagents and protocols supplied by Miltenyi Biotech. Similar magnetic separation systems are also available from other sources (e.g., Dynal Biotech, Qiagen, BD Biosciences, and StemCell Technologies). Before starting, verify the presence of LECs in the endothelial cell cultures using standard immunofluorescence methods and appropriate LEC-specific antibodies. Confirm the suitablility and titer the optimal concentration of the antibody, e.g., by FACS analysis. All steps should be done using sterile equipment and solutions.

Cell staining and magnetic labeling

1. Detach the cells with trypsin-EDTA. Use a minimal amount of time to avoid excessive digestion and irreversible damage of the cells. If using antibodies against a protease-sensitive antigen, you may have to use another method for detaching the cells, such as EDTA treatment.

2. Resuspend the cells in culture medium.

3. Centrifuge the cells at 300 × g for 5 min.

4. Remove the supernatant by suction and wash the cells with cold PBS.

5. Repeat centrifugation and resuspend the cells in cold PBS.

6. Count the cells.

7. Centrifuge the cells again and resuspend in cold PBS-BE (PBS + 0.5 % BSA + 2 mM EDTA). Use a small volume of PBS-BE (i.e., 100 µl per 10^7 cells). Add the appropriate antibody. The concentration of each antibody has to be titered for the approximate final staining volume and cell concentration.

8. Incubate on ice for 10–15 min, mix occasionally by flicking the tube.

9. Wash twice with cold PBS-BE (>20 × vol).

10. Resuspend in cold PBS-BE and add the appropriate antibody-conjugated magnetic microbeads according to the instructions of the manufacturer, e.g., resuspend in 80 µl of buffer per 10^7 total cells and add 20 µl of magnetic beads.

11. Incubate in a refrigerator (6–12 °C) for 15 min, mix occasionally by flicking the tube.

12. Wash the cells twice with cold PBS-BE.

13. Resuspend the cells in 1 ml of cold PBS-BE.

Magnetic separation

a) Depletion column LD

1. Assemble the column into MidiMACS separator and place a preseparation filter on top of the column.

2. Apply 2 ml of cold PBS-BE on the column and pre-separation filter, and let it run through. Discard the effluent.

3. Add the cells (1 ml).

4. Wash the column with 2 × 1 ml of cold PBS-BE and collect the total effluent as depleted fraction (= **BECs**).

5. Remove the column from the magnet and place it on a suitable collection tube. Add 3 ml of cold PBS-BE and flush out the cells with a plunger.

6. Repeat the elution.

7. Centrifuge the cells (6 ml) and resuspend in 500 µl of cold PBS-BE.

b) Positive selection column MS

1. Assemble the column into MiniMACS separator.

2. Apply 500 µl of cold PBS-BE on the column and let it run through. Discard the effluent.

3. Add the cells (500 µl).

4. Wash the column with 3 × 500 µl of cold PBS-BE and collect the total effluent (=mixture of BECs and LECs).

5. Remove the column from the magnet and place it on a suitable collection tube. Add 1 ml of cold PBS-BE and flush out the cells (= **LECs**) with a plunger.

6. Centrifuge the cells, resuspend in complete culture medium and count.

Optional

If the main interest is to get pure LECs whereas the purity of the BEC population is not essential, you may use two MS-positive selection columns instead of a combination of LD depletion and MS-positive selection columns. After labeling, resuspend the cells in 500 µl of the PBS-BE buffer and use the protocol for magnetic separation using the MS-positive selection column. After elution, apply the positive cell fraction into a new MS column to increase the purity of the LEC population. This procedure gives a pure population of the LECs, but the negative BEC cell population obtained as a flow-through usually contains up to 10–20 % of contaminating LECs.

Plating and culture

After separation, LECs adhere and proliferate better when cultured in the presence of extracellular matrix components, such as fibronectin or collagen. For comparison, you may need to culture the BECs in similar conditions. In addition, for the optimal growth of the LECs, add VEGF-C (or VEGF-D) to the culture medium.

1. Coat the cell culture plates with fibronectin:
   Dilute fibronectin in PBS (1 μg/ml) and leave on culture plates for 10 min at 37 °C. Remove by suction.

2. Plate the cells in a density of approx. $8-10 \times 10^3$ cells/cm^2.

3. For optimal growth of LECs, add 10–50 ng/ml of recombinant VEGF-C (or VEGF-D) in the culture medium.

4. Change the culture medium one day after separation. After this point, replace the medium every second day.

5. When the cells are about to reach confluence, they can be split at a 1:2 – 1:3 ratio. Detach the cells with trypsin-EDTA. Use a minimal amount of time to avoid excessive digestion and irreversible damage of the cells. Resuspend in culture medium and centifuge at $300 \times g$ for 5 min. Remove supernatant, resuspend in fresh culture medium and plate.

▨ Expected Results

We have used commercially available human dermal microvascular endothelial cells isolated from foreskin as a source for subsequent separation of the BECs and LECs. Staining for a pan-endothelial cell marker CD31/ PECAM-1 has been used to confirm the lack of contaminating non-endothelial cells in the cultures. The proportion of the LECs in these microvascular endothelial cell cultures varies, typically being 30–60 % (see Fig. 1).

We have usually carried out the separation procedure using $5-20 \times 10^6$ cells as a starting material. We have used both podoplanin (Breiteneder-Geleff et al. 1999; kindly provided by Dr. D. Kerjaschki, Vienna, Austria) and VEGFR-3 antibodies (Mäkinen et al. 2001) for the immunomagnetic isolation of the LECs. High antigen density and exclusive expression on LECs, as well as the resistance against trypsin digestion make podoplanin a particularly suitable target molecule.

After immunomagnetic separation, the resulting BEC and LEC populations are typically >99 % pure, which can be verified by immunofluorescence staining (Fig. 1) or FACS analysis. However, this has to be confirmed after

microvascular EC culture (CD31+)

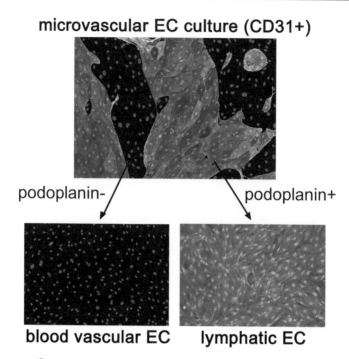

podoplanin-   podoplanin+

blood vascular EC   lymphatic EC

Fig. 1. Immunofluorescence staining of human dermal microvascular endothelial cells as well as immunomagnetically separated lymphatic and blood vascular endothelial cells using antibodies against a lymphatic specific marker, podoplanin. The nuclei are stained with Hoechst fluorochrome

each separation. If impure cell populations are obtained, the specificity and suitability of the antibody used has to be carefully evaluated, and the optimal antibody concentration should be titered. The effluent obtained from the MS-positive selection column is a mixture of BECs and LECs. These cells can be used as a control for subsequent experiments, or they can be used later on for another round of magnetic separation.

The proportions of the separated BECs and LECs of the total number of retrieved cells should reflect the ratio detected in the original cell culture used as a starting material. In our experience, the loss of the cells during the magnetic labeling and separation procedure is 10–15 %. From the retrieved cells, the proportions of BECs and LECs are typically within a range of 30–50 %, and the effluent from the MS-positive selection column is usually 15–20 % of the total cell number.

After plating, most of the cells (typically >90 %) should adhere, spread out on the bottom of the culture dish, and start to proliferate within 48 h after the separation. At this point, they grow as single cells or in small clusters. At confluence they should display a typical endothelial cell-like cobbelstone morphology. However, in the presence of VEGF-C, LECs display an elongated or spindle-shaped morphology (see Fig. 1).

Troubleshooting

Purity of the separated cell populations

The purity of the separated cell populations has to be verified after each isolation. If contaminating cells are found in either cell population, titer and readjust the antibody concentration carefully for an optimal result (see below). In addition, make sure that you do not exceed the capacity of the columns, which may lead to inefficient separation (see below). If contaminating BECs are detected in the LEC population, you may also increase the number of washing steps. Using higher temperatures or longer incubation times may lead to unspecific labeling.

Antibody

The magnetic labeling of the cells depends on the quality of the primary antibody. Verify the specificity and suitability of the antibody first using e.g. FACS analysis. The concentration of each antibody has to be titered for the approximate final staining volume and cell concentration. Too high concentrations or unspecific binding of the antibody can lead to impure LEC populations, while too low concentrations can result in insufficient labeling and contamination of BECs with unlabeled LECs. Depending on the antibody used, the antigenic epitope may be sensitive to trypsin digestion, in which case another cell detachment method, such as EDTA treatment, has to be used.

Columns

The material to be separated should be well suspended and should not contain clumps or aggregates, which may reduce the flow rate and therefore result in inefficient separation. To avoid this, use a 30 µm nylon mesh to filter the cell suspension before magnetic separation. Make sure that you follow the manufacturer's instructions with regard to the optimal cell concentration and the maximum capacity of the columns. Overloading of the columns may compromise efficient separation.

References

Alitalo K, Carmeliet P (2002) Molecular mechanisms of lymphangiogenesis in health and disease. Cancer Cell 1:219–227

Banerji S, Ni J, Wang SX, Clasper S, Su J, Tammi R, Jones M, Jackson DG (1999) LYVE-1, a new homologue of the CD44 glycoprotein, is a lymph-specific receptor for hyaluronan. J Cell Biol 144:789–801

Breiteneder-Geleff S, Soleiman A, Kowalski H, Horvat R, Amann G, Kriehuber E, Diem K, Weninger W, Tschachler E, Alitalo K, Kerjaschki D (1999) Angiosarcomas express mixed endothelial phenotypes of blood and lymphatic capillaries: podoplanin as a specific marker for lymphatic endothelium. Am J Pathol 154:385–394

Hirakawa S, Hong YK, Harvey N, Schacht V, Matsuda K, Libermann T, Detmar M (2003) Identification of vascular lineage-specific genes by transcriptional profiling of isolated blood vascular and lymphatic endothelial cells. Am J Pathol 162:575–586

Kaipainen A, Korhonen J, Mustonen T, van Hinsbergh VW, Fang GH, Dumont D, Breitman M, Alitalo K (1995) Expression of the fms-like tyrosine kinase 4 gene becomes restricted to lymphatic endothelium during development. Proc Natl Acad Sci U S A 92:3566–3570

Kriehuber E, Breiteneder-Geleff S, Groeger M, Soleiman A, Schoppmann SF, Stingl G, Kerjaschki D, Maurer D (2001) Isolation and characterization of dermal lymphatic and blood endothelial cells reveal stable and functionally specialized cell lineages. J Exp Med 194:797–808

Mäkinen T, Veikkola T, Mustjoki S, Karpanen T, Wise L, Mercer A, Catimel B, Nice EC, Kowalski H, Kerjaschki D, Stacker SA, Achen MG, Alitalo K (2001) Isolated lymphatic endothelial cells transduce growth, survival and migratory signals via the VEGF-C/D receptor VEGFR-3. EMBO J 20:4762–4773

Petrova TV, Mäkinen T, Mäkelä TP, Saarela J, Virtanen I, Ferrell RE, Finegold DN, Kerjaschki D, Ylä-Herttuala S, Alitalo K (2002) Lymphatic endothelial reprogramming of vascular endothelial cells by the Prox-1 homeobox transcription factor. EMBO J 21:4593–4599

Wigle JT, Oliver G (1999) Prox1 function is required for the development of the murine lymphatic system. Cell 98:769–778.

Chapter 6

Three-Dimensional Spheroid Culture
of Endothelial Cells

Thomas Korff

▨ Introduction

Endothelial cells (EC) line the inside of all blood vessels as a monolayer and contact smooth muscle cells (SMC) or pericytes (PC) through breaks in the underlying basement membrane (BM). EC do not proliferate under these conditions and are referred to as "quiescent". Primary EC populations are usually cultured in vitro as two-dimensional flat monolayers which reflect some of the properties of the endothelium in vivo (Pauly et al. 1992). Nevertheless, EC cultured in monolayer configuration are never completely quiescent and up to 10 % of EC proliferate even in a confluent monolayer. Furthermore, monolayer EC gradually dedifferentiate and lose their distinct organ-specific and caliber-specific morphological and functional characteristics. For example, CD34, which is constitutively expressed by most EC in vivo, is rapidly downregulated upon transfer into tissue culture (Delia et al. 1993). Similarly, cultured brain EC lose their tight-junction-dependent blood–brain barrier characteristics (Wolburg et al. 1994).

Three-dimensional cell culture models of tumor cells have greatly improved the understanding of tumor cell functions (Jacks and Weinberg 2002). Modifying techniques established for tumor cell spheroids (Müller-Klieser 1994, Kunz-Schughart and Müller-Klieser 2000), we have developed a spheroid-based culture model for EC (Korff and Augustin 1998) and a coculture model of EC and SMC which mimics the three-dimensional assembly of the normal vessel wall with a polarized surface monolayer of EC and a multilayered assembly of smooth muscle cells (SMC) (Korff et al. 2001). Three-dimensional culture of EC in spheroids leads to the formation of a completely quiescent surface monolayer and enhances the cells' differentiated phenotype, particularly if the cells are co-cultured with SMC. This chapter summarizes the techniques to generate cell number- and size-defined EC spheroids as well as EC/SMC coculture spheroids. Details of the procedures are also outlined at www.spherogenex.de.

Springer Lab Manual
H. Augustin (Ed.)
Methods in Endothelial Cell Biology
© Springer-Verlag Berlin Heidelberg 2004

Materials

Methyl cellulose stock solution

The preparation of methyl cellulose stock solution (methocel) is critical. If the concentration of methyl cellulose is too low or the solution contains methyl cellulose debris, single cells will stick to the wall of the culture plate and several small spheroids are formed in each well instead of a single uniform spheroid. For preparation of methocel, methyl cellulose with high viscosity should be used (methyl cellulose from Sigma, Cat. No. m-0512, 4,000 centipoises [cpi], www.sigmaaldrich.com):

1. Autoclave the pure powder (6 g) in a 500 ml flask containing a clean magnetic stirrer.

2. Dissolve the autoclaved methyl cellulose in preheated 250 ml basal medium (60 °C) for 20 min (using the magnetic stirrer).

3. Add 250 ml basal medium (at RT) to a final volume of 500 ml.

4. Mix the solution for 1–2 h (4 °C). Aliquot the final stock solution (50 ml aliquots) and clear by centrifugation (2,500 × g, 2 h, at RT).

5. Use only the clear, highly viscous supernatant for spheroid culture.

Non-adhesive 96-well round bottom plates

EC spheroids are generated in 96-well non-adhesive round bottom tissue culture dishes. The assay has been optimized for use with 96-well plates for suspension culture from Greiner (Cat. No. 650185, www.greiner-lab.com).

Cells

EC monolayers are trypsinized and the cells are suspended in corresponding culture medium containing 20 % methyl cellulose stock solution (see above) for the generation of cell number and size-defined EC spheroids. Make sure that the final suspension of EC is devoid of clustered cells and that the cells are homogenously suspended. The following cell types have been tested and found suitable for spheroid culture (other EC populations may be similarly suitable): HUVEC (human umbilical vein EC), HUAEC (human umbilical arterial EC), HAoEC (human aortic EC), HSaVEC (human saphenous vein EC), HDMVEC (human dermal microvascular EC), BAEC (bovine aortic EC), and PAEC (porcine aortic EC), as well as HUASMC (human umbilical artery SMC), HAoSMC (human aortic SMC), PASMC (porcine aortic SMC), or combinations of EC and SMC. Because of their endogenous FGF-2 production (BAEC, PAEC) or embryonic origin (HUVEC, HUAEC), these EC form more robust spheroids than ECs from adult vascular tissue (HSaVEC, HAoEC, HDMVEC).

Procedure

Depending on the specific application, the spheroids should contain different numbers of cells: for morphological/biochemical analyses (paraffin-embedding, Western blot, RNA analysis), each spheroid should contain 2250 to 3000 cells. Spheroids to be embedded into collagen gels for functional analysis (i.e., sprouting assays) should contain 400 to 750 cells each.

EC spheroids

Protocol to generate 4×96 spheroids containing 3,000 EC/spheroid:

1. Trypsinize a confluent EC monolayer.

2. Suspend these cells in 10 ml methocel containing medium (20 % methocel stock solution (see above), 80 % culture medium [supplements/FCS: cell type-dependent]).

3. Count the cells and suspend approx. 1.2×10^6 (4 plates $\times$ 96 wells $\times$ 3000 cells/well) in 40 ml (4 plates $\times$ 96 wells $\times$ 100 µl/well) methocel containing medium.

4. Distribute the cell suspension into four 96-well plates using a twelve-channel pipette (100 µl/well).

5. Incubate the plates at 37 °C (5 % CO_2, 100 % humidity) for at least 24 h.

EC/SMC coculture spheroids

Protocol to generate 4×96 spheroids containing 1,500 HUVEC and 1500 HUASMC per spheroid:

1. Trypsinize the smooth muscle cells and the endothelial cells.

2. Suspend both cell populations in 10 ml methocel containing medium (20 % methocel stock solution (see above), 80 % culture medium [supplements/FCS: cell type-dependent]).

3. Count the cells and suspend approx. 6×10^5 EC and 6×10^5 SMC (4 plates $\times$ 96 wells $\times$ 1,500 EC and 1,500 SMC/well) in 40 ml (4 plates $\times$ 96 wells $\times$ 100 µl/well) methocel-containing medium.

4. Distribute the cell suspension into four 96-well plates using a 12-channel pipette (100 µl/well).

5. Incubate the plates at 37 °C (5 % CO_2, 100 % humidity) for at least 24 h.

Harvesting of spheroids

Spheroids can be harvested from 96-well plates using standard pipette tips (1 ml) transferring them into 15-ml or 50-ml tubes. To ensure that the spheroids are not sheared during transfer, it is advisable to widen the tip hole by cutting away 1–2 mm of the pipette tip. Spheroids are collected by centrifugation ($200 \times g$, 2–3 min).

Processing of spheroids for paraffin sectioning

1. Harvest spheroids as described above in 15-ml tubes (at least 2×96 spheroids per experimental group).

2. Centrifuge ($500 \times g$, 3 min) and remove the supernatant

3. Fix the spheroid pellet for 12–24 h (4 °C) in freshly prepared buffered paraformaldehyde (4 %) -containing fixative (dependent on the cell type and specific application, different fixatives may be tested).

4. Centrifuge the spheroids and remove the fixative.

5. Wash the spheroids in water (10 ml per tube) for 30 min.

6. Centrifuge again, remove the supernatant, suspend the spheroids in 70 % ethanol and incubate for 45–60 min at RT.

7. Repeat this step with 85 % ethanol, 96 % ethanol, and collect the spheroids in 1.5 ml isopropanol.

8. Carefully transfer the spheroid pellet together with 1 ml isopropanol in a 1.5-ml safe-lock cup. Allow the spheroids to sediment and remove the supernatant.

9. Fill the cup with 1.4 ml 65 °C low melting paraffin (melting temperature: 44–48 °C), close the cup and place it into an oven (65 °C) for 15 min.

10. Invert the cup several times to suspend the spheroids in the paraffin. Make sure that the tip of the cup is oriented towards the bottom and incubate it for 12 h at 65 °C.

11. Place the tip of the cup into ice-cold water until the paraffin becomes solid (the spheroids will be trapped in the solid phase) and discard the supernatant (note that only a small amount of low melting paraffin (ca. 20–50 µl) should remain in the cup. Otherwise it will disturb the solidification of high melting paraffin).

12. Fill the cup with 1.4 ml of 65 °C high melting paraffin (paraplast, melting temperature: 56–58 °C) and incubate for 15 min at 65 °C. Invert the cup

several times to suspend the spheroids. Make sure that all spheroids are suspended in the paraffin as the high melting paraffin may rapidly solidify during this procedure.

13. Allow the paraffin to cool down to room temperature.

14. Cut the tip of the cup (5–7 mm) with a scalpel. The paraffin plug can easily be removed with tweezers. The spheroids are located at the bottom of the paraffin plug.

15. Embed the paraffin plug in a paraffin block in such a way that the spheroids are located at the cutting edge. Since the spheroids are located only in a thin layer inside the upper part of the paraffin block, the paraffin sections should be monitored microscopically during sectioning to ensure that the spheroids are not cut away.

Processing of spheroids for cryostat sectioning

1. Add 500 µl of silicone solution (i.e., Sigmacote [SL-2] from Sigma) to a 15-ml tube and vortex for 5 s. Remove the solution and place the tube upside down for 15 min. Make sure that the tube is completely dry prior to use.

2. Harvest spheroids as described above in the siliconized 15-ml tube (at least 2 × 96 spheroids per experimental group).

3. Centrifuge (200 × g, 3 min) and remove the supernatant.

4. Add 250 µl of a cryo-freezing solution [1 ml PBS stained with Methylene Green (e.g., Methylene Green [M-7766] from Sigma) + 1 ml Tissue Freezing Medium (e.g., TissueTek)] to the spheroid pellet.

5. Vortex the tube briefly (1 s). Make sure that the spheroids stay in their position at the bottom of the tube.

6. Freeze the pellet in liquid nitrogen.

7. Carefully remove the frozen plug by knocking the tube against the table.

8. Embed the frozen plug in tissue freezing medium in such a way that the spheroids are located at the cutting edge. During sectioning, the green color of the frozen plug should indicate the position of the spheroids. Since the spheroids are located only in a thin layer inside the upper part of the cryo block, the cryo sections should be monitored microscopically during sectioning to ensure that the spheroids are not cut away.

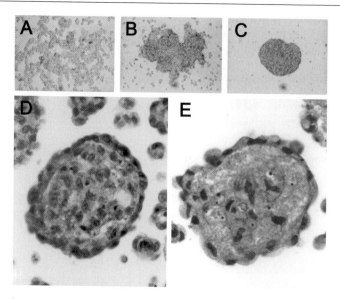

Fig. 1. A-C Phase contrast microscopy of the formation of 3 D EC spheroids. A defined number of single EC is suspended into 96-well non-adhesive round bottom tissue culture plates (**A**). The cells gradually sediment (**B**, 6h) and form a uniform compact spheroid within 24 h (**C**). **D** Paraffin cross-section of an early BAEC spheroid (hematoxylin staining) with a core of unorganized cells and a surface layer of differentiated EC. **E** Paraffin cross section of a later (4 days) hematoxylin-stained BAEC spheroid. Most center EC have undergone apoptosis. A well-differentiated quiescent monolayer of EC is at the surface of the spheroids

Expected Results

ECs suspended in 96-well round dishes will organize to form one single rounded spheroid per well (Fig. 1A–C). Depending on the type of cell, EC cultured in spheroids may spontaneously organize to develop a two compartment system consisting of a surface monolayer and a core of unorganized cells (Fig. 1D) which undergo apoptosis and may disappear over time (Fig. 1E).

EC/SMC coculture spheroids will organize to form a core of SMC and a surface layer of EC mimicking the three-dimensional assembly of the normal vessel wall (in an inside-out orientation) with a polarized monolayer of endothelial cells and an underlying multicellular arrangement of smooth muscle cells. Figure 2A shows a cross section of an EC/SMC coculture spheroid with a surface layer of CD31-positive EC. Depending on the number and relative amount of EC and SMC used to generate EC/SMC coculture spheroids, some EC may be trapped in the spheroid center (Fig. 2B).

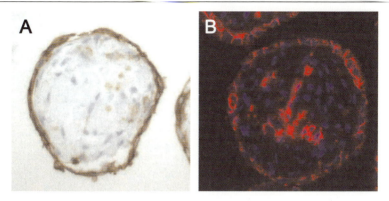

Fig. 2A,B. Representative example of a coculture spheroid of EC and SMC. **A** Paraffin cross section of an EC/SMC coculture spheroid with a CD31-positive surface layer of EC and a core of SMC. **B** Fluorescent image of a CD31-stained cryo-sectioned EC/SMC spheroid with a quiescent EC surface monolayer, CD31-negative SMC in the spheroid center, and few trapped CD31-positive EC with cluster in the spheroid center

Troubleshooting

▶ **Problem:** Formation of more than one spheroid per well.
Check the quality of the methocel-containing medium. Sometimes non-dissolved debris from the methyl cellulose prevents the formation of regular spheroids. Clear the stock solution by centrifugation.

▶ **Problem:** Stimulation of multiple spheroids with minimal amounts of reagents.
Harvest the spheroids as described before and collect them into 2.0-ml cups. Spinning these cups during the stimulation will prevent the spheroids from sticking together. Spheroids can be stimulated for up to 4 days using miniaturized spinner cultures.

References

Delia D, Lampugnani MG, Resnati M, Dejana E, Aiello A, Fontanella E, Soligo D, Pierotti MA, Greaves MF (1993) CD34 expression is regulated reciprocally with adhesion molecules in vascular endothelial cells in vitro. Blood 81:1001–1008

Jacks T and Weinberg RA (2002) Taking the study of cancer cell survival to a new dimension. Cell 111:923–925.

Korff T, Augustin HG (1998) Integration of endothelial cells in multicellular spheroids prevents apoptosis and induces differentiation. J Cell Biol 143:1341–1352

Korff T, Kimmina S, Martiny-Baron G, Augustin HG (2001) Blood vessel maturation in a 3-dimensional spheroidal co-culture model: direct contact with smooth muscle cells regulates endothelial cell quiescence and abrogates VEGF responsiveness. FASEB J 15:447–57

Kunz-Schughart L, Müller-Klieser W (2000) Three-dimensional culture (Chap. 5). In: Animal cell culture: a practical approach. 3rd Edition, John RW Masters (ed), IRL Press at Oxford University Press, Oxford, New York, Tokyo, pp. 123–148

Müller-Klieser W (1994) Multicellular Spheroids. In: Cell and tissue culture: laboratory procedures. Doyle A, Griffiths JB, Newell DG (eds). Wiley & Sons, Chichester, England, pp. 5B8.1–5B8.5

Pauly RR, Passaniti A, Crow M, Kinsella JL, Papadopoulos N, Monticone R, Lakatta EG, Martin GR (1992) Experimental models that mimic the differentiation and dedifferentiation of vascular cells. Circulation 86:III68–73

Wolburg H, Neuhaus J, Kniesel U, Krauss B, Schmid EM, Ocalan M, Farrell C, Risau W (1994) Modulation of tight junction structure in blood-brain barrier endothelial cells. Effects of tissue culture, second messengers and co-cultured astrocytes. J Cell Sci 107:1347–1357

Chapter 7

Immortalization of Endothelial Cells

Yvonne Reiss and Friedemann Kiefer

Introduction

The past decade has witnessed tremendous progress in the understanding of endothelial cell biology. Specifically acting growth factors and their cognate receptors on endothelial cells have been identified and the analysis of signaling pathways regulating endothelial behavior, growth, differentiation, and morphogenesis has taken center stage. Yet, isolation and maintenance of primary mouse endothelial cells is time-, cost- and labor-intensive and the limited proliferative potential of primary endothelial cultures renders extensive biochemical and cell biological analyses tedious. Frequently observed contamination of endothelial cultures with fibroblasts and pericytes, as well as the fact that primary endothelium is largely resistant to transfection with calcium phosphate or lipid-based methods substantially add to the inconvenience. Permanently growing endothelial cell lines that have been immortalized by the action of dominant oncogenes can provide an elegant solution to these limitations that are hampering the analysis of endothelial cells.

DNA tumor viruses frequently usurp essential cellular key signaling mechanisms to stimulate DNA replication and ensure virus propagation resulting in oncogenic transformation. Simian Vacuolating Virus 40 (SV40) induces tumors in newborn hamsters and is known to transform a wide variety of cell types. The SV40 large T (LT) gene product is required for establishment and maintenance of transformation and has been utilized for the generation of permanently growing endothelial cell lines (Noguchi et al. 2002, Lassalle et al. 1992, O'Connell and Edidin 1990). SV40 LT binds and functionally neutralizes the cellular key tumor suppressors p53 and Rb. The Adenovirus gene product E1A also binds and inactivates the non-phosphorylated form of Rb, resulting in cellular proliferation, and has been successfully employed to immortalize endothelial cells (Roux et al., 1994). More recently, overexpression of the catalytic subunit of human telomerase has been employed to enhance oncogene-mediated transformation of endothelium (MacKenzie et al. 2002, O'Hare et al. 2001). In either case, immortaliza-

Springer Lab Manual
H. Augustin (Ed.)
Methods in Endothelial Cell Biology
© Springer-Verlag Berlin Heidelberg 2004

tion by SV40 LT or adenovirus E1A accumulation of spontaneous mutations during graft tumor progression is frequently observed.

Polyomavirus middle T antigen (PymT), the principal oncoprotein of murine polyomavirus, rapidly transforms immature endothelial cells in embryos and newborn mice as well as endothelial progenitors with an apparently single rate limiting step (Kiefer et al. 1994a,b, Ong et al. 2001). This property of PymT can be exploited to derive immortalized endothelial cell lines, which are referred to as endothelioma cells. Taking advantage of the exquisite susceptibility of embryonic and neonate cells to retroviral transduction, retroviral gene transfer is the method of choice to induce expression of PymT. The ensuing transforming/oncogenic process is extremely rapid. After infection of neonates with a PymT-transducing recombinant retrovirus, cystic endothelial tumors or hemangiomas appear within 10 days in 50 % of all inoculated newborns (Fig. 1). Similarly, charac-

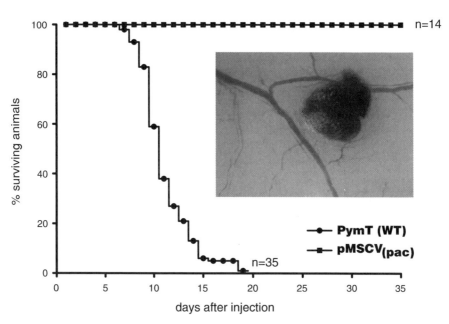

Fig. 1. Transformation of endothelial cells in newborn mice by intraperitoneal inoculation with a recombinant retrovirus transducing PymT. Newborn CD1 mice (n=35) were inoculated within 48h after birth with a single dose of the PymT-transducing virus MSCV-PymT. Approximately 5×10^4 c.f.u. in 200 µl of cell-free supernatant were injected intraperitoneally. Eight days after the inoculation the first animal succumbed to multiple intraperitoneal hemorrhages, more littermates were affected, with 50 % lethality being observed after 12 days and a complete loss of all pups after 20 days (*closed circles*). Animals infected with a control virus (n=14) survived unaffected (*closed squares*). *Inset* Depiction of a subcutaneous PymT-induced hemangioma which formed in the course of these experiments. From explants of such lesions PymT-transformed endothelioma cell lines are readily established in culture

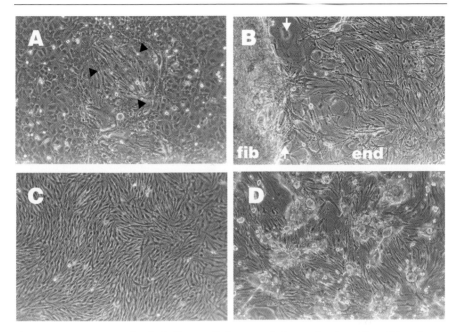

Fig. 2A-D. Generation of endothelioma cells by retroviral transduction of PymT in primary embryonic explant. A primary culture derived from an E11.5 embryo was infected by three subsequent 12 h inoculations with 3×10^6 c.f.u. MSCV-PymT. **A** Within the crude cell mixture the appearance of a first islet of endothelial morphology can be noted (demarked by *black arrowheads*). **B** Endothelioma cells progressively displace fibroblastic regions of the culture (*fib* denotes the fibroblastic components, *end* the expanding endothelioma culture). **C** Endothelioma culture with homogeneous morphological appearance. **D** Non-endothelial cells are repelled from a mixed culture and float off into the medium

teristic islands of endothelial morphology, which indicate PymT-driven expansion of endothelioma cells, can be observed as early as 2 weeks after inoculation of primary cultures (Fig. 2A).

PymT lacks discernible intrinsic enzymatic activity; however, it acts as a platform at the cell membrane resulting in the recruitment and activation of a number of key cellular signaling molecules. The apparent specificity of PymT for immature/proliferating endothelial cells reflects an intrinsic permissiveness of endothelium of embryonic and newborn mice for the cellular signaling components activated by PymT. Serendipitously, PymT appears to impinge on a combination of pathways, which are naturally active in immature endothelium. Consequently, during the subsequent process of oncogenic transformation, there is no need for the accumulation of secondary mutations in tumor suppressor genes, resulting in a fast, single-step oncogenesis. Mature endothelium is largely refractory to PymT-induced transformation, indicating a change in pathway permissiveness at later developmental stages.

The PymT transforming complex contains Src family kinases (Src, Fyn, or Yes), phosphatidyl inositol 3'-kinase (PI 3-K), the adaptor molecule ShcA, phospholipase Cγ-1 (PLCγ-1), phosphatase 2A (PP2A), and a member of the 14-3-3 family of proteins. Each of these molecules directly interacts with PymT. Additional interaction partners like the docking protein Gab2 are recruited via these proteins, giving rise to a large multicomponent complex which mimics one or more constitutively activated cell surface RTKs. Binding of PymT to c-Src results in constitutive activation of the Src tyrosine kinase and subsequent phosphorylation of specific tyrosines on PymT providing docking sites for the SH2 and PTB-domain-driven interactions with PI 3-K, ShcA, and PLCγ-1 (Fig. 3).

Current retroviral gene transfer technology is based on the usage of packaging cell lines and retroviral expression vectors. After introduction of proviral DNA, packaging cells secrete infectious, replication-incompetent viral particles. Packaging lines provide the necessary proteins for genomic RNA and particle formation through stably integrated versions of the viral gag, pol, and env genes, optimally at distinct chromosomal locations, ensuring the production of replication-incompetent virus for retroviral gene transfer. The viral env gene, expressed by the packaging cell line, encodes the envelope protein, which determines the range of infectivity (tropism) of the packaged virus. Viral envelopes are classified according to the receptors used to enter host cells. For example, the retroviral packaging cell line GP+E86 described in the following protocol (Markowitz et al. 1988) is an ecotropic producer line that secretes retroviruses which recognize a receptor present only on mouse and rat cells (for further information on Retroviral Gene

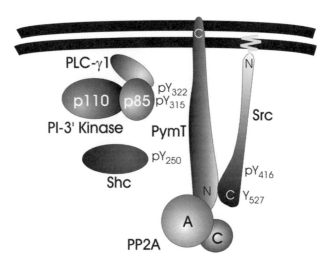

Fig. 3. Transforming complex formed by PymT at the cell membrane (see text for details)

Transfer see www.bdbiosciences.com). We will describe here the derivation of PymT-transformed primary mouse endothelial cells using a PymT-trans-ducing variant of the recombinant retrovirus MSCV. The Murine Stem Cell Virus (MSCV) vectors were derived by Robert Hawley and coworkers to achieve stable, high-level gene expression in hematopoietic and embryonic stem cells through a specifically designed 5'-long terminal repeat (LTR) (Hawley et al. 1994). This LTR differs from the MoMuLV LTR used in other retroviral vectors resulting in changes that enhance transcriptional activity and prevent transcriptional suppression in embryonic stem and embryonal carcinoma cells. As a result, the MSCV drives high-level constitutive expres-sion of a target gene in stem cells or other mammalian cell lines.

Materials

Cells

- Primary mouse endothelial cells (see Chap. 2 of this Vol. by Brandes et al. or follow the protocols described by Engelhardt et al. 1997 and Risau et al. 1990)
- Producer cell line GP+E MSCV-PymT (Ong et al., 2001)
- Murine fibroblasts cell line, e.g., NIH 3T3 (ATCC CRL-1658) or Rat-1A (ATCC CRL-2786)

Specimen

- Primary explants of newborn mice (Ong et al. 2001)

Reagents

- Phosphate-buffered saline (PBS)
- 0.1 % gelatin-coated culture flasks
- G418 (working concentration 0.5–1 mg/ml, a 100 mg/ml stock can be prepared in PBS and kept frozen for several month or at 4°C up to 4 weeks) or puromycin (working concentration 1–2 µg/ml, a 1 mg/ml stock can be prepared in PBS. Puromycin dilutions should be kept frozen, be thawed only once and used within 72 h)
- GP+E medium: DMEM complete medium (high glucose 4500 g/l, Gibco BRL, www.gibcobrl.com) with Glutamax and sodium pyruvate (both Gibco BRL) supplemented with 10 % FBS, 0.001 % monothioglycerol (prepare 1 % stock solution in PBS, Sigma, www.sigmaaldrich.com), and 1 % penicillin/streptomycin. Make sure FBS is heat inactivated at 56°C for 30 min
- Endothelial cell medium: DMEM complete medium supplemented with endothelial cell growth supplement (ECGS, 50 µg/ml, Sigma)

- Polybrene (1,5-dimethyl-1,5-diazaundecamethylene polymethobromide, hexadimethrine bromide, working concentration 8 µg/ml, prepare sterile 1 mg/ml stock in PBS, Sigma)
- Trypsin-EDTA (Gibco BRL)
- Diff Quick staining kit (Dade Behring, www.dadebehring.com)

Equipment

- Class II laminar flow hood and biosafety equipped tissue culture facility (although exclusively ecotropic recombinant murine retroviruses are used in this protocol, we recommend special care as retroviruses transducing an activated oncogene are handled. Use of lab coats and gloves throughout all procedures should be mandatory)
- Autoclave for waste decontamination
- Pipettes and filtered tips
- Centrifuge
- Incubator
- −80 °C freezer

Supplies

- Tissue culture plates and flasks
- 0.45 µm low protein binding filter
- 10 ml syringe

▨ Procedure

Virus production and infection of target cells

1. Thaw packaging cell line GP+E MSCV-PymT and culture cells in T75 culture flasks coated with 0.1 % gelatine in PBS.

2. Passage cells at least 1–2 times (1:5–1:10), apply fresh medium 16 h prior to harvesting cell supernatant for infection or titer determination, allow cells to grow only to subconfluency.

3. Collect spent medium and gently pass supernatant through a low protein binding 0.45 µm filter (do not use nitrocellulose filters, because they bind viral surface proteins and greatly reduce the available titer; avoid excessive shear forces during filtration which also destroy viral particles). Virus-containing media may be stored at −80°C. However, each freeze - thaw cycle will reduce the effective titer by 30–50 %.

4. Add Polybrene to the conditioned supernatant at a final concentration of 5–8 µg/ml. Polybrene is a positively charged polymer that reduces repulsion between the viral and cellular membranes.

5. Target cells may comprise isolated primary endothelial cells, crude embryonic primary explants, or primary explants of newborn mice. Note that the susceptibility to PymT of endothelia prepared from mice older than 2 weeks is dramatically reduced. Depending on the target, the protocol for infection may be adjusted slightly. Isolated primary endothelium should be grown in 35-mm culture flasks for 24–48 h, primary embryonic or newborn cultures will necessitate growth in 60–100 mm dishes for at least 48 h.

6. Wash target cells with PBS and apply virus-containing supernatant for 4 h twice. Alternatively, infect the cells overnight. Multiple rounds of infection can improve results by increasing the number of infected cells. For crude embryonic cultures we repeat infection for as long as 1 week, replenishing the virus-containing supernatant daily. Replace virus-containing medium with fresh endothelial medium after the last round of infection.

7. For crude primary explants, we usually do not apply antibiotic selection. The selectivity of PymT-driven transformation suffices to drive only the expansion of endothelioma cells from mixed cultures. For isolated endothelial cells start selection of infected cells with respective antibiotic (G418: 0.5–1 mg/ml; puromycin: 1–2 µg/ml) 48 h after infection.

8. Small islands of transduced endothelial cells (endothelioma cells) may be visible as early as 3–6 days after application of antibiotic selection. Typical endothelioma islands tend to appear 10–20 days after infection of crude primary cultures (Fig. 2A). Endothelioma cells spread underneath contaminating cell types (e.g., fibroblasts) and repel other cell types (Fig. 2B and D) after forming a continuous monolayer.

9. Culture the cells until they appear homogenously pure, i.e., no contamination with pericytes and fibroblasts as judged by morphological appearance (Fig. 2C). Alternatively, endothelioma cells can be subcloned using cloning cylinders (note: passaging small numbers of mouse endothelioma cells is challenging as the cells generally do not survive high dilutions).

Determination of virus titer

1. Determine susceptibility of a suitable murine fibroblast line (e.g., NIH 3T3 (ATCC CRL-1658) or Rat-1A (ATCC CRL-2786)) for the retrovirally transduced antibiotic resistance. Use lowest concentration at which cells are effectively killed. G418 may be applied at a concentration range of 0.5–1 mg/ml and selection should be finished within 7 days. Puromycin should be applied at 1–2 µg/ml and selection should be finished within 3

days. Puromycin dilutions should be kept frozen, be thawed only once and used within 72 h.

2. Seed between 5×10^4 and 1×10^5 cells in a 60 mm dish 24 h prior to infection for titer determination. The number of fibroblasts plated depends on the actual growth rate of the cells. They must not reach confluency within the next 72 h.

3. Prepare 100-, 1,000-, and 10,000-fold dilutions of 0.45 µm filtered virus supernatant in GP+E medium containing 8 µg/ml Polybrene.

4. Thoroughly aspirate medium and replace with 500 µl of the virus-containing dilutions. Each dilution should be tested in duplicate. Allow infection to proceed for 2 h, gently swirl the dishes every 15 min to ensure equal distribution of the virus dilution.

5. Add 4 ml of normal growth medium after 2 h and incubate for 48 h.

6. Start antibiotic selection 2 days after infection at the determined concentration. Apply fresh medium every 2 days.

7. Inspect cells daily and evaluate results after 7–10 days. Selection should be completed, i.e., no more dead cells floating, and vital, rapidly expanding clones should be present. One of the virus dilutions should produce 10 to 100 colonies per dish.

8. For evaluation, colonies can be counted microscopically or dishes may be fixed with ice-cold methanol and colonies are stained with a rapid staining protocol like the Diff-Quick staining kit. We prefer microscopic evaluation over the counting of stained colonies, as smaller colonies may be overlooked and morphological alterations of the infected cells are noticed.

9. The viral titer (average colony number per dish × dilution × 2) is determined as the number of colony-forming units (c.f.u.)/ml. Viral titers may range from 5×10^6 c.f.u./ml to 3×10^3 c.f.u./ml. Viral titers should be in the order of 10^5 c.f.u./ml to allow for a sufficient multiplicity of infection (1:1 to 10:1).

Expected Results

Immortalization of primary endothelial cells or extracts of embryos can be efficiently achieved employing PymT (or other retroviral oncogenes) as outlined in this chapter and illustrated in Figs. 1–3. Resulting endothelioma (End) cell lines retain endothelial cell characteristics (see Chap. 15) and provide an excellent experimental system to study endothelial cell functions in vitro, e.g., by using a variety of gene-targeted mouse models.

Troubleshooting and General Considerations

- Note that the susceptibility of endothelial cells prepared from mice older than 2 weeks to PymT is dramatically reduced as mature endothelium is largely refractory to PymT-induced transformation.
- Multiple rounds of infection can improve results by increasing the percentage of infected cells as well as by increasing the number of proviral integrations per cell.
- The number of target cells may need to be optimized if poor infection is observed. A multiplicity of infection of 5 viral particles per target cell or higher is desirable. Make sure target cells are not too confluent (plate cells to 60–80 % confluency), as successful proviral integration requires active replication of cells. Premature application of antibiotic selection may be another reason for poor results (wait at least 48 h post infection for maximal gene expression). If low virus production of the packaging cell lines is a problem, subcloning and screening for high-titer clones may be necessary.
- Poor viability of target cells during or after infection may be due to the packaging cell line-conditioned medium that is affecting cell growth: dilute viral medium or shorten exposure time to viral supernatant.
- Avoid continued freeze-thaw cycles of the virus-containing media as it will reduce the effective titer by 30–50 %.
- Excessive exposure to Polybrene (> 24 h) may be toxic to cells.
- Culturing and expanding infected cells may be challenging as mouse endothelial cells generally do not survive high dilutions after passaging. Therefore, it is important to start with sufficient target cell numbers prior to virus infection in order to achieve good results.

References

Engelhardt B, Vestweber D, Hallmann R, Schulz M (1997) E- and P-Selectin are not involved in the recruitment of inflammatory cells across the blood-brain barrier in experimental autoimmune encephalomyelitis. Blood 90(11):4459–4472

Hawley RG, Lieu FH, Fong AZ, Hawley TS (1994) Versatile retroviral vectors for potential use in gene therapy. Gene Ther 1:136–138

Kiefer F, Anhauser I, Soriano P, Aguzzi A, Courtneidge SA, Wagner EF (1994a) Endothelial cell transformation by polyomavirus middle T antigen in mice lacking Src-related kinases. Curr Biol 4:100–109

Kiefer F, Courtneidge SA, Wagner EF (1994b) Oncogenic properties of the middle T antigens of polyomaviruses. Adv Cancer Res 64:125–157

Lassalle P, LaGrou C, Delneste Y, Sanceau J, Coll J, Torpier G, Wietzerbin J, Stehelin D, Tonnel AB, Capron A (1992) Human endothelial cells transfected by SV40 T antigens: characterization and potential use as a source of normal endothelial factors. Eur J Immunol 22:425–431

MacKenzie KL, Franco S, Naiyer AJ, May C, Sadelain M, Rafii S, Moore MA (2002) Multiple stages of malignant transformation of human endothelial cells modelled by co-expression of telomerase reverse transcriptase, SV40 T antigen and oncogenic N-ras. Oncogene 21:4200–4211

Markowitz D, Goff S, Bank A (1988) A safe packaging line for gene transfer: separating viral genes on two different plasmids. J Virol 62:1120–1124

Noguchi H, Kobayashi N, Westerman KA, Sakaguchi M, Okitsu T, Totsugawa T, Watanabe T, Matsumura T, Fujiwara T, Ueda, T, Miyazaki M, Tanaka N, Leboulch P (2002) Controlled expansion of human endothelial cell populations by Cre-loxP-based reversible immortalization. Hum Gene Ther 13:321–334

O'Connell KA, Edidin M (1990) A mouse lymphoid endothelial cell line immortalized by simian virus 40 binds lymphocytes and retains functional characteristics of normal endothelial cells. J Immunol 144:521–525

O'Hare MJ, Bond J, Clarke C, Takeuchi Y, Atherton AJ, Berry C, Moody J, Silver AR, Davies DC, Alsop AE, Neville AM, Jat PS (2001) Conditional immortalization of freshly isolated human mammary fibroblasts and endothelial cells. Proc Natl Acad Sci USA 98:646–651

Ong SH, Dilworth S, Hauck-Schmalenberger I, Pawson T, Kiefer F (2001) ShcA and Grb2 mediate polyoma middle T antigen-induced endothelial transformation and Gab1 tyrosine phosphorylation. EMBO J 20:6327–6336

Risau W, Engelhardt B, Wekerle H (1990) Immune function of the blood-brain barrier: incomplete presentation of protein (auto-)antigenes by rat brain microvascular endothelium in vitro. J Cell Biol 110:1757

Roux F, Durieu-Trautmann O, Chaverot N, Claire M, Mailly P, Bourre JM, Strosberg AD Couraud PO (1994) Regulation of gamma-glutamyl transpeptidase and alkaline phosphatase activities in immortalized rat brain microvessel endothelial cells. J Cell Physiol 159:101–113

Chapter 8

Viral Gene Transfer Into Endothelial Cells

CHRISTOPH VOLPERS and STEFAN KOCHANEK

Introduction

Viral vectors are being employed to transfer genes into endothelial cells for experimental purposes in basic research as well as for gene therapeutic approaches for cardiovascular and oncological disorders. Therapeutic gene transfer strategies in the field of cardiovascular disorders are directed towards the stimulation of angiogenesis for ischemic heart disease, the inhibition of vascular smooth muscle cell proliferation, and functional protection of endothelial cells in order to prevent restenosis after angioplasty, the reduction of cholesterol blood levels for prevention of atherosclerosis, and the vasodilation or blockade of vasoconstriction in systemic and pulmonary hypertension (Khurana et al. 2001, Ehsan et al. 2000). The inhibition of endothelial cell proliferation and neoangiogenesis in tumors is a promising approach in oncological viral gene therapy developed in recent years.

The transduction efficacy of retroviral and lentiviral vectors as well as adeno-associated viral (AAV) vectors in quiescent endothelial cells, especially in vivo, is low; retroviral vectors can be used for gene transfer into cultured, stimulated cardiovascular cells in vitro (Ehsan et al. 2000, Shichinohe et al. 2001). Adenoviral vectors are the most widely used viral vectors for gene transfer into endothelial cells, in vitro and in vivo. Although endothelial cells express considerably lower levels of the primary adenoviral receptor than epithelial cells, they can be transduced with reasonable efficacy by adenoviral vectors at higher MOIs (multiplicity of infection, infectious vector particles per cell). Therefore, the protocols outlined in this chapter will focus on the use of adenoviral vectors.

The DNA genome of adenoviral vectors remains episomal and does not integrate into the cellular chromosomal DNA after delivery into the nucleus, potentially a safety advantage for in vivo application. In the commonly used first-generation adenoviral vectors, the transgene to be transfered is inserted

Springer Lab Manual
H. Augustin (Ed.)
Methods in Endothelial Cell Biology
© Springer-Verlag Berlin Heidelberg 2004

in the region of the deleted E1 gene, which is essential for viral replication and is provided in trans by the producer cell lines used for vector propagation (Danthinne and Imperiale 2000). As the E1 function is not present in the transduced endothelial cells, the vector DNA does not replicate in these cells. First-generation adenoviral vectors can accommodate up to 8 kb of foreign DNA. In order to increase the transgene capacity and to reduce toxicity of the vectors, some additional viral genes (second-generation adenoviral vectors) or all viral genes (high-capacity or "gutless" adenoviral vectors) have been removed from the vector genome. High-capacity adenoviral vectors can accommodate large transgenes of up to 36 kb in size and are especially suitable for in vivo applications due to their low toxicity and immunogenicity (Kochanek 1999, Kochanek et al. 2001, Schiedner et al. 2002).

The transgene can be inserted into the vector genome by cloning, bacterial recombination or transposition, or by recombination in mammalian cells (for review, see Danthinne and Imperiale 2000). Straightforward, kit-based systems to produce recombinant vector genomes are provided by several commercial suppliers (e.g., AdEasy™-System by QBIOgene [www.qbiogene.com], Adeno-X™ by BD Biosciences [www.bdbiosciences.com], RAPAd™-System by ViraQuest Inc. [www.viraquest.com], AdHTS^R-System by Neurogenex Inc. [www.neurogenex.com], and Helper Dependent Adenovirus Vector Kit by Microbix [www.microbix.com]). The recombined vector genome is transfected into one of the available E1-complementing producer cell lines, HEK293 cells (Graham et al. 1977), N52.E6 cells (Schiedner et al. 2000), 911 cells (Fallaux et al. 1996) or PER.C6 cells (Fallaux et al. 1998), and the vector is rescued and amplified from these cells. In Protocol A, amplification and preparation of high-titer vector stocks is described; Protocol B provides step-by-step guidance for endothelial cell transduction.

Materials

Cells

- Endothelial cells: e.g., Human umbilical vein endothelial cells (HUVEC), Microvascular endothelial cells (MVEC)
- Producer cell line: N52.E6 (Schiedner et al. 2000) or 293 cells

Reagents

- Tris-buffered saline (TBS) (Sambrook and Russell 2001)
- Phosphate-buffered saline (PBS)
- Trypsin/EDTA (Life Technologies)
- Fetal bovine sera (FBS) (Life Technologies)
- 100× Penicillin/Streptomycin (Life Technologies)

- Endothelial cell media (ECM)
  (suitable for, e.g., HUVEC, primary human choroidal and retinal endothelial cells):
  M199 Media with Earle's salts (Life Technologies) with 20 % FBS, 60 µg/ml ECGS [Endothelial Cell Growth Supplement (Sigma, www.sigma aldrich. com)], 0.1 mg/ml heparin (Sigma), 1× Penicillin/Streptomycin
- Alpha-Minimal Essential Medium (Alpha-MEM, Life Technologies) with 10 % FBS, 1× Penicillin/Streptomycin

Equipment

- Micropipettor and filter-plugged tips
- 1.5 ml-Safe-lock tubes (Eppendorf, www.eppendorf.com)
- Sterile 15 ml and 50 ml centrifuge tubes
- Autoclaved 250 ml centrifuge beakers
- Ultracentrifuge tubes (Beckman Ultraclear, 14 × 89 mm)
- 6 cm and 15 cm cell culture dishes
- 24-well cell culture microplate
- 2 ml syringe, 18G needles
- Desalting columns (Amersham Pharmacia Column PD-10 with Sephadex G-25 M, Amersham Biosciences, www.amershambiosciences. com)
- Low-speed centrifuge
- Ultracentrifuge
- SW41 rotor
- Incubator with CO_2 atmosphere
- −80 °C freezer
- Autoclave
- Safety level laboratory for all adenoviral vector works

▓ Procedure

Protocol A: Amplification and preparation of first-generation adenoviral vector stocks

This protocol is suited for preparing a large stock of adenoviral vector from small aliquots or from material isolated ("picked") from plaques in agarose overlays of producer cells transfected with recombinant vector DNA ("plaque rescue", Hitt et al. 1995).

1. Seed 2×10^6 N52.E6 or 293 producer cells in complete Alpha-MEM in 6-cm dish.
   The next day, remove the medium, add 2 ml fresh medium, pipet vector inoculum (about 1×10^7 infectious units or half of the material isolated

from a plaque) into the media and gently move the plate to distribute inoculum.

Fill up to total volume of 5 ml after 2 h and incubate at 37 °C.

2. Harvest culture when all cells have rounded up and detached from the bottom of the plate (cytopathic effect, CPE).
   Lyse cells by 3 cycles of freezing in liquid nitrogen and rapid thawing in a 37 °C water bath.

3. Seed 1.5×10^7 N52.E6 or 293 cells in complete Alpha-MEM in 15 cm dish. Next day, remove the medium, add 10 ml of fresh medium, add half of the cell lysate from step 2, and distribute evenly. Fill up to total volume of 30 ml after 2 h and incubate at 37 °C.

4. Harvest culture when all cells show CPE (usually after 48 h).
   Lyse cells by 3 cycles of freeze/thaw.

5. Seed 1.5×10^7 N52.E6 or 293 cells/dish in ten 15 cm dishes.
   Next day, infect the cells with 1/10 of the lysate from step 4 per dish as described in step 3.

6. Harvest culture when all cells show CPE (usually after 48 h). Spin culture in sterile centrifuge beakers for 15 min at 500 × g. Discard supernatants, resuspend cell pellets in a total volume of 10 ml sterile TBS. Lyse cells by 3 cycles of freeze/thaw.

7. Spin cell debris down for 10 min at 500 × g.
   Take supernatant and fill up with TBS to a total volume of 20 ml. Add 10 g CsCl and mix carefully. Fill 11 ml of the suspension each into two SW41 ultracentrifuge tubes.

8. Spin tubes in ultracentrifuge for 20 h at 32,000 rpm, 4 °C.
   Puncture the side of the tube using a 2 ml syringe with an 18G needle slightly below the band. Collect vector band in 1–1.5 ml volume. (Vector band should be at about the upper third of the gradient, incomplete viral particles might be visible as faint additional bands at the top of the gradient). Pool aspirated vector from both centrifuge tubes.

9. Fill up pooled vector preparation to a total volume of 11 ml with CsCl solution (0.5 g/ml CsCl in TBS) into a fresh ultracentrifuge tube. Repeat ultracentrifugation as in step 8.

10. Remove CsCl by quick one-step-chromatography over PD-10 desalting column according to the manufacturer's instruction.
    Pool vector containing column fractions #2 and #3, add 10 % sterile glycerol, and store at –80 °C.

Protocol B: Transduction of endothelial cells in vitro

1. Seed isolated endothelial cells in ECM at 70–80 % confluence the day before transduction (e.g., 5–7 × 10⁴ HUVEC cells per well in 24-well cell culture plates).

2. Remove media, add small volume of ECM containing 5 % FBS well covering the cell monolayer (e.g., 200 μl per well for 24-well cell culture plate).

3. Dilute vector for infection with desired MOI (multiplicity of infection; infectious vector particles per cell) in TBS to a minimum volume of 10 μl, pipet vector dilution onto cells without touching the monolayer and gently move the dish to distribute vector.

4. Incubate cells for 2 h at 37 °C.

5. Fill up with complete ECM to regular volume (e.g., 1 ml per well for 24-well plate). (Or remove inoculum and add fresh complete ECM if transgene product is secreted into the cell culture supernatant).

6. Incubate cells for 48 h or longer and harvest cells or supernatant for detection, analysis or purification of the transgene product.

For assessment of viral transduction efficacy in the respective experiment, a lacZ reporter gene containing vector can be used at the same MOI in parallel infections with the vector containing the gene of interest. Transduced cells are identified after 48 h by fixation and staining of the cell monolayer according to standard procedures (Sambrook and Russell 2001).

Example

Human umbilical vein endothelial cells (HUVEC) were seeded in 24-well plates at 3 × 10⁴, 5 × 10⁴, and 7 × 10⁴ cells per well the day before transduction, corresponding to a cell culture confluence of <50 %, 75 %, and >90 %, respectively. A first-generation β-galactosidase reporter vector (Ad-β-Gal) was used for transduction of the cells with 10, 30, 100, and 300 MOI according to Protocol B. After 48 h, cells were stained with X-Gal substrate and the percentage of blue cells was determined by counting 5 representative microscopic fields each.

As shown in Table 1, the gene transfer efficacy with Ad-β-Gal in HUVEC depends on cell culture confluence and the vector dosage. About 1–4 % of cells were found to be transduced with 10 MOI of vector, and all cells of the almost confluent culture could be transduced with 300 MOI of vector.

Table 1. Transduction of HUVEC with different amounts of a first-generation adenoviral vector (AdβGal) expressing bacterial β-galactosidase as reporter protein (Values given are percent transduced cells of total cell number)

Confluence of culture[a]	10 MOI[b]	30 MOI	100 MOI	300 MOI
<50 %	0.2–1 %	2–5 %	20–30 %	65–70 %
75 %	1–3 %	10–15 %	40–50 %	80–90 %
>90 %	2–4 %	15–20 %	65–75 %	95–100 %

[a] HUVEC cells were seeded at 3×10^4, 5×10^4, and 7×10^4 cells/well, respectively, in a 24-well plate the day before transduction.
[b] MOI, multiplicity of infection, infectious vector particles per cell.

▨ Troubleshooting and Practical Tips

Adenoviral vector handling

Safety glasses, gloves, and lab coat should always be worn when doing freeze/thaw cycles during vector amplification. Adenoviral vector stocks should be aliquoted and stored with 10 % glycerol at –80 °C. Repeated freezing and thawing of aliquots should be avoided as it will cause a gradual loss of vector titer. Adenoviral vector preparations should be kept on ice during transduction experiments and should be resuspended after thawing by snipping of the Eppendorf tube with a finger tip or short, mild vortexing (twice for 1–2 s). Sterile filter-plugged tips should always be used for pipetting of vector suspension.

Vector titration

Adenoviral vector preparations contain immature and uninfectious particles. An infectious:total particle ratio of 1:10 to 1:20 is well within the usual range. Infectious particle titers can be determined by different methods including plaque assay (Hitt et al. 1995), limiting dilution assay (also called tissue culture infectious dose 50 method (TCID50) (Summers and Smith 1987), slot blot (Kreppel et al. 2002) and real-time PCR (Ma et al. 2001). Of note, results of these methods are not readily comparable, e.g., determination by plaque assay usually results in lower titers than other methods. A quick and rough estimate of infectious titers of first-generation vectors can be obtained by infection of 6 cm dishes plated with about 2×10^6 cells of the producer cell line used with serial dilutions of the vector preparation. About 5 MOI (multiplicity of infection, ratio of infectious particles per cell) of the vector, i.e., 1×10^7 infectious units, will yield a cytopathic effect (CPE) in all cells after 48 h (CPE of the cells is characterized by rounding up and detaching from the bottom of the plate).

Total particle titers can be determined by the following method based on the spectrophotometric measurement of DNA content in the vector particles: An aliquot of the vector preparation is diluted 1:15 in PBS containing SDS in a final concentration of 0.1 %. The diluted preparation is shortly vortexed, incubated for 10 min, and OD(260) is determined. Total particle per ml can be calculated by OD(260) $\times$ dilution factor (i.e., 15) $\times$ 1 $\times$ 10^{12}. This method, however, will yield reliable results only when vector preparations are used which had been highly purified over two CsCl gradients because contaminating free viral and cellular nucleic acids will be assessed as well.

Replication-competent adenovirus (RCA)

During amplification of first-generation adenoviral vectors in 293 cells, replication-competent adenovirus (RCA) can arise due to homologous recombination between overlapping sequences in the viral E1-containing fragment integrated in the cellular genome and the left end of the vector genome. If transgene expression is low or lacking in spite of high infectious titers of a vector preparation, this possibility should be considered. RCA in a vector preparation can be detected by serial passage on HeLa cells, since RCA, but not E1-deleted vector will replicate in these cells (Hitt et al. 1995). The generation of RCA can be avoided by use of more recently developed producer cell lines such as N52.E6 (Schiedner et al. 2000) which contain clearly defined E1 fragments in the cellular genome without any sequence overlap with the vector.

Endothelial cell density

Low transduction efficacy with adenoviral vectors in cultured endothelial cells may be due to low cell density. The expression level of the primary cellular adenovirus receptor, CAR (Coxsackie-and Adenovirus Receptor), in cultured endothelial cells depends on the cell density: CAR expression is increased with culture confluence (Carson et al. 1999). This explains the variations in transduction efficacy observed in our Example experiment with HUVEC. We routinely transduce cultures with 70–90 % confluence.

β-Gal staining of transduced endothelial cells

If a β-Gal reporter vector is used in parallel with the vector containing the gene of interest for quantitative assessment of transduction efficacies, the staining should be performed over night, since many endothelial cell types in culture stain only weakly and dot-like. However, do not stain for longer than 20 h, because crystal aggregates will form and make counting of blue cells difficult.

Comments and Modifications

Endothelial-specific promoters

Particularly for in vivo applications, the use of endothelial-specific promoters may be of interest to reduce background expression in other tissues. The von Willebrand factor promoter and the vascular endothelial growth factor receptor type-I (flt-1) promoter have successfully been used in the context of adenoviral vectors for endothelial-specific expression (Nicklin et al. 2001a).

Capsid-modified adenoviral vectors for higher transduction efficacy

Capsid-modified adenoviral vectors are available which transduce endothelial cells with higher efficacy. Adenoviral vectors with short peptide sequences incorporated into the fiber capsid protein (e.g., SIGYPLP, CDCRGDCFC, Poly-K peptides) binding to cell surface molecules on endothelial cells are considerably more efficient in gene delivery to these cells as compared to unmodified vector (Nicklin et al. 2001b, Biermann et al. 2001, Wickham et al. 1997), thus reducing required vector doses. The same effect has been achieved by complexing adenoviral vector particles with bispecific antibody or scFv constructs composed of a capsid-binding and an endothelial-binding moiety (Wickham et al. 1996).

Adenoviral gene transfer in vivo

Their transduction efficiency for quiescent cells and high titers make adenoviral vectors attractive for cardiovascular gene transfer in vivo and for transfer to arterial and venous grafts ex vivo. The routes of delivery of the vector to the cardiovascular system are endoluminal, e.g., using an injection needle for tail vein injection of mice or using a catheter during angioplasty, periadventitial, or direct injection into the myocardium, coronary artery or ventricle (Khurana et al. 2001, Ooboshi et al. 1997). For arterial grafts, high adenoviral vector doses of 5×10^9 plaque-forming units (PFU) per ml and an instillation pressure of 80 to 100 mm Hg have been found to increase the transduction efficacy (Brevetti et al. 2002).

Activation of kinase cascades by adenoviral vectors

Cellular binding and internalization of adenoviral vectors has been shown to trigger certain kinase cascades, e.g. phosphatidylinositol 3-OH kinase (PI3K) and MAP kinase. If adenoviral vectors are to be used for transfer of genes encoding signal transduction components into endothelial cells, control experiments including vector without the transgene are important.

Biological safety level for adenoviral gene transfer experiments

Production and gene transfer applications of adenoviral vectors have to be performed in safety level laboratories and with the required regulatory approval for the respective project. Prior to planning and performing gene transfer experiments employing adenoviral vectors, the relevant national regulatory authorities should be contacted for advice on safety standards and approval of experimental procedures.

References

Biermann V, Volpers C, Hussmann S, Stock A, Kewes H, Schiedner G, Herrmann A, Kochanek S (2001) Targeting of high-capacity adenoviral vectors. Hum Gene Ther 12:1757–1769

Brevetti LS, Chang DS, Sarkar R, Messina LM (2002) Effect of adenoviral titer and instillation pressure on gene transfer efficiency to arterial and venous grafts ex-vivo. J Vasc Surg 36:263–270

Carson SD, Hobbs JT, Tracy SM, Chapman NM (1999) Expression of the Coxsackievirus and Adenovirus receptor in cultured human umbilical vein endothelial cells: regulation in response to cell density. J Virol 73:7077–7079

Danthinne X, Imperiale MJ (2000) Production of first generation adenovirus vectors: a review. Gene Ther 7:1707–1714

Ehsan A, Mann MJ, Dzau VJ (2000) Gene therapy for cardiovascular disease and vascular grafts, pp 421–438. In: NS Templeton and DD Lasic (eds), Gene therapy. Therapeutic mechanisms and strategies, Marcel Dekker Inc., New York

Fallaux FJ, Bout A, van der Velde I, van den Wollenberg DJ, Hekir KM, Keegan J, Auger C, Cramer SJ, van Ormondt H, van der Eb AJ, Valerio D, Hoeben RC (1998) New helper cells and matched early region 1-deleted adenovirus vectors prevent generation of replication competent adenoviruses. Hum Gene Ther 9:1909–1917

Fallaux FJ, Kranenburg O, Cramer SJ, Houweling A, van Ormondt H, Hoeben RC, van der Eb AJ (1996) Characterization of 911: A new helper cell line for the titration and propagation of early region 1-deleted adenoviral vectors. Hum Gene Ther 7:215–222

Graham FL, Smiley J, Russell WC, Nairn R (1977) Characteristics of a human cell line transformed by DNA from human adenovirus type 5. J Gen Virol 36:59–74

Hitt M, Bett AJ, Addison CL, Prevec L, Graham FL (1995) Techniques for human adenovirus vector construction and characterization. Methods in Molecular Genetics 7.13–30

Khurana R, Martin JF, Zachary I (2001) Gene therapy for cardiovascular disease. Hypertension 38:1210–1216

Kochanek S (1999) High-capacity adenoviral vectors for gene transfer and somatic gene therapy. Hum Gene Ther 10:2451–2459

Kochanek S, Schiedner G, Volpers C (2001) High-capacity "gutless" adenoviral vectors. Curr Opin Mol Ther 3:454–463

Kreppel F, Biermann V, Kochanek S, Schiedner G (2002) A DNA-based method to assay total and infectious particle contents and helper virus contamination in high-capacity adenoviral vector preparations. Hum Gene Ther 13:1151–1160

Ma L, Bluyssen HAR, De Raeymaeker M, Laurysens V, Van der Beek N, Pavliska H, Van Zonneveld AJ, Tomme P, Van Es HHG (2001) Rapid determination of adenoviral vector titers by quantitative real-time PCR. J Virol Methods 93:181–188

Nicklin SA, Reynolds PN, Brosnan MJ, White SJ, Curiel DT, Dominiczak AF, Baker AH (2001a) Analysis of cell-specific promoters for viral gene therapy targeted at the vascular endothelium. Hypertension 38:65–70

Nicklin SA, von Seggern DJ, Work LM, Pek DCK, Dominiczak AF, Nemerow GR, Baker AH (2001b) Ablating adenovirus type 5 fiber-CAR binding and HI loop insertion of the SIGYPLP peptide generate an endothelial cell-selective adenovirus. Mol Ther 4:534–542

Ooboshi H, Rios CD, Heistad DD (1997) Novel methods for adenovirus-mediated gene transfer to blood vessels in vivo. Mol Cell Biochem 172:37–46

Sambrook J, Russell DW (eds) Molecular Cloning: A Laboratory Manual. 3rd edition 2001. Cold Spring Harbor Laboratory Press, Cold Spring Harbor, New York

Schiedner G, Clemens PR, Volpers C, Kochanek S (2002) High-capacity "gutless" adenoviral vectors: Technical aspects and applications, pp 429–446. In: DT Curiel and JT Douglas (eds), Adenoviral vectors for gene therapy, Academic Press, London

Schiedner G, Hertel S, Kochanek S (2000) Efficient transformation of primary human amniocytes by E1 functions of Ad5: Generation of new cell lines for adenoviral vector production. Hum Gene Ther 11:2105–2116

Shichinohe T, Bochner BH, Mizutani K, Nishida M, Hegerich-Gilliam S, Naldini L, Kasahara N (2001) Development of lentiviral vectors for antiangiogenic gene delivery. Cancer Gene Ther 8:879–889

Summers MD, Smith GE (1987) A manual of methods for baculovirus vectors and insect cell culture procedures. Texas Agricultural Experiment Station Bulletin. Tex Agric Exp Stn Vol. 1555. College Station Press, Texas

Wickham TJ, Segal DM, Roelvink PW, Carrion ME, Lizonova A, Lee GM, Kovesdi I (1996) Targeted adenovirus gene transfer to endothelial and smooth muscle cells by using bispecific antibodies. J Virol 70:6831–6838

Wickham TJ, Tzeng E, Shears LL, Roelvink PW, Li Y, Lee GM, Brough DE, Lizonova A, Kovesdi I (1997) Increased in vitro and in vivo gene transfer by adenovirus vectors containing chimeric fiber proteins. J Virol 71:8221–8229

Part II:
Functional Assays in Vitro

Chapter 9

Leukocyte Adhesion to Endothelium in Vitro Under Shear Stress

OLAF ZILLES, MARKUS HAMMEL, and RUPERT HALLMANN

Introduction

Endothelial cells separate the blood from the environment of local tissues, but also mediate signals between these two compartments and facilitate directed leukocyte migration. Leukocyte transmigration through the blood vessel wall is part of the recirculation pathway of mononuclear leukocytes between the blood and the lymphatic system, and it is essential for leukocyte extravasation into sites of inflammation. Leukocyte adhesion to the endothelial cells of the blood vessel wall is the first step of leukocyte extravasation. In vitro cell–cell adhesion assays have contributed to the identification and functional analysis of cell adhesion molecules which are important for the interaction of leukocytes with endothelium. Different leukocyte subsets accumulate outside the vascular wall, dependent on the type of inflammatory insult. In vitro assays allow the study of the individual adhesion characteristics of distinct leukocyte subsets like polymorphonuclear granulocytes (PMN), eosinophilic granulocytes, T and B lymphocytes, or monocytes under defined conditions. Endothelial cell adhesion molecules relevant for leukocyte adhesion include members of the selectin family (E-selectin and P-selectin) as well as immunoglobulin superfamily molecules ICAM-1 and ICAM-2, VCAM-1, the JAM family, and members of the mucin-type cell surface molecules like CD34, GlyCAM-1, or MADCAM-1 (Butcher and Picker 1996, Johnston and Butcher 2002, Patel et al. 2002, Vestweber and Blanks 1999). These endothelial cell surface adhesion molecules are recognized by their leukocyte counterparts, L-selectin, integrins like alpha4 beta1 and alpha4beta7, and sialylated sugar moieties which are detected by the endothelial cell selectins, e.g., PSGL-1, ESL-1, and other sLex-containing glycoproteins.

Basic leukocyte adhesion assays are used in many research laboratories. The main distinction between the two assay systems utilized here is the different application of shear forces during coculturing by which the shear of

Springer Lab Manual
H. Augustin (Ed.)
Methods in Endothelial Cell Biology
© Springer-Verlag Berlin Heidelberg 2004

the blood flow along the endothelial cell surface is mimicked. The application of shear forces has greatly increased our ability to analyze the role of different adhesion molecules in the leukocyte extravasation cascade.

Materials

Cells

- Different endothelial cell populations can be employed. For the analysis of murine leukocyte adhesion to murine endothelial cells, we routinely use the mouse microvascular endothelial cell lines bEND.3 or mlEND.1 Other endothelioma cells can be passaged up to 75 times without loss of differentiation. The cells should be analyzed regularly by flow cytometry for differentiation markers such as MECA-32 and, after IL-1 activation for 4 h, for expression of CD62E (E-selectin).
- Leukocytes: Mouse primary leukocytes are isolated from bone marrow with a 25-g needle, washed through a mesh100 nylon mesh and are further separated into subsets by negative magnetic sorting if necessary. Mouse bone marrow has about 80 % differentiated neutrophilic granulocytes (PMN). Isolation of bone marrow B lymphocytes can be performed by negative sorting with Miltenyi magnetobeads for GR-1 to segregate the PMN population. Human U937 monocytic cells are obtained from ATCC and are cultured by standard procedures. Human peripheral blood leukocytes are isolated by standard procedure with Histopaque 1077 or 1119 density gradient centrifugation.

Reagents

- DMEM (Gibco) with 10 % FCS (Biochrom) and 2 mM Glutamax (Gibco, www.gibcobrl.com)
- 5 mM EDTA in PBS
- Trypsin-EDTA
- Phosphate-buffered saline (PBS)
- Interleukin-1 (recombinant mouse IL-1, Genzyme)
- Histopaque (Sigma, www.sigmaaldrich.com) or similar

Equipment

- Cell culture facility with CO_2-incubator

Random shear system

- Labtek chamber slides with glass bottom (Nalge Nunc International, www.nalgenunc.com)

- Horizontal shaker with excenter larger than 15 mm
- Cold room

Parallel shear system

- Focht Live-Cell Chamber system FCS2 with pump and controller (Bioptechs, www.bioptechs.com)
- Inverted microscope with heated chamber (Zeiss, www.zeiss.com) , Bioptechs stage adapter, and camera (digital still or video)
- Imaging system (e.g., OpenLab, www.openlab.ch or Improvision, www.improvision.com)

Procedure

1. Culture bEND.3 or mlEND.1 endothelial cells in T75 culture flasks in DMEM + Glutamax, 10 % FCS at 37 °C and 7,5 % CO_2 at saturating humidity until cell monolayer is confluent.

2. Briefly wash cells in 5 mM EDTA in PBS followed by short (less than 3 min) trypsinization to produce single cells. All cell centrifugations are at $200 \times g$ for 10 min.

3. Count cells and seed them into glass Labtek chamber slides (40,000/well) or onto Bioptechs coverslips for the BioptechsFCS2 system (200,000/coverslip in 6-well plate).

4. Grow cells to confluence for 24–48 h (END monolayer).

For random shear system

1. Labtek END will be activated according to experimental design with 10–20 ng/ml IL-1 (or TNF) for 4 h for optimal E-selectin cell surface expression.

2. Labtek plastic domes are removed by cutting off silicon connection points – without removing the silicon ring surrounding each of the eight wells.

3. END monolayer is washed with plain DMEM to remove the cytokine and 50 µl of plain DMEM is added to each well.

4. As described above, mouse or human granulocytes are freshly isolated from mouse bone marrow or from human peripheral blood, respectively. U937 cells are washed once in DMEM .

5. Leukocytes are added at 10^6 leukocytes/50 μl DMEM without FCS to the END monolayer under constant rotation at 75 rpm on a horizontal shaker with an excenter of 20 mm.

6. The two cell types are cocultured for 20 min at either 4–9 °C, 21 °C or at 37 °C, depending on which adhesion molecule is at the center of interest: for integrin function, RT or 37 °C is appropriate, with the exception of PMN adhesion studies which should be performed at 4 °C to avoid artificial activation of beta2-integrins. The selectin-mediated binding becomes dominant in vitro.

7. Only under coldroom conditions, Labteks are washed twice in DMEM, followed by fixation in 2.5 % glutardialdehyde in DMEM for at least 2 h or maximally overnight.

8. Analyze the cells by imaging four random fields under a phase contrast microscope (define field size; in our experimental setup 1.69 mm²). Still images are processed by an image analysis system, e.g., Openlab (Improvision). Density slicing of the still images allows the automatic counting of adherent leukocytes. Assays are usually done in quadruplicates. Assays can be counted manually with the help of an ocular grid in the microscope in the absence of an automated image analysis system.

For parallel flow studies

1. The cover slip with confluent endothelial cell monolayer is inserted into the chamber system as described by the manufacturer. Medium is pumped across the monolayer for 30 min to produce an equilibrium. Care should be taken to have the whole system at a uniform temperature prior to injecting the purified leukocytes (see above) into the system.

2. Reagents and cells are added into the open circuit system by a needle injection port.

3. The interaction of moving leukocytes on the endothelial cell monolayer is monitored by phase contrast microscopy on an inverted microscope equipped with a video camera and connected to an image capture and analysis software (Improvision). Individual cells are tracked while they are in the field of vision and their behavior and migration paths are recorded.

▪ Expected Results

PMN adhesion is strongly increased when the endothelial cells are preactivated with proinflammatory cytokines like IL-1. A typical example of the

Fig. 1A,B. Adhesion of polymorphonuclear granulocytes (PMN) to mouse
bEND.3 endothelial cells. Mouse bone
marrow PMN were cocultured at 6 °C
under random shear stress with either
bEND.3 cells without cytokine activation (**A**), or with bend.3 cells activated
with 10 ng/ml IL-1 for 4 h (**B**). The
cytokine was washed out prior to the
application of the PMN

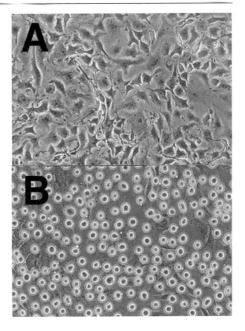

change of adhesion under random shear is given in Fig. 1. Not activated
endothelial cell monolayers show hardly any adhesion of PMN under random shear at 4 °C. In contrast, IL-1-activated bend.3 cells bind up to 6,000
PMN per mm^2, making the endothelial cell monolayer nearly invisible.
Under these conditions, leukocytes remain spherical in shape and do not
spread on the endothelial cell monolayer. Spreading occurs at 21 °C or at 37
°C assay temperature, and the evaluation of the experiment will be much
more difficult without labeling of the leukocyte, because the leukocyte can
no longer be distinguished from the endothelial cells. Adhesion to mouse E-
selectin expressed on the endothelial cell surface can also be studied with
human U937 cells, a monocyte cell line, to avoid the purification of primary
leukocytes. U937 cell adhesion at 4 °C can be blocked completely by prein-
cubation of the endothelium with a functional blocking antibody directed
against mouse and human E-selectin, UZ4 (Hammel et al. 2001). We find that
inhibition of cell-cell adhesion is most efficient when the inhibiting agent is
applied before the coculture of leukocytes with endothelium, and is present
also during the coculture period.

The parallel flow system allows the application of defined shear forces,
and allows the observation and the tracking of individual cells. The reaction
of endothelial cells and leukocytes to cytokines or other reagents can be
observed morphometrically, but also functionally in a changed adhesion
behavior. The physical parameters have been described extensively
(Lawrence and Springer 1991, Ley et al. 1989)

Troubleshooting

Poor adhesion

▶ Mouse END endothelial cells can be cultured for up to 75 passages. Regularly analyze for differentiation markers such as CD31, MECA-32 (Hallmann et al. 1995) and, after IL-1 activation for 4 h, for CD62E.

▶ Do not trypsinize END for more than 3 min – this causes dedifferentiation.

▶ Check integrity of the cell monolayer before starting the assay – the endothelial cell monolayer should be confluent, even after the wash out the cytokine.

▶ Do not change activation time – Selectin expression is narrowly regulated (expression level will change within 2 h).

No specific adhesion

▶ Leukocytes (especially PMN) already activated? – not kept on ice during preparation?

▶ END monolayer dried during application of the leukocytes?

Poor standard errors

▶ Check integrity of the cell monolayer.

▶ Avoid taking images from corners or edges of the well.

Shear stress consideration

The amount of shear utilized in in vitro assays is dependent on calculations of in vivo shear forces. The basic physical equations for this calculation are insufficient because they assume an ideal fluid, which is obviously not true for blood with its particles and viscosity. Nevertheless, the application of shear forces has greatly increased our ability to analyze the role of different adhesion molecules in the leukocyte extravasation cascade.

References

Butcher EC, Picker LJ (1996) Lymphocyte homing and homeostasis. Science 272:60–66

Hahne M, Jäger U, Isenmann S, Hallmann R, Vestweber D (1993) Five TNF-inducible cell adhesion mechanisms on the surface of mouse endothelioma cells mediate the binding of leukocytes. J Cell Biol 121:655–664

Hallmann R, Mayer D, Broermann R, Berg E, Butcher EC (1995) Novel endothelial cell marker is suppressed during formation of the blood brain barrier. Dev Dynamics 202:325–332

Hammel M, Weitz-Schmidt G, Krause A, Moll T, Vestweber D, Zerwes H, Hallmann R (2001) Species-specific and conserved epitopes on mouse and human E-selectin important for leukocyte adhesion. Exp Cell Res 269:266–274

Johnston B, Butcher EC (2002) Chemokines in rapid leukocyte adhesion triggering and migration. Sem Immunol 14:83–92

Lawrence MB, Springer TA (1991) Leukocytes roll on a selectin at physiologic flow rates: distinction from and prerequisite for adhesion through integrins. Cell 65:859–873

Ley K, Lundgren E, Berger E, Arfors KE (1989) Shear-dependent inhibition of granulocyte adhesion to cultured endothelium by dextran sulfate. Blood 73:1324–1330

Patel KD, Cuvelier S, Wiehler S (2002) Selectins: critical mediators of leukocyte recruitment. Sem Immunol 14:73–81

Vestweber D, Blanks J (1999) Mechanisms that regulate the function of the selectins and their ligands. Physiol Rev 79:181–213

Chapter 10

Application of Biomechanical Forces on Endothelial Cells

ANDREAS SCHUBERT and HENNIG MORAWIETZ

Introduction

Endothelial cells in situ are permanently exposed to biomechanical forces. Important mechanical forces are wall shear stress and cyclic strain resulting in deformation of the endothelial cells. Biomechanical forces can markedly influence structure, growth, and function of endothelial cells, resulting in a variety of intracellular molecular changes (Fisher et al. 2001, Frangos et al. 2001). Increasing evidence supports an important role of mechanical forces in the maintenance of vascular integrity and the development of vascular diseases. The localization of atherosclerotic lesions to arterial geometry associated with disturbed flow patterns suggests an important role for local hemodynamic forces in atherogenesis (Gimbrone et al. 2000). Unbranched arterial segments that are exposed to more uniform laminar flow appear relatively protected from atherosclerotic lesion development (Traub and Berk 1998). In contrast, atherosclerotic lesions first develop in the vicinity of arterial branch points and areas of major curvature, regions associated with non-laminar flow, flow reversal, and roaming stagnation points. The mean physiological shear stress acting on endothelial cells is higher in arterial vessels (approximately 15–30 dyn/cm^2) compared to venous vessels (approximately 1 dyn/cm^2). This difference in shear stress affects the shape and differentiation of endothelial cells in arteries and veins. Endothelial cells align their shape and reorganize their cytoskeleton in response to the direction and degree of shear stress. Furthermore, elongation of cells by 6% to 22% in response to mechanical strains is observed in certain intact blood vessels (Osol 1995).

Several devices have been developed to apply biomechanical forces on cultured endothelial cells in vitro. Shear stress is usually applied using cone-and-plate viscometers or parallel flow chambers. Cyclic mechanical strain can be applied by cyclic deformation of cells cultured on distensible elastic membranes. Biological compression conditions can be simulated using pressure to deform cells cultured in vitro. We will focus in this chapter on the

Springer Lab Manual
H. Augustin (Ed.)
Methods in Endothelial Cell Biology
© Springer-Verlag Berlin Heidelberg 2004

application of shear stress on endothelial cells by a cone-and-plate viscometer and on the application of mechanical strain by cyclic deformation of flexible membranes.

Materials and Equipment

Cells

- Confluent endothelial cells of various sources

Shear stress

- Cells on tissue culture dishes (e.g., 60 × 15 mm or 94 × 16 mm plates; Greiner, www.greiner-lab.com)
- Parallel plate flow chambers or circular flow chambers, (e.g., Glycotech, www.glycotech.com; Flexcell)
- Cone-and-plate viscometer (Fig. 1), for details please contact Dr. H. Lehnich, Martin Luther University Halle-Wittenberg, Germany (www.medizin.uni-halle.de/zmg/lehnich; email: holger.lehnich@medizin.uni-halle.de)

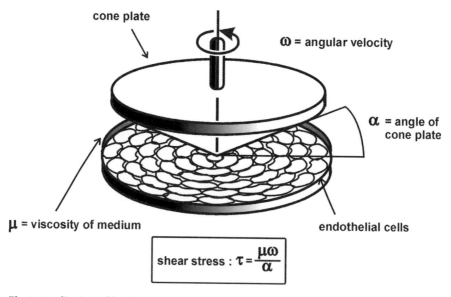

cone plate

ω = angular velocity

α = angle of cone plate

μ = viscosity of medium

endothelial cells

$$\text{shear stress} : \tau = \frac{\mu\omega}{\alpha}$$

Fig. 1. Application of laminar shear stress on human endothelial cells. Cells are exposed to laminar shear stress in a cone-and-plate viscometer. The apparatus consists of a cone with an angle of 0.5° rotating on top of a cell culture dish. The degree of shear stress τ depends on the viscosity of the medium μ, the angular velocity ω, and the angle of the cone α

Strain

- Cells on Bioflex culture plates (Flexcell, www.flexcellint.com)
- Flexcell Tension PlusT System (Flexcell)
  The stress unit is a modification of the unit originally described by Banes et al. (1985) and consists of a computer controlled vacuum unit and a baseplate to hold the culture dishes. Vacuum is repetitively applied to the rubber-bottomed dishes via the baseplate which is placed in a humidified incubator with 5 % CO_2 at 37 °C. The computer system controls the frequency of deformation and the negative pressure applied to the culture plates (Fig. 2).

Cell culture medium

- Standard medium: e.g., medium M199 with 1.25 mg/ml sodium bicarbonate, 100 µg/ml L-glutamine (Invitrogen, www.invitrogen.com), supplemented with 20 % FCS, 15 mmol/l HEPES, 100 U/ml penicillin, 100 µg/ml streptomycin, 250 ng/ml fungizone (Invitrogen), and 16.7 ng/ml endothelial cell growth supplement (C.C. Pro, www.c-c-pro.com)
- Medium with dextran: Dissolve 5 % dextran (MW 71.400; Sigma, www.sigmaaldrich.com) in standard medium in order to increase the viscosity and sterilize by passing through a 0.45 µm filter.
- Phosphate-buffered saline (PBS)

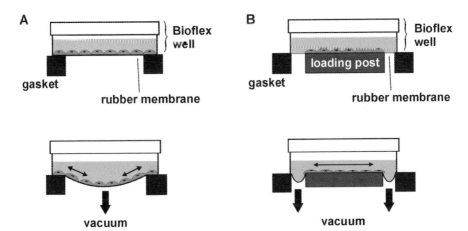

Fig. 2A,B. Application of mechanical strain on endothelial cells. Endothelial cells are cultured on distensible membranes under static conditions (*top*) or during cyclic application of vacuum causing mechanical deformation of membranes and attached cells (*bottom*). The originally described apparatus (**A**) allows the application of mechanical strain of 30 % and higher, but has the disadvantage of inhomogeneous cellular deformation (regions of high strain at the periphery and regions of lower strain in the center of the plate). The modified new strain device (**B**) is limited by the maximally applicable strain, but it serves as an uniform cellular deformation device

▨ Procedure

Shear stress

1. Confluent cultures of endothelial cells (e.g., approximately 1×10^6 cells/60-mm dish) are supplemented with fresh medium and transferred to a CO_2 incubator containing a cone-and-plate apparatus. A control dish with endothelial cells from the same preparation is treated in the same way.

2. After 2 h, remove the lid of one dish and place the culture dish in the cone-and-plate apparatus.

3. Start the rotation of the cone by adjusting the speed (e.g., 40 rpm).

4. Slowly layer the rotating cone on the culture dish and stop immediately before the cone is touching the cells. This can be monitored on the corresponding scale (e.g., in 10-µm steps).

5. Adjust rotational speed as desired and apply shear stress for appropriate time. The control dish is incubating under static conditions for the same time.

6. During long term application of shear stress, e.g., 24 h, water evaporating from the medium has to be replaced. Routinely, we slowly add up to 1 ml sterile water twice to keep the volume of the medium constant.

7. After application of shear stress, the cone is lifted, the culture dish is transferred to a sterile flow box, and the cells are washed 2 times with PBS. Subsequently, cells can be characterized by standard molecular techniques including RNA or protein preparation, immunofluorescence, etc.

Mechanical strain

1. Confluent cultures of endothelial cells on Bioflex culture plates (Flexcell) are supplemented with fresh medium and transferred to a CO_2 incubator containing the baseplate of the Flexcell Tension PlusT System (Flexcell). For application of uniform strain, use the FX-4000T Version (or higher). A control dish with endothelial cells from the same preparation is treated in the same way.

2. Follow the Quickstart instructions of the manufacturer. In brief, place four Loading Stations into the four baseplate wells and apply lubricant to the tops of the bottons. Place four gaskets on four Bioflex plates and place a Bioflex plate and gasket in each baseplate well.

3. Turn the FlexLink and afterwards the computer system on. Double click on the corresponding "FX-4000" icon starting the software.

4. Click on the "User" button and register as a user. Afterwards, quit and return to the main window.

5. Using the "Regimen" button, create a regimen by entering values into the appropriate spaces. When completed, save step and then save regimen. Exit the regimen editor.

6. With the "Assign" button, choose the desired baseplate, regimen, and user, and click on assign at the button of the window. Confirm the regimen.

7. If the program is downloaded, you may simulate the regimen and confirm by stop.

8. Turn on vacuum system. Click on start to run the regimen. The cells on the plates should begin flexing.

9. Check the water trap to see if water is accumulating. Empty the trap before it is more than half full.

10. You can pause or stop the regimen with the corresponding buttons at any time. Usually, the program will stop when the regime is complete.

11. After application of mechanical strain, the Bioflex culture plates are transferred to a sterile flow box and the cells are washed two times with PBS. Subsequently, cells can be characterized by standard molecular techniques.

Results

Cells align in the direction of flow after long-term application of laminar shear stress (Fig. 3). In contrast, turbulent shear stress does not result in alignment of endothelial cells (Fig. 3). Application of laminar shear stress using the cone-and-plate viscometer does not increase the temperature of the cell culture medium (Schubert et al. 2000). The achievement of equal degrees of shear stress at lower rotational speed by using additional dextran has been shown to yield similar results (Malek and Izumo 1992) and does not affect cell viability, detachment, or increased release of lactate dehydrogenase (LDH) into the medium as an indicator of cell integrity. In addition, dextran had no effect on control genes like glyceraldehyde 3-phosphate dehydrogenase mRNA expression (Schubert et al. 2000).

For application of mechanical strain, the previously used Flexcell apparatus allows the application of mechanical strain of more than 30 %, but has the disadvantage of inhomogeneous cellular deformation (regions of high strain at the periphery and regions of lower strain in the center of the plate) (Fig. 2A; (Gilbert et al. 1994). The modified new strain device (e.g., version

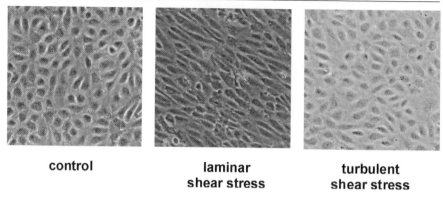

control   **laminar shear stress**   **turbulent shear stress**

Fig. 3. Adaptation of endothelial phenotype in response to shear stress. Human endothelial cells were cultured under standard static conditions (control) or were subjected to laminar or turbulent shear stress of 30 dyn/cm² in a cone-and-plate viscometer for 24 h. Cells align in the direction of flow after application of laminar shear stress. In contrast, turbulent shear stress does not result in alignment of endothelial cells

FX-4000 and higher, Fig. 2B) is limited by the maximally applicable strain, but it serves as a uniform cellular deformation device.

The expression of many endothelial cells genes is affected by shear stress. For example, long-term application of laminar shear stress induces endothelial nitric oxide synthase (Nishida et al. 1992) and downregulates pre-pro endothelin-1 (Malek and Izumo 1992, Morawietz et al. 2000). Furthermore, laminar shear stress protects endothelial cells from apoptosis (Dimmeler et al. 1996), e.g., by downregulation of proapoptotic genes and upregulation of antiapoptotic genes (Bartling et al. 2000).

Mechanical strain can increase pre-pro-endothelin-1 (Lauth et al. 2000) and apoptosis (Cattaruzza et al. 2000). Furthermore, transient application of shear stress or strain might induce immediate-early genes (Du et al. 1995, Shyy et al. 1995; Morawietz et al. 1999).

Modifications

The degree of shear stress $\tau = \mu \cdot \omega/\alpha$ (τ: shear stress, μ: viscosity of medium, ω: angular velocity, α: angle of cone plate) can be adjusted by varying the rotational speed of the cone. The value n (in rpm) can be calculated in the following way: $n = \tau \cdot \alpha/2 \cdot \pi \cdot \mu$. As an example, using a cone of 0.5° ($\alpha = 2 \cdot \pi \cdot 0.5°/360° = 0.0087$ rad) and our standard medium ($\mu = 0.007$ dyn $\cdot$ s/cm²), the value n in order to achieve a shear stress of 1 dyn/cm² (1 dyn/cm² = 0.1 N/m², low or venous shear stress) is 12 rpm. In order to keep the cell culture medium volume constant and to avoid spillover of the medium even at high rotational speed, for application of higher arterial levels of shear stress (>15 dyn/cm^2), 5 % dextran is added to the cell culture medium to increase the viscosity of the standard medium 2.95-fold to 0.02065 $dyn \cdot s/cm^2$. A level of shear stress of 15 dyn/cm^2 (15 dyn/cm^2 = 1.5 N/m^2) using medium with dextran can be achieved with $n = 60$ rpm. In this case, the control cells receive medium with dextran as well.

The flow conditions using a cone with an angle of 0.5° are laminar, because the parameter R ($r^2 \cdot \omega \cdot \alpha^2/12 \cdot v$) (Sdougos et al. 1984) is in each case smaller than 4 (e.g., $R_{1dyn/cm}^2$: 0.006; $R_{15dyn/cm}^2$: 0.03; $R_{30dyn/cm}^2$: 0.06). A cone with an angle of 5° can be used in order to achieve turbulent shear stress. Using this approach, the cone has to rotate with higher speed, compared to the same degree of laminar shear stress, and the parameter R increases to values higher than 4 resulting in turbulent shear stress (Sdougos et al. 1984).

Alternatively, shear stress can be applied on endothelial cells grown on plain or matrix covered glass cover slips using parallel plate flow chambers or circular flow chambers, e.g., by Glycotech or by the Flexcell Shear Stress Device or Flexcell Streamer. The Flexcell Shear Stress Device allows even the combination of shear stress and mechanical strain.

Defined pressure can be applied on endothelial cells using the Flexcell Compression PlusT System.

Troubleshooting

In order to minimize variations of primary cultures, isolated endothelial cells should be pooled, subsequently split, and grown in standard medium under identical conditions. Note: The growth status and the time of incubation is accompanied by an internal control in each experiment and at every time point to account for the release of growth factor and signaling molecules which may vary with cell preparation. We recommend standardizing the procedure of adding fresh medium prior to starting the experiment. Be aware that even changing the medium can induce a variety of genes (e.g., immediate-early genes).

The use of dextran to enhance the viscosity of the medium and hence shear stress is feasible, but could comprise the problem of changing the osmotic pressure as well. Therefore, some laboratories employ polyvinyl-pyrrolidone instead which does not pose this problem and at 3 % attains a much higher increase in viscosity (typically tenfold) than dextran.

Be careful to avoid bubbles when layering the rotating cone on the culture dish and try to avoid touching the cells in the device. Otherwise, the cells could be scraped off from the plate. In order to avoid direct contact of the rotating cone with the edge of the tissue culture dish, a circular area of 91 % of the cultured cells is exposed to the indicated amount of laminar shear stress. Therefore, the measured effect of shear stress in the cells harvested for standard biochemical methods might be even slightly underestimated.

In experiments applying turbulent shear stress, the parameter R depends on the radius of the cone due to the geometry of the cone-and-plate viscometer. It therefore needs to be considered that a small area in the center of the plate is stimulated with laminar shear stress, while the majority of cells will receive turbulent flow.

Genetic polymorphism like the -786C/T polymorphism of the endothelial NO synthase gene might affect the responsiveness of the gene to changes in shear stress and, thus, the NO-synthesizing capacity of endothelial cells.

Be sure that the Bioflex plate and the baseplate are in tight contact during application of mechanical strain to ensure the desired deformation by vacuum.

References

Banes AJ, Gilbert J, Taylor D, Monbureau O (1985) A new vacuum-operated stress-providing instrument that applies static or variable duration cyclic tension or compression to cells in vitro. J Cell Sci 75:35–42

Bartling B, Tostlebe H, Darmer D, Holtz J, Silber RE, Morawietz H (2000) Shear stress-dependent expression of apoptosis-regulating genes in endothelial cells. Biochem Biophys Res Commun 278:740–746

Cattaruzza M, Dimigen C, Ehrenreich H, Hecker M (2000) Stretch-induced endothelin B receptor-mediated apoptosis in vascular smooth muscle cells. FASEB J 14:991–998

Dimmeler S, Haendeler J, Rippmann V, Nehls M, Zeiher AM (1996) Shear stress inhibits apoptosis of human endothelial cells. FEBS Lett 399:71–74

Du W, Mills I, Sumpio BE (1995) Cyclic strain causes heterogeneous induction of transcription factors, AP-1, CRE binding protein and NF-kB, in endothelial cells: species and vascular bed diversity. J Biomech 28:1485–1491

Fisher AB, Chien S, Barakat AI, Nerem RM (2001) Endothelial cellular response to altered shear stress. Am J Physiol Lung Cell Mol Physiol 281:L529–533

Frangos SG, Knox R, Yano Y, Chen E, Di Luozzo G, Chen AH, Sumpio BE (2001) The integrin-mediated cyclic strain-induced signaling pathway in vascular endothelial cells. Endothelium 8:1–10

Gilbert JA, Weinhold PS, Banes AJ, Link GW, Jones GL (1994) Strain profiles for circular cell culture plates containing flexible surfaces employed to mechanically deform cells in vitro. J Biomech 27:1169–1177

Gimbrone MA, Jr., Topper JN, Nagel T, Anderson KR, Garcia-Cardena G (2000) Endothelial dysfunction, hemodynamic forces, and atherogenesis. Ann N Y Acad Sci 902:230–240

Lauth M, Wagner AH, Cattaruzza M, Orzechowski HD, Paul M, Hecker M (2000) Transcriptional control of deformation-induced preproendothelin-1 gene expression in endothelial cells. J Mol Med 78:441–450

Malek A, Izumo S (1992) Physiological fluid shear stress causes downregulation of endothelin-1 mRNA in bovine aortic endothelium. Am J Physiol 263:C389–396

Morawietz H, Ma YH, Vives F, Wilson E, Sukhatme VP, Holtz J, Ives HE (1999) Rapid induction and translocation of Egr-1 in response to mechanical strain in vascular smooth muscle cells. Circ Res 84:678–687

Morawietz H, Talanow R, Szibor M, Rueckschloss U, Schubert A, Bartling B, Darmer D, Holtz J (2000) Regulation of the endothelin system by shear stress in human endothelial cells. J Physiol (Lond) 525:761–770

Nishida K, Harrison DG, Navas JP, Fisher AA, Dockery SP, Uematsu M, Nerem RM, Alexander RW, Murphy TJ (1992) Molecular cloning and characterization of the constitutive bovine aortic endothelial cell nitric oxide synthase. J Clin Invest 90:2092–2096

Osol G (1995) Mechanotransduction by vascular smooth muscle. J Vasc Res 32:275–292

Schubert A, Cattaruzza M, Hecker M, Darmer D, Holtz J, Morawietz H (2000) Shear stress-dependent regulation of the human β-tubulin folding cofactor D gene. Circ Res 87:1188–1194

Sdougos HP, Bussolari SR, Dewey CFJ (1984) Secondary flow and turbulence in a cone-plate device. J Fluid Mech 138:379–404

Shyy JY, Li YS, Lin MC, Chen W, Yuan S, Usami S, Chien S (1995) Multiple cis-elements mediate shear stress-induced gene expression. J Biomech 28:1451–1457

Traub O, Berk BC (1998) Laminar shear stress: mechanisms by which endothelial cells transduce an atheroprotective force. Arterioscler Thromb Vasc Biol 18:677–685

Chapter 11

Endothelial Cell Permeability Assays in Culture

Maria Grazia Lampugnani and Elisabetta Dejana

Introduction

One of the most specific functions of the endothelial layer is to act as a selective barrier between blood/lymph and tissues. A standardized assay able to measure permeability of in vitro cultured endothelial cells is therefore highly desirable. In principle such an assay measures the passage of an easily detectable tracer between two compartments separated by an endothelial layer. The assay of trans-monolayer electrical resistance, which measures the passage of ions through the cell layer, while very effective for testing barrier properties of epithelial cells which form very tight layers in culture (Madara 1998), is less suitable for endothelial cells. With the exception of arterial or brain microvascular endothelium, most types of endothelial cells in vitro form layers that present low electrical resistance, which makes it difficult to pick up a further decrease.

A conceptual aspect is the definition of the route followed by the tracer to pass through the endothelial layer. That is the contribution of the intercellular and transcellular pathways to the overall value of permeability measured (Van Hinsbergh and van Nieuw Amerongen 2002). In this chapter, we will describe a method that gives an overall measure of endothelial permeability. This method detects different organization of cell-cell contacts, as determined by the presence or absence of junctional molecules, or by treatment with antibodies to these structures. We will also describe a variation of the method and a pharmacological tool which can be applied to evaluate the contribution of transcellular transport.

Materials

Cells

- Endothelial cells in culture; depending on the experimental aim, endothelial cells of different species and origin as well as carrier cells

Springer Lab Manual
H. Augustin (Ed.)
Methods in Endothelial Cell Biology

transfected for endothelial specific proteins can be used (see Sect. "Expected Results")

Reagents

- Culture medium appropriate for the specific endothelial type under test, PBS without Ca and Mg and Trypsin-EDTA.
- Fibronectin (e.g., fibronectin from human plasma, Sigma, Cat. No. F2006, www.sigmaaldrich.com); fibronectin is supplied lyophilized and is dissolved in sterile PBS with Ca and Mg at a concentration of 200 µg/ml; aliquots of 50–100 µl are then separated and stored at –20 °C. Note: all handlings should be performed under sterile conditions.
- Fluorescein isothiocianate (FITC)-conjugated dextran (e.g., MW 40,000, Sigma, Cat. No. FD-40S); as an alternative to Sigma, Molecular Probes (www.probes.com) offers a wide choice of dextran conjugates (see also Sect. "Procedure").

Supplies

- 24-well plate (e.g., Falcon, www.bdbiosciences.com or Costar, www.lab-pages.com)
- Transwell cell culture insert (e.g., polycarbonate membrane, pore size 0.4 µm, insert membrane growth area 0.33 cm², Costar, Cat. No. 3413)
- Forceps, about 15 cm long
- Micropipettes (e.g., p100, p200 and p1000, Gilson, www.gilson.com)

Equipment

- Cell culture facilities
- Gas burner
- Fluorimeter (e.g., plate fluorimeter CytoFluor 2350, Millipore, www.millipore.com)

Procedure

First day: cell seeding on Transwell membrane

Choice of Transwell insert membrane

A wide variety of membrane types is available from Costar that vary in chemical composition and pore size. In our experience, the best suitable membrane for measuring permeability of endothelial cells of different origin is the polycarbonate one with a pore size of 0.4 µm. The only drawback of this membrane is the fact that it is not clear and cells can be observed only after staining (see below Sect. "Control of monolayer integrity"). Transwell clear polyester membrane allow observation of the cell layer during culture, but the sensitivity of tracer detection is strongly decreased, which may be due to the lower density of pores (4×10^6 pores/cm^2 vs. 1×10^8 pores/cm^2 for polycarbonate membrane).

Coating of Transwell insert membrane

1. Set each Transwell unit in a well of a 24-well plate. This and all following handlings of Transwells during cell culture require a laminar flow cabinet and sterile conditions. To this purpose, the Transwell insert is lifted and transferred using sterile forceps. It is useful to havve a gas burner and a tube with alcohol available under the hood to sterilize the forceps with the flame if necessary. Please consider at least triplicates for each observation, which is particularly needed in this method that can present some tricky steps as indicated below. Consider also three extra wells that will remain free of cells and will serve as blank reference (see below Sect. "Adding the tracer").

2. Prepare a solution of fibronectin (7 µg/ml in PBS with Ca and Mg). Add 50 µl solution with a micropipette to each filter. Be warned not to touch the filter with the pipette tip. This is critical at each step of the procedure. You have to avoid damaging the filter at this stage and later the cell layer, as this will result in unspecific increase of tracer passage.

3. Incubate fibronectin on filters for at least 1 h at RT under the hood.

4. Meanwhile, prepare the cell suspension.

Preparation of cell suspension

1. Trypsin-EDTA is generally the best choice to detach cells from culture vessels. Endothelial cells, in particular of murine origin, can be difficult to detach. It is very effective to apply the following steps to facilitate the process: Wash the cell layer twice with PBS without Ca and Mg (e.g., 5 ml for 25 cm^2 flask), then incubate in PBS for 15–30 min at RT. Wash quickly with Trypsin-EDTA (e.g., 1 ml for 25 cm^2 flask), which is discarded, and add Trypsin-EDTA which will be maintained on cells until they round up. Observe the cells frequently under the microscope throughout the procedure to control their status. Then, add culture medium containing serum and growth factors. The requirement of medium, serum, as well as growth factor type and concentration depend on the specific cell type under test. The general criterion to be followed is that the culture condition has to be applied, which allows best cell growth and survival at confluence.

2. Count the cells and dilute the suspension to the required cell concentration (generally between 15×10^4 cells/ml and 30×10^4 cells/ml). The seeding density can vary depending on the cell dimension and on the aim of the experiment. In general, a cell number at seeding is chosen which gives an almost confluent culture when the cells adhere to the membrane and spread. The best seeding density for the specific cell type can be calculated from the cell density at confluence. 100 µl cell suspension is seeded into each Transwell insert. The growth area is 0.33 cm², which corresponds to the area of a well in a 96-well plate.

Cell culture on Transwell membranes: cell seeding and medium change

1. Remove fibronectin: To this purpose, lift the Transwell insert with forceps (held with left hand for right handed). Bend it a little towards yourself to see the bottom. Position the tip at the lower corner (not against the membrane), aspirate fibronectin with micropipette (held with right hand, set it to 70 µl, to be sure to aspirate all the volume). Again be very careful not to damage the membrane. It is preferable to leave some liquid than to risk damaging the membrane. In any case, the residual liquid will be diluted during the washing step. Set the Transwell in a well of a 24-well plate and add 100 µl serum free culture medium. Again, remove liquid as described for fibronectin (set the micropipette to 120 µl).

2. Set the Transwell in a well of a 24-well plate containing 600 µl complete culture medium (lower compartment). Before seeding cells, check that no air bubbles are entrapped under the Transwell membrane. If air bubbles are present, they can be easily removed lifting the Transwell insert and repositioning it back. Load each Transwell with 100 µl cell suspension (upper compartment). Do not mix the plate. The volume added is high enough to avoid formation of a meniscus and to ensure homogeneous distribution of the seeded cells.

3. Place the plate in an incubator at 37 °C and 5 % CO_2 and let the cells adhere and form a tight monolayer.

4. Culture the cells for 72 h prior to testing permeability. If you want to measure the permeability of a tightly confluent monolayer, this is indicated in most cases. For specific aims and cell types, this culture time can be prolonged. However, you have to take into consideration that some endothelial cells, for example HUVEC, cannot remain confluent for many days. After reaching maximal saturation density, they may start to detach. This will unspecifically increase permeability. A detailed knowledge of the behavior of the specific cell type used will help in the planning of the seeding density and length of culture.

5. If culturing has to be prolonged for more than three days, it is advisable to change the medium. Transwells are handled as described above and set in a well with 600 μl fresh medium. The medium from the upper compartment is removed and 100 μl fresh medium is carefully added to the upper compartment.

Assay day

Adding the tracer

1. Add 5 μl FITC-dextran (20 mg/ml in PBS without Ca and Mg, the final concentration in the upper compartment is 1 mg/ml) to the upper compartment (see also below Sect.'Tracer choice'). This can be followed by a stimulus to increase permeability, if required by the experiment. As described above, be careful not to damage the cell layer. To this aim, position the tip of the micropipette just under the meniscus.

2. Gently rotate the plate and incubate for the required time.

To measure maximum permeability, two types of controls can be introduced. a) Treatment of cells with EGTA (5 mM EGTA in serum free medium. Note that the lower compartment also has to contain serum free medium) will induce severe cell retraction within few minutes and maximal increase of permeability. b) Another control of maximum permeability consists in measuring the permeation of the probe across an empty Transwell filter.

Tracer choice

The characteristics of the tracer determine the functional interpretation of the measurements. Fluorescent dye-conjugated dextrans offer the advantage that they are available in a wide range of MW (from MW 4,000 to 150,000, either from Sigma or from Molecular Probes) and therefore may serve to asses the 'tightness' of the monolayer. Dextrans are mostly membrane impermeant and can be assumed to give a measure of the barrier activity of cell-cell junctions. However, the contribution of endocytosis to the total permeability measurements cannot be assessed if non-neutral dextran is used. The relative contribution of transcellular transport can be measured (Balda et al, 1996). Additionally, the activity of junctional proteins modifies the passage of FITC-dextran through the endothelial monolayer in many different experimental conditions. This suggests that the assay described can measure paracellular permeability (see Sect. "Expected Results").

Probes labeled with fluorescent dyes offer good sensitivity of detection and are easy to handle. Their only disadvantage is the need of a fluorimeter for the measurement. If a fluorimeter is not available, radio-labeled tracers can be used. However, there are no commercially available radio-labeled dextrans and, consequently, different molecules have to be applied. Mannitol (MW 184) and inulin (MW 5,200), which are tracers of paracellular flux, are small molecules that may not be suitable to discriminate permeability of relatively leaky monolayers. Albumin (about 60 kD) can also be transported across the cell. Therefore, the contribution of transcellular transport has to be evaluated (see above), if a specific measurement of paracellular permeability is required. The calveolae inhibitor filipin can be used if the transport through caveolae has to be evaluated (Schnitzer et al. 1994). However, the cellular mechanisms controlling transcellular permeability such as caveolae or fenestrae are usually not maintained in cultured endothelial cells.

Collection of samples

1. Culture media (50 µl) are sampled from the lower compartment. The tip of the micropipette is introduced in one of the openings along the wall of the Transwell insert to have access to the lower compartment without removing the upper unit.

2. To maintain hydrostatic equilibrium, 50 µl of fresh culture medium is added to the lower compartment.

3. Sampling can start after few minutes and continue for some hours depending on the experimental condition (for example, short acting stimuli are histamine or thrombin, a long acting one VEGF). The comparison can also be done between different lines of transfected cells expressing or not a specific protein in wild type or mutated form, see Sect. "Expected Results". In this case, permeability can be checked consecutively over several hours). Samples can be stored at 4 °C for at least 24 h prior to testing the concentration of the tracer.

Measurement of tracer concentration

FITC-conjugated probes can be measured easily using a fluorimeter, either the classical cuvette-type or the plate reader-type. Absorbance and emission wavelengths are set at 492 nm and 520 nm, respectively. Samples are in general diluted 1:20 in PBS without Ca and Mg prior to measurement. This ratio may change for specific cell types or stimuli and may need to be determined in preliminary experiments. Using a plate fluorimeter, a volume of 100 µl sample is set in each well of a 96-well plate.

Results can be expressed as 'Fluorescence units' directly using the reading values obtained by the fluorimeter. Alternatively, concentrations can be calculated using a titration curve of known concentrations of the same probe.

Control of monolayer integrity

The cell layer on the Transwell membrane can be stained after collection of the last sample to control the integrity of the monolayer. Transfer the Transwell insert to a new well, using a micropipette to carefully remove the medium from the upper compartment. Add 100 µl Diff-Quick fixative (2 mg/l Fast Green in methanol) for 5 min, remove it and add Crystal Violet (0.5 % in 20 % methanol) for 3 min. Wash twice with distilled water. Fixative, stain, and water can also be added to the lower compartment. Filters are allowed to dry upside down and inserts are placed in a 24-well plate and observed under the microscope for altered monolayer integrity. This observation can be very useful to interpret the results of tracer permeation, e.g., to eventually eliminate extreme results which have resulted from lesions to the monolayer. In most cases, monolayer damage results from the manual manipulation. Therefore, the monolayer may appear regular, but in some areas scratches or holes are visible. If the entire monolayer appears discontinuous, the culture conditions most likely do not suit the needs of the cells and have to be changed.

For a more detailed morphological analysis, it is also possible to stain cells on Transwell membranes for immunofluorescence microscopy. Cells can be fixed and stained on the membrane in the Transwell insert. The membrane can be detached from the plastic frame after staining, using a fine needle or a razor blade. It is then fixed on a glass slide in mounting medium (e.g., Mowiol 4-88, Calbiochem, www.calbiochem.com) under a glass coverslip.

▧ Expected Results

Some typical agents and cell types tested are summarized in Table 1.

▧ Troubleshooting

The most critical aspect of the method is the handling of the Transwell insert and of the cell layer cultured on it in a way that maintains the integrity of the membrane and the cell layer. Many suggestions have been introduced throughout the description of the procedure, which should help to get a continuous cell layer and an undamaged membrane.

Non-endothelial cells on filter may be unable to form a continuous monolayer, especially if these cells are used as carrier of endothelial genes. For example, CHO do not adjust to fibronectin coating, which results in poor

Table 1. Summary of some typical agents and cell types tested in the described permeability assay

Permeability modifying agent	Cell type	Outcome	Reference
Thrombin (1–10 U/ml)	HUVEC	increased permeability	Rabiet et al. 1996 van Nieuw Amerongen et al. 1998
Histamine (1×10^{-6}–1×10^{-4} M)	HDMVEC HUVEC	increased permeability	Andriopoulou et al. 1999 van Nieuw Amerongen et al. 1998
VEGF (1–100 ng/ml)	HUVEC	increased permeability	Esser et al. 1998 Breslin et al. 2003
TNF-α (0.1–10 µg/ml)	HUVEC	increased permeability	Tiruppathi et al. 2001 Friedl et al. 2003
MAb to VE-cadherin (50 µg/ml)	HUVEC, murine endothelial cells	increased permeability Fig. 1	Corada et al. 2001
absence of VE-cadherin	whole embryo and ES-derived VE-cadherin null endothelial cells	increased permeability	Fig. 1
transfer of wildtype VE-cadherin	CHO VE-cadherin null endothelial cells	decreased permeability	Breviario et al. 1995
Catenin-binding domain-truncated VE-cadherin	CHO	no effect on permeability	Navarro et al. 1995
Absence of JAM	lung endothelial cells from JAM null mice	no effect on permeability	Fig. 1

HUVEC, human umbilical vein endothelial cells
HDMVEC, human dermal microvascular endothelial cells
CHO, Chinese hamster ovary cells

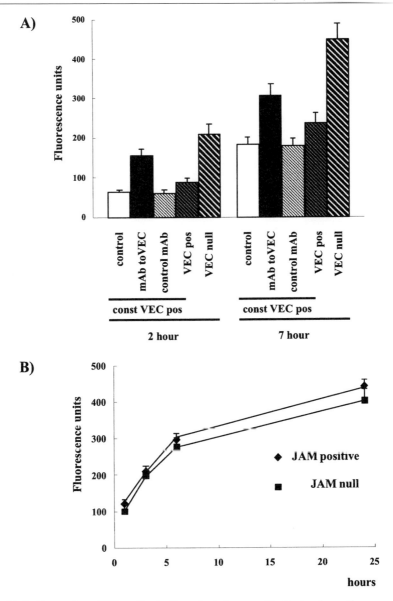

Fig. 1A,B. Expression of VE-cadherin (transfected or constitutive) reduces the permeability of murine endothelial monolayer. Endothelial cells from VE-cadherin null embryo (VEC null) were transfected to express wild type VE-cadherin (VEC positive, Lampugnani et al. 2002). Endothelial cells from wild type littermate embryo, which constitutively express VE-cadherin (ConstVEC positive), were also tested (**A**). In comparison, endothelial cells derived from the lungs of animals expressing or not the junctional molecule JAM (JAM positive and JAM null, respectively) have a comparable permeability (**B**). mAb to the EC1 extracellular repeat of VE-cadherin (mAb to VEC, BV13) increases monolayer permeability (**A**), whereas a mAb to VE-cadherin EC4 extracellular repeat (control mAb, BV14) does not (Corada et al. 2002). Permeability was measured as described in the text using FITC-dextran (MW 40,000, Sigma) as a tracer

spreading. The optimal culture condition has to be considered for each specific cell type.

References

Andriopoulou P, Navarro P, Zanetti A, Lampugnani MG, Dejana E (1999) Histamine induces tyrosine phosphorylation of endothelial cell-to-cell adherens junctions. Arterioscler Thromb Vasc Biol 19:2286–2297

Balda MS, Whitney JA, Flores C, Gonzalez S, Cereijido M, Matter K (1996) Functional dissociation of paracellular permeability and transepithelial electrical resistance and disruption of the apical-basolateral intramembrane diffusion barrier by expression of a mutant tight junction membrane protein. J Cell Biol 134:1031–1049

Breslin JW, Pappas PJ, Cerveira JJ, Hobson RW 2nd, Duran WN (2003) VEGF increases endothelial permeability by separate signaling pathways involving ERK-1/2 and nitric oxide. Am J Physiol Heart Circ Physiol 284:H92–H100

Breviario F, Caveda L, Corada M, Martin-Padura I, Navarro P, Golay J, Introna M, Gulino D, Lampugnani MG, Dejana E (1995) Functional properties of human vascular endothelial cadherin (7B4/cadherin-5), an endothelium-specific cadherin. Arterioscler Thromb Vasc Biol 15:1229–1239

Corada M, Liao F, Lindgren M, Lampugnani MG, Breviario F, Frank R, Muller WA, Hicklin DJ, Bohlen P, Dejana E (2001) Monoclonal antibodies directed to different regions of vascular endothelial cadherin extracellular domain affect adhesion and clustering of the protein and modulate endothelial permeability. Blood 97:1679–1684

Corada M, Mariotti M, Thurston G, Smith K, Kunkel R, Brockhaus M, Lampugnani MG, Martin-Padura I, Stoppacciaro A, Ruco L, McDonald DM, Ward PA, Dejana E (1999) Vascular endothelial-cadherin is an important determinant of microvascular integrity in vivo. Proc Natl Acad Sci U S A 96:9815–9820

Corada M, Zanetta L, Orsenigo F, Breviario F, Lampugnani MG, Bernasconi S, Liao F, Hicklin DJ, Bohlen P, Dejana E (2002) A monoclonal antibody to vascular endothelial-cadherin inhibits tumor angiogenesis without side effects on endothelial permeability. Blood 100:905–911

Esser S, Lampugnani MG, Corada M, Dejana E, Risau W (1998) Vascular endothelial growth factor induces VE-cadherin tyrosine phosphorylation in endothelial cells. J Cell Sci 111(Pt 13):1853–1865

Friedl J, Puhlmann M, Bartlett DL, Libutti SK, Turner EN, Gnant MF, Alexander HR (2002) Induction of permeability across endothelial cell monolayers by tumor necrosis factor (TNF) occurs via a tissue factor-dependent mechanism: relationship between the procoagulant and permeability effects of TNF. Blood 100:1334–1339

Lampugnani MG, Zanetti A, Breviario F, Balconi G, Orsenigo F, Corada M, Spagnuolo R, Betson M, Braga V, Dejana E (2002) VE-cadherin regulates endothelial actin activating Rac and increasing membrane association of Tiam. Mol Biol Cell 13:1175–1189

Madara JL (1998) Regulation of the movement of solutes across tight junctions. Annu Rev Physiol 60:143–159

Navarro P, Caveda L, Breviario F, Mandoteanu I, Lampugnani MG, Dejana E (1995) Catenin-dependent and -independent functions of vascular endothelial cadherin. J Biol Chem 270:30965–30972

Rabiet MJ, Plantier JL, Rival Y, Genoux Y, Lampugnani MG, Dejana E (1996) Thrombin-induced increase in endothelial permeability is associated with changes in cell-to-cell junction organization. Arterioscler Thromb Vasc Biol 16:488–496

Schnitzer JE, Oh P, Pinney E, Allard J (1994) Filipin-sensitive caveolae-mediated transport in endothelium: reduced transcytosis, scavenger endocytosis, and capillary permeability of select macromolecules. J Cell Biol 127:1217–1232

Tiruppathi C, Naqvi T, Sandoval R, Mehta D, Malik AB (2001) Synergistic effects of tumor necrosis factor-alpha and thrombin in increasing endothelial permeability. Am J Physiol Lung Cell Mol Physiol 281:L958–968

van Hinsbergh VW and van Nieuw Amerongen GP (2002) Intracellular signalling involved in modulating human endothelial barrier function. J Anat 200:549–560

van Nieuw Amerongen GP, Draijer R, Vermeer MA, van Hinsbergh VW (1998) Transient and prolonged increase in endothelial permeability induced by histamine and thrombin: role of protein kinases, calcium, and RhoA. Circ Res 83:1115–1123

Chapter 12

Three-Dimensional in Vitro Angiogenesis Assays

THOMAS KORFF

Introduction

Stimulatory events such as ovarian corpus luteum angiogenesis (physiological condition) or tumor angiogenesis (pathological condition) activate endothelial cells (EC) to degrade their underlying basal membrane, to migrate into the surrounding matrix, to proliferate, and to establish new anastomosing networks which mature by recruiting mural cells (pericytes and smooth muscle cells [SMC]). A number of two-dimensional in vitro assays such as lateral sheet migration and proliferation assays have been utilized to study some of the functional and phenotypic properties of endothelial cells during angiogenesis (Sato and Rifkin 1988, Pepper et al. 1989, Augustin and Pauli 1992). Correspondingly, three-dimensional assay systems with EC invading fibrin or collagen matrices have been developed in order to more closely mimic the angiogenic process and to study specific matrix-dependent steps of the angiogenic cascade (Montesano and Orci 1985, Madri et al. 1988). Embedding of single EC at high density in a three-dimensional matrix has been exploited as an endothelial cell alignment assay (Ilan et al. 1998). Alignment of EC reflects some morphogenetic properties of EC, but is not necessarily an appropriate in vitro surrogate of the angiogenic cascade. EC dispersed at lower density as single cells in a matrix will undergo apoptosis, even in the presence of survival factors (Pollman et al. 1999, Satake et al. 1998). We have observed that spheroidal culture of EC stabilizes them and induces their differentiation. Based on these observations, we have used EC spheroids as stabilized delivery devices for the focal embedding of EC in a collagen matrix, establishing a highly reproducible and robust three-dimensional angiogenesis assay system (Korff and Augustin 1999). This chapter outlines the details of this assay with the embedding of EC spheroids in collagen gels, their stimulation, and the subsequent quantitative analysis of capillary sprout formation originating from stimulated spheroid embedded EC.

Springer Lab Manual
H. Augustin (Ed.)
Methods in Endothelial Cell Biology
© Springer-Verlag Berlin Heidelberg 2004

Materials

Methyl cellulose stock solution

The preparation of methyl cellulose stock solution (methocel) is critical. If the concentration of methyl cellulose is too low or the solution contains methyl cellulose debris, single cells will stick to the wall of the culture plate and several small spheroids are formed in each well instead of a single uniform spheroid. For preparation of methocel, methyl cellulose with high viscosity should be used (Methyl cellulose from Sigma, Cat. No. m-0512, 4,000 centipoises [cpi], www.sigmaaldrich.com).

1. Autoclave the pure powder (6 g) in a 500-ml flask containing a clean magnetic stirrer (the methyl cellulose powder is resistant to this procedure).

2. Dissolve the autoclaved methyl cellulose in preheated 250 ml basal medium (60 °C) for 20 min (using the magnetic stirrer).

3. Add 250 ml basal medium (room temperature) to a final volume of 500 ml.

4. Mix the solution for 1–2 h (4 °C). Aliquot the final stock solution (50-ml aliquots) and clear by centrifugation (5,000 × g, 2 h, RT).

5. Use only the clear, highly viscous supernatant for spheroid culture.

Collagen stock solution

The following protocol is for the preparation of a collagen stock solution from two rat tails. Use of commercially available collagen is possible but may extend polymerization times.

1. Place two rat tails for 20 min in 70 % ethanol.

2. Remove the skin by cutting it from the tail root to the tip using scissors and pull it away it from the tail corpus.

3. Wash the scalped tails in ethanol.

4. Break each second (depending on the length of the tails) vertebra and extract the tendons. Make sure to isolate tendons without any attached connective tissue.

5. Collect the tendons and place them in 500 ml ethanol for 20 min.

6. Dry the tendons under a lamina air flow for 20–30 min.

7. Put the tendons in 250 ml 0.1 % acetic acid (v/v in H_2O) and place them in the refrigerator for 48 h shaking the solution at least twice each day.

8. Aliquot the final solution in autoclaved tubes and centrifuge them at 4 °C at 17,000 g for 1 h.

9. Collect the clear supernatant. Be sure not to collect any debris. The clear acid extracted collagen stock solution can be stored in the refrigerator for at least 6 month.

Collagen equilibration

For equilibration, mix 4 ml of the collagen stock solution with 0.5 ml of ten-fold concentrated medium (M199, DMEM or as required) and keep the mixed solution on ice for at least 15 min. The collagen concentration of the stock solution will be too high for most applications and the mixed solution may solidify. This can be avoided by diluting (1:2, 1:3, 1:4) the collagen stock solution with 0.1 % sterile acetic acid. Mix the diluted stock solution with tenfold-concentrated medium as described above until the final solution remains liquid. The collagen stock solution is ready to use. To keep endothelial baseline sprouting levels low, it is advisable to prepare the stock solution at least 4 weeks prior to use. Freshly prepared collagen will lead to higher absolute sprouting levels, and may cause a higher interexperimental variation.

Neutralization solution

Sterile 0.2 N NaOH solution aliquoted in 15-ml tubes.

Tissue culture plates

The spheroid-based in gel angiogenesis assay is performed in 24-well non-adhesive tissue culture plastic. The assay has been optimized for use with 24-well plates for suspension culture from Greiner (Cat. No. 662102; Frickenhausen, Germany). For the generation of size and cell number-defined spheroids, 96-well non-adhesive round bottom tissue culture dishes are used [96-well plates for suspension culture from Greiner (Cat. No. 650185, www.greiner-lab.com)].

Fixative

Aqueous solution of 10 % paraformaldehyde.

■ Procedure

At least 48 spheroids per collagen gel (1 ml) should be used. For analysis of EC network formation 400, spheroids per collagen gel should be used to analyze the sprouting angiogenic activity of a test compound. A higher number of spheroids (400 per collagen gel) is necessary for network formation experiments. The following protocol is for the preparation of 8 collagen gels containing 48 spheroids per gel.

Generation and harvesting of EC spheroids

Protocol to generate 4×96 spheroids containing 400 EC/spheroid:

1. Trypsinize a confluent EC monolayer.

2. Suspend these cells in 10 ml methocel-containing medium [20 % methocel stock solution (see above), 80 % culture medium (supplements/FCS: cell type dependent)].

3. Count the cells and suspend approx. 1.6×10^5 (4 plates $\times$ 96 wells $\times$ 400 cells/well) in 40 ml (4 plates $\times$ 96 wells $\times$ 100 µl/well) methocel containing medium.

4. Distribute the cell suspension into four 96-well plates using a 12-channel pipette (100 µl/well) and incubate the plates overnight at 37 °C (5 % CO_2, 100 % humidity).

5. Next morning: Harvest the spheroids from 96-well plates using standard pipette tips (1 ml) transferring them into 15-ml or 50-ml tubes. To avoid shearing (shear stress) of the spheroids during transfer, it is advisable to widen the tip hole by cutting away 1–2 mm (not more) of the pipette tip. Spheroids are collected by centrifugation (500 $\times$ g, 2–3 min).

3D spheroid-based angiogenesis assay

1. Fill the outer wells of a 24-well plate with 0.5–1.0 ml sterile PBS and place the plate in the incubator (37 °C).

2. Carefully mix 0.5 ml 10-fold-concentrated medium (cell type-dependent) and 4 ml collagen stock solution in a 15-ml tube and place the final collagen-medium on ice.

3. Place the neutralization solution on ice.

4. Harvest 4×96 spheroids (400 EC/spheroid) in one 50-ml tube.

5. Centrifuge the spheroids (3 min, 500 $\times$ g).

6. Carefully remove the supernatant and shortly scratch the tube over a rough underground to loosen the pellet (do not let the pellet stay for longer than 15–30 min, otherwise the spheroids will stick together).

7. Overlay (do NOT mix) the pellet with 4.5 ml FCS-containing methocel stock solution (10–40 % FCS, 90–60 % methocel stock solution; FCS con-

tent depending on the cell type; note that the final FCS content in the gel is only the half of the concentration in the methocel solution).

8. Immediately prior to use, neutralize the collagen medium (see Step 2) solution using ice cold neutralization solution (the phenol red pH indicator in the collagen medium should change the color from yellow to flesh/red-colored). Mix immediately by carefully inverting the tube to avoid polymerization. The neutralized collagen solution should be clear.

9. Keep the neutralized collagen medium on ice (it should take 3–5 min until the solution polymerizes).

10. Mix 4.5 ml neutralized collagen medium with the 4.5 ml spheroid containing methocel solution (again: do it FAST and CAREFULLY and avoid bubbles!). Homogenous mixing of all the components is most critical for the formation of a clear and sprout-supporting collagen matrix and for the homogenous dispersion of the spheroids in collagen gels.

11. Divide the spheroid containing collagen/methocel solution on the PRE-WARMED 24-well plate (8 × 1 ml).

12. Immediately place the plate into an incubator (37 °C, 100 % humidity).

13. Let the collagen gels polymerize for at least 30 min.

14. For stimulation, overlay the collagen gels with 100 µl medium containing compounds (10fold-concentrated) to be tested. It takes approx. 1–2 h for the compounds to penetrate the collagen gel.

15. Incubate the collagen gels for 24–48 h.

16. Add 1 ml fixative (see above) to stop the assay. Assay plates can be stored for up to 4 weeks at 4 °C.

Quantitative analysis endothelial cell sprouting

The pro- or antiangiogenic potency of a compound is reflected by its capacity to increase/decrease the number AND the length of the sprouts originating from the spheroids (Fig. 1A). For this reason, the cumulative length of all the sprouts originating from one spheroid is a sensitive parameter which is calculated as cumulative sprout length ([CSL], Fig. 1B), measuring the CSL of at least 10 randomly selected spheroids per data point.

1. Let the 24-well plate adjust to room temperature and remove 1 ml of the fixative-containing supernatant.

2. Place the plate under an inverted microscope.

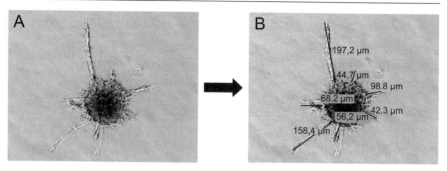

Fig. 1. A Phase contrast image of a collagen gel embedded HUVEC spheroid sprouting for 24 h. **B** Automated computerized image analysis quantitating the cumulative sprout length of all sprouts originating from the spheroid

3. Randomly select 10 individual spheroids and measure their CSL (e.g., with an ocular grid or preferably by digital image analysis; Fig. 1B). Do not analyze spheroids at the bottom of the well (EC originating from these spheroids do not form sprouts but spread as monolayer).

Processing of the collagen gels for paraffin sectioning

Note that fixation of collagen gels with formaldehyde-containing fixatives makes the collagen more fragile and that – depending on the number of embedded spheroids – several sections may be prepared for cross sections of sprouting endothelial cells.

1. Carefully remove untreated collagen gels from the 24-well plate.

2. Place each gel in 50 ml of buffered fixative containing 4 % freshly prepared paraformaldehyde for 24 h at 4 °C.

3. Replace the fixative with 50 ml water and let it penetrate the gel for 2 h at RT.

4. Remove the water and incubate the gels in 70 % ethanol for 24 h.

5. Repeat this step with 85 %, 99 % ethanol, and isopropanol.

6. Remove the isopropanol and incubate the gels in low melting paraffin (melting temperature: 44–48 °C) for 24 h at 65–70 °C.

7. Replace the paraffin by high melting paraffin (paraplast, melting temperature: 56–58 °C) and incubate for 48 h at 65–70 °C .

8. Embed the collagen gel for paraffin sectioning

Expected Results

The sprouting activity is strongly dependent on the type of endothelial cell, the concentration of FCS, and the quality of the collagen gel. Usually, HUVEC spheroids in collagen gels containing 10 % FCS show low baseline sprouting activity (Fig. 2A,B). Capillary sprout formation can be significantly stimulated by VEGF within 24h (Fig. 2A,C) down to concentrations below 1 ng/ml (Fig. 2A,D). If HUVEC spheroids are used in the angiogenesis assay to analyze the angiogenic activity of different compounds, the assay should be stopped after 24–48 h. The collagen gels should be overlaid with up to 0.5 ml growth factor-containing medium for prolonged incubation times (e.g., for the study of capillary network formation) and the medium should be changed every other day. BAEC, which produce endogenous FGF-2, perform well in long-term angiogenesis assays (Korff and Augustin, 1999).

Compared to VEGF, FGF-2 induces the formation of capillary like sprouts with a more continuous morphology (Fig. 3). Depending on the cell type and stimulus, sprouting of EC originating from collagen-embedded spheroids is accompanied by lumen formation, which is most prominent for BAEC.

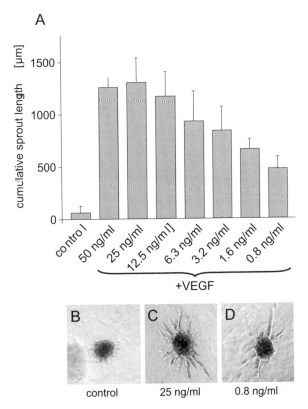

Fig. 2A-D. Representative example of a VEGF titration experiment demonstrating the sensitivity and robustness of the assay. A Titration of the angiogenic stimulus VEGF with a dilution factor 2 leads to a gradual declining dose-response curve with a half maximal stimulation of approx. 1 ng/ml. B Control HUVEC spheroid with very little baseline sprouting activity. C Maximal stimulation of sprouting activity with 25 ng/ml VEGF. D Induction of significant sprouting activity with as little as 0.78 ng/ml VEGF

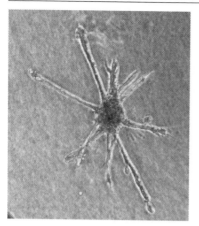

Fig. 3. Induction of in gel sprouting angiogenesis with FGF-2, which causes a robust angiogenic response with solid, continuous capillary-like structures. In contrast, VEGF stimulation leads to a different morphophenotype with loosely structured capillaries and individual endothelial cell scattering into the gel (compare Fig. 2)

HUVEC spheroid
+ FGF-2 (36h)

Troubleshooting

▶ **Problem:** Spheroids sink to the bottom of the well:
The polymerization of the collagen takes too long and the spheroids are not homogenously distributed throughout the collagen gel. This may be due to inadequate mixing of the collagen gel components or to a low collagen concentration in the collagen stock solution. After mixing all the components, polymerization of the collagen gels should take not more than 2 min at 37 °C.

▶ **Problem:** High baseline sprouting:
To reduce baseline sprouting, HUVEC monolayers and HUVEC spheroids should be prepared in medium containing all supplements (ECGS, FCS). To keep baseline sprouting low, the FCS content in collagen gels should be the same as in the culture medium. The collagen gel should contain 10 % FCS if the culture medium contains 10 % FCS. Supplements should not be given into the collagen gel, as this may dramatically increase baseline sprouting activity. If baseline sprouting activity is still too high, the spheroids should be cultured for 48 h (instead of 24 h) prior to use or the FCS concentration should be reduced (not below 5 %). Note that some cell types produce endogenous FGF-2 (e.g., BAEC) and show a high baseline sprouting which can be decreased by the addition of FGF-2-neutralizing antibodies. Furthermore, freshly prepared collagen sometimes causes increased baseline sprouting activity increasing the inter-experimental variability of the assay. This can be prevented by using collagen stock solution that is produced 4 weeks prior to use.

▶ **Problem:** Apoptosis of sprouting EC:
Some endothelial cell types are more "survival factor"-dependent (e.g., HDMVEC) than other cells (e.g., BAEC). Furthermore, depending on the amount of spheroids embedded in a collagen gel, growth/survival factors are rapidly depleted (FGF-2 has generally proven to be more stable than VEGF in collagen gels with large amounts of spheroids (400 HUVEC spheroids for 48 h). If EC undergo apoptosis, the concentration of growth factors and/or FCS in the collagen gels can be increased. Changing the medium including the growth factors every other day may also prevent early apoptosis of EC.

▶ **Problem:** Insufficient quality of RNA isolated from collagen embedded sprouting EC:
Homogenize collagen gels containing at least 400 spheroids per gel in RNA isolation reagent (i.e., TRI-reagent BD, Sigma) which supports the isolation of RNA from liquid samples. The resulting RNA should be cleaned at least once by standard phenol/chloroform extraction (larger amounts of spheroids than 400/ml may cause high proteolytic activity which may destabilize the gel, lead to EC apoptosis and RNA degradation).

References

Augustin HG, Pauli BU (1992) Quantitative analysis of autocrine-regulated, matrix-induced, and tumor cell-stimulated endothelial cell migration using a silicon template compartmentalization technique. Exp Cell Res 198:221–227

Ilan N, Mahooti S, Madri JA (1998) Distinct signal transduction pathways are utilized during the tube formation and survival phases of in vitro angiogenesis. J Cell Sci 111:3621–3631

Korff T, Augustin HG (1999) Tensional forces in fibrillar extracellular matrices control directional capillary sprouting. J Cell Sci 112:3249–3258

Madri JA, Pratt BM, Tucker AM (1988) Phenotypic modulation of endothelial cells by transforming growth factor-beta depends upon the composition and organization of the extracellular matrix. J Cell Biol 106:1375–1384

Montesano R, Orci L (1985) Tumor-promoting phorbol esters induce angiogenesis in vitro. Cell 42:469–477.

Pepper MS, Spray DC, Chanson M, Montesano R, Orci L, Meda P (1989) Junctional communication is induced in migrating capillary endothelial cells. J Cell Biol 109:3027–3038

Pollman MJ, Naumovski L, Gibbons GH (1999) Endothelial cell apoptosis in capillary network remodeling. J Cell Physiol 178:359–370

Satake S, Kuzuya M, Ramos MA, Kanda S, Iguchi A (1998) Angiogenic stimuli are essential for survival of vascular endothelial cells in three-dimensional collagen lattice. Biochem Biophys Res Commun 244:642–646

Sato Y, Rifkin DB (1988) Autocrine activities of basic fibroblast growth factor: regulation of endothelial cell movement, plasminogen activator synthesis and DNA synthesis. J Cell Biol 107:1199–1205.

Chapter 13

Rat Aortic Ring Assay of Angiogenesis

ROBERTO F. NICOSIA and WEN-HUI ZHU

▓ Introduction

Angiogenesis, the formation of new blood vessels from preexisting vessels, plays a critical role in embryonal development, fetal growth, the female menstrual cycle, and wound healing. Newly formed blood vessels promote collateral blood flow in coronary artery disease and other ischemic vascular disorders, but also contribute to the progression of cancer, rheumatoid arthritis, and diabetic retinopathy. Among the many methods currently used to test the activity of pro-angiogenic and anti-angiogenic factors (Jain et al. 1997), the rat aortic ring model, first described by our group in the early 1980s (Nicosia et al. 1982), has proven to be a practical and cost effective assay of angiogenesis. The purpose of this chapter is to review in detail protocols currently used in our laboratory to study angiogenesis in this system.

Aortic ring cultures: general overview of methodological principles

The aortic ring model is an ex vivo assay of angiogenesis that combines advantages of both in vivo and in vitro models. Angiogenesis occurs in a chemically defined culture environment, which can be easily adapted to different experimental conditions. Since cells of the aortic outgrowth have not been modified by repeated passages in culture, microvessels developed in this system are essentially indistinguishable from microvessels formed during angiogenesis in vivo. Both the growth medium and the collagen gel in which microvessels grow can be modified to test the activity of soluble factors and extracellular matrix molecules (Nicosia and Ottinetti 1990, Nicosia et al. 1993, Nicosia et al. 1994A, Nicosia et al. 1994B). The effect of angiogenesis agonists and antagonists can be measured in the absence of serum molecules which may otherwise bind, inactivate or simulate the action of the substances being tested (Kawasaki et al. 1989, Nicosia and Ottinetti 1990). The aortic ring model can be used to study not only angiogenesis but also vascular regression because neovessels are eventually reabsorbed during the second and third week of culture (Nicosia and Ottinetti 1990, Zhu et al. 2000).

Springer Lab Manual
H. Augustin (Ed.)
Methods in Endothelial Cell Biology
© Springer-Verlag Berlin Heidelberg 2004

We describe here two types of aortic ring assay: (1) standard assay with thick collagen gels, and (2) modified assay with thin collagen gels. The standard aortic ring assay with thick gels was the first method developed in our laboratory to study angiogenesis in serum-free medium (Nicosia and Ottinetti 1990). More recently, we introduced the thin prep modification of the assay (Zhu and Nicosia 2002), which significantly facilitates the preparation and subsequent immunostaining of the cultures (see below). Gel thickness in the modified assay is reduced by decreasing the amount of collagen per culture to 1/10 of its original volume. This results in thin, wafer-like gels that are easily penetrated by labeling reagents used for immunohistochemistry. Cultures with thick collagen gels are, however, preferable to study vascular regression, which occurs more slowly and can be more effectively modulated with reagents in a thick gel. Because they float, thick collagen gels with angiogenic outgrowths can also be co-cultured with different cell types seeded on the bottom of the culture dish. The procedure for the preparation of the aortic rings is the same for both thick and thin prep assays. Collagen can be replaced with fibrin, which is an optimal matrix for angiogenesis. We prefer collagen or fibrin to Matrigel, which may confound the interpretation of the assay because non-endothelial cells including fibroblasts have been shown to organize into networks in this matrix (Donovan et al. 2001). Collagen and fibrin represent instead provisional matrices that better simulate the solid-phase milieu in which vessels develop during physiologic and pathologic processes.

Materials

Animals

- 1–2-month-old Fischer 344 male rats (Harlan Sprague Dawley, www.harlan.com)

Tissue culture facility

- Tissue culture room with HEPA-filtered air

Equipment

- Dissecting microscope (e.g., American Optical Corp. New York, USA)
- 25 × 53 × 90 cm custom-made rectangular wooden hood with glass top and front opening
- Laminar flow hood (e.g., The Baker Company, www.bakercompany.com)
- CO_2 incubator (e.g., Forma Scientific, www.forma.com)
- Inverted and standard light microscopes (e.g., Scientific Instrument Co., www.simicroscopes.com)

- Laser scanning microscope (e.g., Leica Microsystems, www.leica-microsystems.com)
- Transmission electron microscope (e.g., JEOL, www.jeol.com)

Supplies for preparation of assay

- Electric hair clipper for small animal (Oster, www.oster.com)
- 4-0 silk sutures (Ethicon, www.ethicon.com)
- 100 % CO_2 (Praxair, www.praxair.com) or sodium pentobarbital (Steris Laboratories, www.steris.com) for euthanasia
- Sterile cotton gauze (Ethicon)
- Dissection instruments including Noyes scissors (Fine Science Tools, www.finescience.com)
- Stainless steel #22 blades (BD, www.bd.com)
- Endothelial Basal Medium (EBM, Clonetics, www.cambrex.com)
- 10 × Minimal Essential Medium (Gibco BRL, www.gibcobrl.com)
- 100x15 Petri and compartmentalized Felsen dishes
- Falcon Blue Max 15 ml conical tubes
- 0.22 µm syringe filters
- Transfer pipettes
- Dialysis tubing (MWCO 6,000-8,000)
- Weighing paper
- Spatulas
- Glacial acetic acid
- Methanol
- 10 % buffered formalin (Fisher Scientific, www.fisherscientific.com)
- Nunclon 4-well multidishes (Nalge Nunc, www.nalgenenunc.com)
- Syringes (BD)
- Collagen solution (see section *Preparation of collagen gel*)
- Sodium bicarbonate (J.T. Baker Chemicals, www.jtbaker.com)
- Sterile bidistilled water and saline solution (Baxter Healthcare Corp., www.baxter.com)
- Type VI-A agarose

Supplies for immunohistochemistry, image analysis and electron microscopy

- Monoclonal mouse anti-α smooth muscle actin antibody
- Biotin-conjugated secondary antibodies and DAB
- Tween 20 (Sigma, www.sigmaaldrich.com)
- Biotinylated or Alexa Fluor 488-conjugated Griffonia Simplicifolia iso-lectin-B4, and Alexa Fluor 568-conjugated goat anti-mouse IgG secondary antibody (Molecular Probes, www.probes.com)
- Vectastain ABC kit (Vector, www.vectorlabs.com)
- Bioquant Image Analysis Software (Bioquant, www.bioquant.com)

- Gelvatol mounting medium (Monsanto, www.monsanto.com)
- Glutaraldehyde, osmium tetroxide (OsO_4) and EPON reagents (Ted Pella, Inc., www.tedpella.com)
- Electron microscopy rubber molds (Ladd Research Industries, www.laddresearch.com)
- Uranyl acetate, lead citrate and propylene oxide (Polysciences Inc., www.polysciences.com)

Protocols

Excision of the rat aorta

Excision of the aorta is performed in an area of the laboratory that is clean and dust-free. A 1–3 months old Fischer 344 male rat is sacrificed by intraperitoneal injection of sodium pentobarbital or by CO_2 asphyxiation. Immediately after the animal has died, the thoracic and abdominal regions are shaved with a hair clipper. The rat is then laid on a rectangular block of styrofoam wrapped in absorbent lab soaker with its legs extended and pinned down with needles. After being wetted and sterilized with 80 % ethanol, the skin of the thoracic and abdominal walls is cut with a 22-gauge scalpel blade in the form of a Y-shaped incision. After the skin has been dissected from the underlying muscle layer, the abdominal cavity is opened using a cross-shaped cut. The ribs and the attached diaphragm are then cut with scissors to obtain a triangular sternal plate still attached to the rib cage via the sternal manubrium. The xyphoid process of the sternum is then clamped with a hemostat and the whole sternal plate is folded over the right side of the animal to expose the thoracic cavity. The intestines, the stomach, the spleen and the liver are displaced to the right side. The diaphragm is then sectioned in a ventral/dorsal direction, paying attention not to cut the diaphragmatic vessels. The thoracic aorta, which is now visible along the vertebral column, is ligated with a 4-0 silk suture distally, above the diaphragm. The suture is passed around the aorta after creating an opening with fine curved microdissection forceps between the aorta and the vertebral column. While being held by the distal suture, the aorta is dissected with fine curved scissors from the posterior mediastinum, cut just below the aortic arch, and transferred into a compartmentalized Felsen dish. This final step is carried out by cutting the aorta just below the suture. The Felsen dish has four compartments each containing 4 ml of serum-free EBM (Knedler and Ham 1987). During excision, attention is paid not to stretch the aorta or let it dry, and not to cut the adjacent veins, to avoid excessive bleeding.

Preparation of aortic rings

Dissection of the aorta including preparation of the aortic rings is carried out under a dissecting microscope on an open bench in a tissue culture room with HEPA-filtered air. We prefer this set-up to a laminar flow hood which keeps the operator too far from the aortic explant.

1. Working under 10× magnification, carefully dissect periaortic fibroadi-pose tissue away from the aortic wall with Noyes scissors and curved microdissection forceps. Do not stretch, cut, or crush the aortic wall. Remove intraluminal blood clots with forceps. Trim stumps of inter-costals arteries.

2. Transfer the aorta to the successive compartments of the Felsen dish to wash the sample. After dissection and washes, the aorta should appear as a clean tube witout any adipose tissue or blood.

3. Perform subsequent steps including cutting and collagen embedding in a wooden hood with glass top and front opening. The glass protects the cultures from possible dust and allows direct vertical view from above. This arrangement greatly facilitates cutting, washing, and collagen embedding, which are otherwise difficult in a laminar flow hood. Alter-natively, carry out these procedures in an open tissue culture hood with positive horizontal airflow.

4. After removing excess medium to keep explant from floating, cross-sec-tion aorta into 1–2 mm long rings with a 22-gauge scalpel blade. Discard proximal and distal 1 mm-long segments.

5. Clean aortic rings of residual blood by thoroughly rinsing them in serum-free medium; this is accomplished by sequentially transferring the rings with microdissection forceps into 8 consecutive compartments of Felsen dishes, each filled with 4 ml medium (Fig. 1 and 2). Wash aortic rings by gently shaking the dish a few times after each transfer. Take care not to crush or otherwise mechanically damage the rings.

Preparation of standard aortic ring assay with thick (floating) collagen gel

Prepare thick collagen gel/aortic ring constructs in cylindrical agarose wells according to the following protocol (Fig. 1). Agarose wells and collagen solu-tion are prepared in advance as described in the sections entitled "Prepara-tion of agarose wells" and "Preparation of collagen gels." Agarose wells are prepared the same day of the experiment. Collagen solution can be stored at 4 °C for several months.

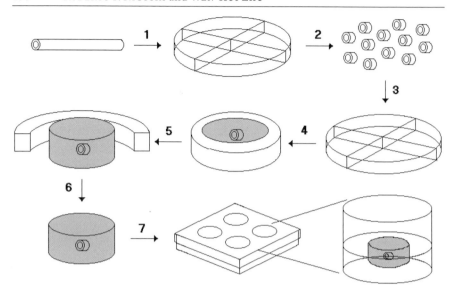

Fig. 1. Steps in the preparation of the standard rat aortic ring assay. Thick collagen gel cultures of rat aorta are prepared in agarose wells, which serve as temporary molds for the gelation of collagen (approx. 300 μl). Once transferred to a 4-well dish, each gel floats in 0.5 ml of serum-free endothelial basal medium with or without pro-angiogenic or anti-angiogenic factors

1. Transfer with a spatula agarose rings into 100 mm dishes (4 per dish) and tap them with the spatula to assure adherence of the agarose to the bottom of the dish. This results in the formation of an agarose well with a plastic bottom.

2. Place 3–4 drops of working collagen solution (for preparation of collagen see section *Preparation of collagen gel*) on the bottom of each agarose well. Transfer dishes into a humidified incubator for 5 min at 35.5–37.0 °C.

3. After the bottom collagen has gelled, place an aortic ring on the edge of the agarose well. Use a transfer pipette to fill the well with collagen solution and let the aortic ring sink to the bottom of the well.

4. Use fine microdissection forceps to position the ring with its luminal axis parallel to the bottom of the dish so that the profiles of its cutting edges are clearly visible.

5. Incubate for 30 min at 35.5–37.0 °C.

6. After the collagen has gelled, cut open the agarose well with a 22-gauge scalpel blade and separate it from the gel with fine microdissection forceps and a spatula.

7. Transfer each collagen/aortic ring construct with a spatula to an 18 mm well containing 0.5 ml of serum-free medium with or without angiogenic agonists or antagonists (4-well NUNC dish). To avoid damaging the delicate gels, perform this step after filling each 100 mm dish with 10 ml of serum-free medium, and keeping the dish slightly tilted towards you to let the gels float. This procedure facilitates the lifting of the gels with the spatula and the transfer of the gels to their final culture well.

8. Transfer cultures to a humidified CO_2 incubator kept at 35.5–37.0 °C. Change growth medium three times a week starting from day 3 of culture. Alternatively, cultures can be maintained in the same medium for the duration of the experiment.

Preparation of agarose wells (standard assay with thick collagen gels)

1. Pour 300 ml of distilled water (sterile irrigation water) into a 500 ml media bottle.

2. Add 4.5 g agarose, Type VI-A.

3. Heat mixture in a microwave oven until agarose goes into solution.

4. Place cap lightly on bottle, and autoclave, liquids cycle. Keep the agarose solution at RT until the bottle is warm to the touch.

5. Move to a laminar flow hood and dispense with a sterile pipette 35 ml of agarose solution into a 100 × 15 mm sterile culture dish (Falcon). Repeat this procedure until the entire stock of agarose is transferred to culture dishes.

6. Allow dishes to cool with the lid slightly ajar in the laminar flow hood.

7. Once the agarose plates have gelled, place the lids on them and stack them in a sealed, airtight container.

8. Agarose is best used within 2–3 days, but it can be stored at 4 °C up to 2–3 weeks. It is important to avoid dehydration of the agarose, which complicates separation of thick collagen gels from agarose wells.

9. Prepare agarose rings before each experiment by punching two concentric circles in the agarose with punchers of 10 and 17 mm diameter (Nicosia and Ottinetti 1990). Remove excess agarose inside and outside the wells with a bent glass pipette.

10. Using a bent spatula, transfer agarose rings to clean 100 × 150 dishes and gently tap them from above with the spatula to make them adhere to the bottom of the dish. This method produces cylindrical agarose wells for the preparation of thick collagen gels.

Preparation of collagen gel

Harvesting rat tails

1. After harvesting the aorta, excise tail using a #22 scalpel blade.

2. Place tail in 100 mm culture dish containing 80 % ethanol, and disinfect skin surface thoroughly for a few seconds.

3. Transfer tail to a dry culture dish and let dry with the lid slightly ajar.

4. Wrap tail in aluminum foil.

5. Store at –20 °C until needed.

Collagen extraction and purification

The following procedure is based on tails obtained from 2–3 month old rats. It is important to work in a clean and dust-free environment. Rat-tail collagen isolation and washings are carried out in 20 × 100 mm disposable sterile culture dishes. This procedure yields mostly interstitial type I collagen.

1. Thaw four tails and disinfect skin surface quickly with 80 % ethanol.

2. Expose collagen fibers by tearing and pulling the skin and subcutaneous tissue of the rat tail with a hemostat. The skin must be incised at 1–2 cm intervals with a scalpel blade before being pulled with the hemostat.

3. Cut collagen fibers with sterile scissors and place them into a dry culture dish.

4. Transfer collagen fibers with sterile forceps into a culture dish containing a saline solution.

5. Using curved microdissection forceps, separate collagen fibers by gently teasing the tendons apart. Collagen fibers are recognizable under a dissecting microscope by their ribbon-like velvety appearance.

6. Remove blood vessels which are pink/red and have a blood-filled lumen. Cut and discard hemorrhagic segments of collagen fibers.

7. Wash collagen fibers in 8 consecutive baths of sterile saline.

8. Transfer collagen fibers to a dish containing 80 % ethanol and keep them submerged for 20 min. This treatment insures sterility of the preparation and denatures possible growth factor contaminants. The collagen fibers become dehydrated and stiff after this treatment.

9. Transfer the collagen to a dish containing distilled water to rehydrate the fibers.

10. Wash collagen fibers in 5 additional baths of distilled water.

11. Transfer collagen fibers with forceps to an Erlenmeyer flask containing 100–120 ml of 0.5 M acetic acid. The acetic acid solution is prepared with sterile distilled water and glacial acetic acid designated for collagen purification only. This is to avoid contamination of collagen with unwanted toxic chemicals which might interfere with the angiogenesis assay.

12. Place flask on a magnetic stirrer at 4 °C.

13. After 48 h filter the collagen solution which has become dense and viscous, through 2–3 thin layers of sterile gauze into sterile 50 ml centrifuge tubes.

14. Centrifuge at 12,000 × g for 1 hour at 4 °C, and transfer supernatant to a sterile glass bottle with a 10 ml glass pipette. Label the stock collagen solution and store at 4 °C.

15. To determine the concentration of collagen, dispense 5 ml of stock collagen solution to an aluminum weighing dish which is weighed before the procedure. Let acetic acid evaporate on a hot plate. As soon as all the liquid has evaporated, let the aluminum dish cool and weigh it again. The difference in weight corresponds to the amount of collagen present in 5 ml. Divide by five to obtain the concentration of collagen per ml. The optimal concentration is 1.25–1.7 mg/ml. Dilute, if necessary, with 0.5 M acetic acid. This solution can be stored at 4 °C for up to one year.

Collagen dialysis

1. Cut approximately 15 inches of dialysis tubing.

2. Wet tubing with distilled water, making sure that the tubing opens and the water goes through its lumen. Squeeze water out of tubing.

3. Boil tubing for 10 min on a Bunsen burner in a tissue culture grade beaker containing 1,000 ml sterile irrigation distilled water. Repeat this procedure two additional times, with fresh water each time, using a sterile pipette to transfer the tubing between washings.

4. Transfer tubing to a sterile 100 mm culture dish and squeeze water out of tubing paying attention not to contaminate the end that will be used to fill the tubing with collagen solution.

5. Allow tubing to cool to RT.

6. In a clean and dust-free environment (if you use a laminar flow hood, be careful not to let the tubing dry), tie two knots at one end of the dialysis tubing.

7. With a 20 ml syringe fill tubing with 25–30 ml of stock collagen solution in 0.5 M acetic acid.

8. Tie top end of tubing leaving an air bubble between the collagen meniscus and the knot.

9. Transfer the collagen containing dialysis tubing to a 4 L autoclaved glass beaker filled with 3.5 L of a 0.1× MEM, pH 4.0, solution prechilled to 4 °C. This solution is prepared by adding 30 ml of 10× Minimal Essential Medium to 2.970 l of sterile distilled water. 2N HCl is used to acidify the solution to pH 4.

10. Place large stirring bar (autoclaved and designated for tissue culture only) on the bottom of the beaker and place collagen-containing tubing in the solution.

11. Place beaker on a magnetic stir plate in 4 °C refrigerator or cold room, and keep overnight.

12. Change dialysis solution the next day and repeat dialysis for another 24 h.

13. On the third day, transfer the tubing to a large sterile culture dish. Disinfect the tubing over the top air bubble with 80 % ethanol using sterile gauze. Nick the tubing with sterile fine-tip scissors and through this opening extract the collagen using a sterile 20 ml syringe. Transfer to a chilled sterile 150 ml glass bottle, and keep on ice until the procedure is completed.

14. Label collagen solution with the appropriate batch number and store at 4 °C.

Collagen gelation

There are three fundamental conditions that regulate collagen fibrillogenesis and gel formation: ionic strength, pH, and temperature. In order to keep collagen in solution, the ionic strength must be low (0.1× MEM), the pH low (pH 4.0) and the temperature low (4 °C). Once any of these conditions is modified, the collagen starts gelling. Optimal gelation occurs when ionic strength is physiologic (1× MEM), pH is 7.0–7.4 and temperature is 35.5–37.0 °C. Gelation is induced by adjusting these parameters at the time of the experiment (see below).

1. Mix in a sterile tube, on ice, 1 volume 10× Eagle's MEM (Gibco) with 1 volume 23.4 mg/ml $NaHCO_3$, and allow for color solution to change orange.

2. To this mixture add 8 volumes of dialyzed collagen in 0.1× MEM, pH 4.0 and mix thoroughly. Avoid making air bubbles. Place tube with working collagen solution on ice.

3. Use this solution to make collagen gels. The working collagen solution will gel at 35.5–37.0 °C. For details about procedural differences in times recommended for collagen gelation in the standard and thin prep modification of the aortic ring assay refer to the sections on the preparation of the respective assays.

Preparation of modified aortic ring assay with thin collagen gel (thin prep assay)

The thin prep modification of the aortic ring model (Fig. 2) simplifies the procedure and allows immunostaining of aortic outgrowths as whole mounts. Reducing collagen gel thickness facilitates penetration of reagents. The thin prep assay can be used to characterize the cellular composition of the outgrowths and localize at different levels of the developing neovasculature expression of proteins implicated in the angiogenic process (Zhu and Nicosia 2002).

1. Place each aortic ring into an 18 mm well (4-well NUNC dishes), with the luminal axis of the ring lying parallel to the bottom of the culture dish. To avoid carrying over medium that could dilute the collagen solution, release excess medium from the ring by gently dragging the ring with forceps along the bottom of a dish.

2. After placing the ring into the culture well, apply 25–30 μl of working collagen solution onto each ring and uniformly spread the collagen with fine microdissection forceps or a pipette tip to form a thin disc of approximately 8 mm in diameter around each explant. Pay attention to suspend the ring within collagen, above the plastic surface of the culture dish.

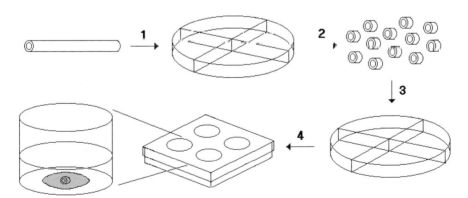

Fig. 2. Steps in the preparation of the thin prep rat aortic ring assay. Aortic rings are suspended in a wafer-thin collagen gel (25–30 μl) on the bottom of a culture well in a 4-well dish, and cultured in 0.5 ml serum-free endothelial basal medium with or without pro-angiogenic or anti-angiogenic factors.

3. After the collagen solution has gelled for 5–10 min at 37 °C, add 0.3–0.5 ml serum-free EBM, with or without angiogenic or anti-angiogenic factors of interest, to each culture. The thin collagen gel remains anchored to the bottom of the culture dish. The aortic ring/collagen gel cultures are maintained in a humidified CO_2 incubator and treated as described above for the standard assay.

Additional protocols for morphologic evaluation of rat aortic ring assay

Immuno- and lectin histochemistry

When the assay was first described, histologic studies including immunohistochemistry were carried out on paraffin sections of thick collagen gel cultures. With the development of the thin prep modification, paraffin-embedding and histologic sectioning, which are labor intensive and require a histology laboratory, are no longer needed. Thin collagen gels can be easily permeated by immunoreactants and immunostained as whole mounts (Zhu and Nicosia 2002). Here we describe how to process thin gels for whole mount immuno- or lectin-histochemistry.

1. Fix thin collagen gel cultures with 10 % neutral buffered formalin for 20–30 min.

2. Detach gels with a transfer pipette, wash them three times with distilled water, and stored them in the last wash at 4 °C for at least 12 h.

3. Transfer each collagen gel to a 35 mm dish with a bent spatula. Perform the following steps on a rotating platform to insure uniform exchange of reagents during the entire staining procedure.

4. For evaluation of intracellular antigens, permeabilize cells with PBS containing 0.1 % Tween 20.

5. Quench endogenous peroxidase with 3 % hydrogen peroxide for 5 min.

6. For antibody staining, block cultures with 5 % normal serum of the same species as the secondary antibody in PBS for 1 h at RT, rinse in PBS (1 × 5 min), react for 1 h with the primary antibody of interest, rinse in PBS (3 × 10 min), incubate for 1 h with biotin-conjugated secondary antibody, and rinse in PBS (3 × 10 min).

7. For lectin histochemistry, react gels with biotinylated *Griffonia simplicifolia* Isolectin B-4 diluted 1:200 for 1 h, and wash in PBS (3 × 10 min).

8. Visualize reactions with the Vectastain ABC kit and DAB, according to the manufacturer's recommendations.

9. Counterstain samples lightly with hematoxylin (less than 30 s), place them on a histology glass slide, and mount them in aqueous mounting medium.

Double immunofluorescence and confocal microscopy

1. Perform fixing, washing, blocking, antibody or lectin reactions, and mounting as described in the previous section.

2. Double stain with a cocktail of antibodies of interest, for example, antibodies against endothelial (anti-CD31, anti-Tie2 or anti-von Willebrand Factor antibodies) and pericyte (anti-α smooth muscle actin antibody) markers. The two cell types are visualized in different colors by fluorescent dyes (Alexa Fluor 488 or 568) conjugated to the appropriate secondary antibodies.

3. For studies with the endothelial marker Griffonia Simplicifolia isolectin-B4, add the Alexa Fluor 488-conjugated lectin at the same time as the Alexa Fluor 568-conjugated secondary antibody against α-smooth muscle actin.

4. Examine gels by confocal microscopy (Fig 3).

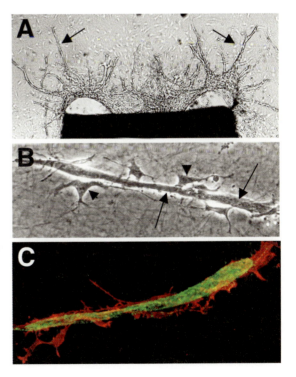

Fig. 3. Morphology of neovessels in aortic ring cultures. A Light micrograph of 7-day-old living collagen gel culture of rat aorta, *arrows* indicate neovessels (×25); B Phase-contrast image of a neovessel composed of endothelial cells (*arrows*) and surrounding pericytes (*arrowheads*) (×200); C Confocal image of neovessel double-stained for endothelial cells with Griffonia Simplicifolia isolectin-B4 (*green*) and pericytes with anti-α-smooth muscle actin antibody (*red*) (×400)

Transmission electron microscopy

Electron microscopic studies are performed to evaluate the ultrastructural characteristics of the vascular outgrowths. First among these is the formation of luminal spaces surrounded by properly polarized endothelial cells.

1. Fix cultures in 2.5 % glutaraldehyde 0.1 % Na Cacodylate buffer, pH 7.4.

2. Change the fixative 3 times per well the first 30 min of fixation to wash out the salts of the growth medium.

3. Store in glutaraldehyde at 4 °C overnight.

4. Identify areas of interest under a dissecting microscope, and cut gels into strips with a razor blade. The strips are sized to fit the rectangular rubber molds used for the final embedding.

5. Transfer gels to glass scintillation vials and rinse with 0.1 % Na Cacodylate buffer pH 7.4, 3 × 10 min. Perform next steps in a fume hood.

6. Postfix in 1 % OsO_4, 1 × 60 min

7. Wash in distilled water, 4 × 10 min

8. Incubate in 1 % uranyl acetate, 1 × 30 min. Cover bottles in tin foil to block out the light (uranyl acetate is a light sensitive compound).

9. Wash in distilled water 2 × 10 min

10. Dehydrate in graded ethanol:

    a) 50 % ethanol – 1 × 20 min

    b) 70 % ethanol – 1 × 20 min

    c) 80 % ethanol – 1 × 20 min

    d) 90 % ethanol – 1 × 20 min

    e) 100 % ethanol – 2 × 20 min

11. Incubate in propylene oxide 3 × 20 min

12. Store in a 1:1 solution of propylene oxide and EPON overnight at RT. Preparation of EPON: Pour 44.3 ml LX 112 Resin into a large plastic beaker containing a stirring bar. Add 31.1 ml NMA and 15.0 ml DDSA, and mix for approximately 15 min. Add 0.8 ml DMP and stir until the mixture becomes light amber in color. The EPON can be stored in 10 ml syringes at –70 °C.

13. The next day, transfer specimens to 100 % EPON, and store overnight.

14. After 24 h, transfer each piece with a wooden spatula to an appropriately labeled flat embedding mold containing EPON. On the bottom of each mold, place a small paper label with the appropriate sample information.

15. Incubate in a 60 °C oven for 48 h.

16. Identify areas of interest on toluidine stained 1 micron thick sections under a light microscope, thin section, mount on copper grids, stain with lead citrate according to standard protocols, and examine with a transmission electron microscope.

Results and Comments

Characterization of angiogenesis

Fibroblasts are the first cells to migrate out of the aortic wall. The angiogenic response starts at days 2–3 and ends at days 7–9. The neovessels arise primarily from the cutting edges of the explants and are composed of endothelial cells and pericytes (Fig. 3). After it has formed, the vascular outgrowth undergoes a process of remodeling and regression. During the maturation/remodeling phase, pericytes increase in number by migrating and proliferating around the endothelium. Vascular regression is associated with the formation of a halo of collagen lysis around the explant, retraction of vessel branches, and fragmentation of neovessels.

Measurement of angiogenesis

The angiogenic response of the rat aorta can be quantitated by visual counts (Nicosia and Ottinetti 1990A, Nicosia and Bonanno 1991). Angiogenic growth curves are obtained by counting microvessels every other day. Cultures are examined under bright-field microscopy using an inverted microscope equipped with 4× and 10× objectives. Optimal contrast and depth of field are obtained by closing the iris diaphragm of the condenser. Angiogenesis is scored according to the following criteria (Fig. 4).

1. Microvessels are distinguished from fibroblasts based on their greater thickness and cohesive pattern of growth.

2. The branching of one microvessel generates two additional microvessels.

3. Each microvascular loop is counted twice because it frequently originates from two converging microvessels.

Although the visual count method is rapid, practical, and reproducible, it requires experience and patience, and becomes difficult to apply when the

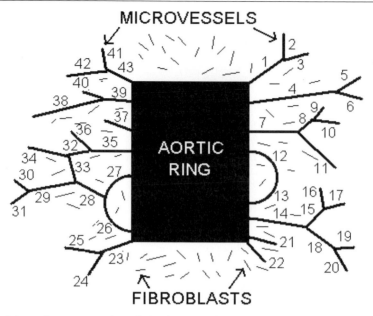

Fig. 4. Schematic representation of visual counts of microvessels in collagen gel culture of rat aorta. The aortic ring, which is portrayed with its luminal axis parallel to the bottom of the dish and therefore not visible, appears as a rectangular square against a white background. Microvessels, which typically develop from the cutting edges of the explant, are counted as marked, according to criteria described in the Section "Measurement of angiogenesis" under "Results and Comments"

cultures have more than 250 microvessels, a condition that occurs, for example, in the presence of saturating concentrations of VEGF. In this case, image analysis approaches provide a more accurate quantitation of the outgrowth. Computer-assisted image analysis of aortic cultures can be performed manually (Nicosia et al. 1993) or automatically using image-processing algorithms and digital filters that separate images into vascular and nonvascular (fibroblasts) compartments based on object size and shape (Nissanov et al. 1995, Blacher et al. 2003).

Induction of quiescence

Aortic rings can be made quiescent by a 10–14 day long incubation step in serum-free medium prior to embedding in collagen gel. This treatment renders the aortic rings unable to sprout spontaneously under serum-free conditions. Addition of VEGF or other angiogenic inducers reinitiates the angiogenic process. This method differentiates molecules that are capable of inducing angiogenesis from molecules that promote the angiogenic process but cannot induce it. For example, VEGF, bFGF, Ang-1, and IGF-1 are all capable of stimulating angiogenesis in cultures of freshly cut aortic rings, but of these angiogenic regulators only VEGF and bFGF can induce angiogenic sprouting from quiescent aortic rings (Nicosia et al. 1994b, Zhu et al. 2002, and unpublished data). To make aortic rings quiescent use the following protocol:

1. Coat wells of a 4-well NUNC dish with a thin layer of agarose to inhibit cell attachment (use same agarose described in the preparation of agarose wells for thick collagen gel cultures).

2. Rinse wells 2× with serum-free EBM, and add 0.5 ml EBM to each well.

3. Place one aortic ring per well. Incubate floating aortic rings at 35.5–37.0 °C for 10–14 days, changing the medium three times a week.

4. After 10–14 days embed aortic rings in collagen gel and culture in the presence or absence of angiogenic regulators.

Recovery experiment after inhibition of angiogenesis

To rule out that the inhibitory effect of a test reagent is due to toxicity use the following protocol:

1. After documenting the inhibitory effect of the reagent, remove inhibitor-containing medium and wash cultures three times with serum-free EBM.

2. Treat cultures with serum-free EBM containing 10 ng/ml VEGF.

3. Follow cultures for 7 days. If they are viable, aortic rings will produce a florid angiogenic response to VEGF after inhibitor withdrawal. No response will be detected if the rings are no longer viable due to prior exposure to a toxic substance.

Vascular survival studies

The effect of molecules of interest on vascular survival can be studied by treating cultures with test reagents after vessels have formed.

1. Culture freshly cut aortic ring in collagen gel with or without angiogenic factors, and feed cultures three times a week.

2. At days 6–7 start treatment with putative neovessel stabilizer.

3. Follow cultures up to three weeks, and count vessels every other day, as per standard protocol.

4. Compare vessel counts at different time points in untreated control with treated cultures, and calculate neovessel survival accounting to the following formula:

$$\% \text{ survival} = \frac{\text{Neovessel counts at day of interest} \times 100}{\text{Neovessel counts at peak of growth}}$$

Troubleshooting

Some degree of variability in the angiogenic response of the aortic rings is to be expected. For example, a 4-culture control group may have the following counts: 70, 90, 55, 80. This result is acceptable, because it is within the normal range of biologic variability. Conversely, the following counts indicate a technical problem with the assay: 110, 10, 85, 5. Usually, this type of variability is due to errors made while performing the assay. Possible solutions to variability problems are listed in Table 1.

Concluding Remarks

The aortic ring model can be used to study the molecular regulation of angiogenesis and to test the activity of angiogenic factors and anti-angiogenic molecules. Like any other assay, it requires attention to detail and judicious use of published methodology. Over the years we have been able to technically improve the model making it more user-friendly. It is our hope that methods and protocols described in this chapter will facilitate the learning process for newcomers to the field.

Approaches reported for the rat aorta model have been used to study the angiogenic response of blood vessels from other species (Masson et al. 2002, Price et al. 2202, Zhu et al. 2003). When this culture system is applied to vessels other than the rat aorta, however, optimal growth conditions are likely to differ. For example, mouse aortic rings, unlike rat aortic rings, are unable to spontaneously produce an angiogenic outgrowth, but respond to angiogenic stimulation by bFGF and VEGF (Zhu et al. 2003). In addition, the speed, extent, and duration of mouse aortic angiogenesis are greatly influenced by the genetic background and age of the animals (Zhu et al. 2003). Vascular organ culture models, therefore, must be tailored to the growth requirements of the blood vessel of interest, taking into account that animal species, strain and age play a critical role in the angiogenic behavior of the explants.

Acknowledgements. This work was supported by the National Institute of Health (HL52585; R.F.N.) and the Medical Research Service, Department of Veterans Affairs (R.F.N).

Table 1. Troubleshooting

Problem	Possible Cause	Solution
Excessive variability of vascular outgrowths	– Aorta mechanically damaged during excision or dissection	– Avoid stretching, crushing, or air drying explant.
	– Collagen gel mechanically damaged	– Position aortic rings within 20–30 s, before collagen solution starts gelling
	– Collagen solution not adequately mixed before gelation	– Always premix 10 × MEM with NaHCO$_3$ before adding stock collagen to this mixture, mix the final working solution thoroughly, and keep on ice
	– Aortic endothelium not in direct contact with collagen gel	– Remove carried over medium from aortic ring before transferring ring into collagen solution. Gently mix collagen solution with a pipette after adding aortic ring and prior to gelation
		– Avoid damaging collagen gel (see above)
Uniformly poor growth of microvessels	– Unsatisfactory reagents	– Use EBM that is not more than 5–6 month old and keep shielded from light
		– Make your own collagen
		– Do not use excessively diluted collagen
	– Toxins	– All reagents must be endotoxin-free and tissue culture grade
		– Any glassware used for experiment should be marked for "tissue culture only"
	– Incubator temperature below 35 °C	– Check incubator temperature regularly
	– Alkaline pH	– Avoid keeping aortic rings in alkaline pH for too long during preparation of the assay

References

Blacher S, Devy L, Noel A, Foidart JM (2003) Quantification of angiogenesis in the rat aortic assay. Image Anal Stereol 22:43–48

Donovan D, Brown NJ, Bishop ET, Lewis CE (2001) Comparison of three in vitro human 'angiogenesis' assays with capillaries formed in vivo. Angiogenesis 4:113–121

Jain RK, Schlenger K, Hockel M, Yuan F (1997) Quantitative angiogenesis assays: progress and problems. Nat Med 11:1203–1208

Kawasaki S, Mori M, Awai M (1989) Capillary growth of rat aortic segments cultured in collagen without serum. Acta Pathol Japonica 39:712–718

Knedler A, Ham RG (1987) Optimized medium for clonal growth of human microvascular endothelial cells with minimal serum. In Vitro Cell Dev Biol 23:481–491

Masson V Ve, Devy L, Grignet-Debrus C, Bernt S, Bajou K, Blacher S, Roland G, Chang Y, Fong T, Carmeliet P, Foidart JM, Noel A (2002) Mouse aortic ring assay: A new approach of the molecular genetics of angiogenesis. Biol Proced Online 4:24–31

Nicosia RF, Ottinetti A (1990) Growth of microvessels in serum-free matrix culture of rat aorta: A quantitative assay of angiogenesis in vitro. Lab Invest 63:115–122

Nicosia RF, Tchao R, Leighton J (1982) Histotypic angiogenesis in vitro: light microscopic, ultrastructural, and radioautographic studies. In Vitro 18:538–549

Nicosia RF, Bonanno E, Smith M (1993) Fibronectin promotes the elongation of microvessels during angiogenesis in vitro. J Cell Physiol 154:654–661

Nicosia RF, Bonanno E, Yurchenco P (1994A) Modulation of angiogenesis in vitro by laminin-entactin complex. Dev Biol 164:197–206

Nicosia RF, Nicosia SV, Smith M (1994B) Vascular endothelial growth factor, platelet derived growth factor, and insulin-like growth factor-1 promote rat aortic angiogenesis in vitro. Am J Pathol 145:1023–1029

Nissanov J, Tuman RW, Gruver LM, Fortunato JM (1995) Automatic vessel segmentation and quantification of the rat aortic ring assay of angiogenesis. Lab Invest 73:734–739

Price DK, Ando Y, Kruger EA, Weiss M, Figg WD (2002) 5'-OH-thalidomide, a metabolite of thalidomide, inhibits angiogenesis. Ther Drug Monit 24:104–110

Zhu WH, Nicosia RF (2002) The thin prep rat aortic ring assay: A modified method for the characterization of angiogenesis in whole mounts. Angiogenesis 5:81–86

Zhu WH, Guo X, Villaschi S, Nicosia RF (2000) Regulation of vascular growth and regression by matrix metalloproteinases in the rat aorta model of angiogenesis. Lab Invest 80:545–555

Zhu WH, MacIntyre A, Nicosia RF (2002) Regulation of angiogenesis by vascular endothelial growth factor and angiopoietin-1 in the rat aorta model: distinct temporal patterns of intracellular signaling correlate with induction of angiogenic sprouting. Am J Pathol 161:823–830

Zhu WH, Iurlaro M, MacIntyre A, Fogel E, Nicosia RF (2003) The mouse aorta model: Influence of genetic background and aging on bFGF- and VEGF-induced angiogenic sprouting. Angiogenesis 6:193–199

Chapter 14

Endothelial Cell Chemotaxis and Chemokinesis Assays

HELLMUT G. AUGUSTIN

▓ Introduction

Endothelial cells form a perfectly quiescent monolayer of cells that lines the inside of all blood vessels. Quiescent endothelial cells do not move and divide only rarely within months to years. Yet, upon angiogenic activation, they can dissociate from the monolayer to acquire a highly motile and invasive phenotype. Endothelial cell invasion and migration are, thus, early key regulatory processes of the angiogenic cascade. Significant effort is consequently aimed at studying the functional and molecular properties of migrating and invading endothelial cells.

A number of two-dimensional and three-dimensional assays have been developed to study the migration of endothelial cells. One of the simplest approaches is the scraping of a confluent monolayer with a razor or scalpel blade (Sato and Rifkin 1989). This allows the endothelial cells at the wounded edge to migrate into the scraped area mimicking the lateral migration of endothelial cells as it, for example, occurs after denudation of a blood vessel in vivo (compare Chap. 18). The scraping technique has been widely used. Yet, razor blade wounding of a monolayer is a non-standardized mechanical manipulation that may lead to scratches in the tissue culture dish and damages the endothelial cells at the migration front. We have consequently developed a standardized assay that allows the non-traumatic release of endothelial cells from growth arrest (Augustin-Voss and Pauli 1992b). In this assay, endothelial cells are grown to confluence within a tissue culture dish which contains a silicon template of defined size that has been inserted into the culture dish prior to the seeding of the cells. The silicon template is removed after the cells are grown to confluence and the cells are allowed to laterally migrate into the area that was previously occupied by the silicon template. The defined area of the silicon template allows for the convenient quantitative microscopic analysis of lateral cell migration over 2 to 3 days.

Lateral endothelial cell migration assays are usually performed as long-term assays studying the effect of pro-migratory or anti-migratory com-

Springer Lab Manual
H. Augustin (Ed.)
Methods in Endothelial Cell Biology
© Springer-Verlag Berlin Heidelberg 2004

pounds over a period of 2 to 4 days. These are classical non-directional chemokinetic experiments aimed at determining the overall migration-regulating effect of a cytokine or pharmaceutical test compound. In contrast, chemotaxis experiments are aimed at determining the migratory effect of a target cell population towards a stimulus that is acting in a gradient-dependent manner. Two compartment vertical chemotaxis systems were pioneered by Stephen Boyden more than 40 years ago (Boyden 1962). The Boyden chamber is comprised of an upper and a lower fluid filled compartment which are separated by a membrane with pores of defined size (typically 5 µm, 8 µm, or 12 µm). A test compound is added to the lower compartment and the cells are seeded on top of the membrane in the upper compartment. The test compound in the lower chamber generates a chemotactic gradient towards which the cells in the upper chamber migrate by passing though the pores in the membrane. The number of migrated cells on the bottom side of the filter is determined as a quantitative readout of the chemotactic agent.

Cytokine gradients can usually not be maintained in culture over longer periods of time. Chemotaxis experiments are therefore short-term experiments performed mostly within a time window of up to 4 h. The published literature is frequently not very specific to explicitly differentiate between chemotaxis and chemokinesis, which requires a carefully performed checkerboard analysis. This is done by generating different gradient levels between the two compartments up to the neutralization of the gradient (i.e., same concentration of test compound in both compartments = chemokinesis).

Originally developed to study the migratory properties of circulating hematopoietic cells, a number of miniaturized multiwell modifications of the original Boyden chamber have been developed over the years and applied to study chemotaxis and chemokinesis of many different cell populations including endothelial cells (Falk et al. 1980). Recently, multiwell chemotaxis chambers have also become available as disposable labware.

This chapter outlines two standard protocols for two-dimensional lateral endothelial cell migration using the silicon template assay and vertical chemotactic migration using a modified Boyden chamber assay. Modifications of the basic protocol are discussed at the end of the chapter along with a troubleshooting guide of common problems.

▨ Materials

Cells

- Endothelial cells (mouse or human, primary or cell lines)

Reagents

- Cell culture media (depending on the specific conditions of the experiment)

Equipment

- For lateral silicon template-based chemokinetic migration assay: 8-well Flexiperm templates (www.vivascience.com/en/cell_products/flexiperm/flexiPERM.shtml; also available from Sigma [product #: Z376647], www.sigmaaldrich.com) (Fig. 1A). The silicon templates are cut apart to be used as insertable walls of defined width (2 mm). We routinely prepare 4 mm long pieces which fit into 48-well plates. Alternatively, the

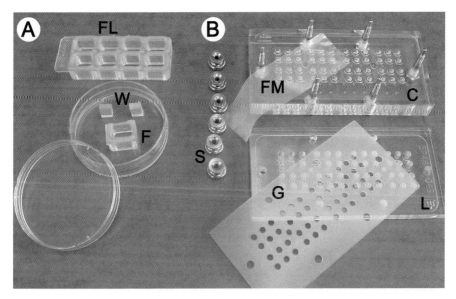

Fig. 1. A Specific equipment for silicon template lateral migration assay; Flexiperm rectangular silicon templates are sold as 8-well manifold (*FL*). The 8-well templates can be cut apart to make individual silicon fences with an inner surface area of 0.9 cm² (*F*) and small 2 mm wide walls (*W*) that can be used to produce a cell-free area of defined surface area. **B** Vertical modified Boyden chemotaxis assay: Components of the 48-well microchemotaxis chamber consisting of a bottom corpus (*C*), to which the filter membrane (*F*) is applied after filling of the bottom chambers followed by the silicon gasket (*G*) and the lid (*L*) which is fixed with 6 screws (*S*)

templates may be cut in rectangular fences defining a growth area of 0.9 cm2 with a culture volume of 500 µl. These fences may be inserted in 12-well plates or any larger culture dish. The silicon templates are reusable. They are soaked immediately in distilled water upon use to avoid drying of serum containing culture medium onto the silicon which may render the template wall adhesive to cells after a while. Upon extensive washing in distilled water, the templates are put in 70 % alcohol for disinfection and allowed to air dry overnight in a tissue culture hood with the UV light turned on. Avoid the use of detergents for the cleaning of the silicon templates. If the templates loose their non-adhesiveness for cultured cells (usually not before at least 40 to 50 times of repeated use), they may briefly be cleaned with detergent after which they are extensively soaked in distilled water.

- For vertical chemotactic experiments with modified Boyden chamber: Nucleopore 48-well microchemotaxis chamber and accessories (Neuroprobe, www.neuroprobe.com) (Fig. 1B). The chamber is reusable. Upon use, all components of the chamber have to be immediately soaked in distilled water to prevent proteins from drying on the instrument. The wells should be soaked dry after cleaning and the chamber is allowed to dry (preferably in the flow of the hood [without UV light]). All components of the chamber should be periodically soaked in Terg-a-Zyme®, an enzyme active detergent that is available from most major laboratory suppliers. Following detergent treatment, the chamber should be intensely rinsed in distilled water.
  The acrylic chamber should be treated with care. Never allow it to get in contact with solvents (e.g., alcohol, acetone, etc.). Don't autoclave the chamber and don't put it in a drying oven. Do not use compressed air containing oil or other contaminants to blow the chamber dry.
- Standard tissue culture equipment (hood, incubator, water bath, microscope, centrifuge) and disposables (pipettes, tubes, Eppendorf cups, etc.)

Procedure

Lateral endothelial cell migration assay

Set-up of silicon templates
The set-up of the silicon template depends very much on the specific aspects of the experiment. For a standard quantitative lateral migration assay, insert one of the 4 mm silicon walls shown in Fig. 1A in the center of a 48-well or 24-well plate. To prepare migrating and adjacent resting endothelial cells for comparative cytochemical applications, a rectangular silicon ring may be inserted into a larger culture dish or on an autoclaved glass slide which is

then inserted into a petri dish. A two-compartment co-culture system can be established by inserting a rectangular silicon template in the center of a 6-well plate. Two different cell populations (e.g., endothelial cells and tumor cells) can be grown in the inner and the outer compartment and the cells can be grown separately in the same dish as long as the template is inserted in the dish. Upon removal of the template, the cells start to migrate towards each other and cellular interactions can be studied under well defined conditions. It is also possible to gel-overlay the cells upon removal of the template to study gradient-induced effects of one cell population on the other (e.g., tumor cell-induced endothelial cell migration).

Lateral migration assay
1. Trypsinize one confluent dish (100 mm or 75 cm^2 flask) of endothelial cells.

2. Collect the cells by centrifugation.

3. Resuspend cells in a volume of medium that corresponds to a 1:3 to 1:4 split.

4. Seed the cells in the dishes with the silicon templates at a density that allows the cells to grow to confluence within approximately 2 days (i.e., usually 1:3–1:4 split).

5. Incubate the dishes in the incubator at standard tissue culture conditions (5 % CO_2, 100 % humidity). Allow the cells to grow to confluence.

6. Remove the silicon templates after the cells have grown to confluence (soak silicon templates immediately in distilled water).

7. Replace medium, wash the cells once in HBSS, and add the test compound containing medium.

8. Inspect the cells under the microscope to confirm straight and undamaged migration starting lines.

9. Return the dishes to the incubator and incubate for the desired period of time (usually 1–2 days). Very rapidly migrating (tumor) cells will close the gap left by the silicon wall within a day. Most endothelial cell populations will not cover the area within 2–3 days.

10. Migration of the cells is measured at the end of the experiment with an inverted phase contrast microscope containing an ocular grid. The silicon templates leave an area of defined width upon removal (2 mm). When starting to work with the silicon template assay as a quantitative migration assay, you first have to standardize the system for your microscope by measuring the width of the empty area immediately after

removing the silicon template. This can be done with an ocular grid or with an automated image analysis system. Results are then processed in ocular grid units or preferably translated into μm. Actual measurements are made by determining the width between the two migration fronts that migrate towards each other using the ocular grid or an image analysis system. Migration is then calculated by subtracting this value from the previously determined standardized starting width. The absolute migration rate of the monolayer cells is 50 % of this value which is preferably expressed in μm per 24 h.

Vertical chemotaxis assay in modified Boyden chamber

The following protocol is for a chemotaxis experiment with HUVEC. Other endothelial cell populations can similarly be used. Yet, some modifications of the culture conditions may be necessary.

Coating of filter (evening before experiment!)
Polyvinylpyrrolidone-free polycarbonate membranes are available at different pore sizes ranging from 2 μm to 12 μm. We recommend 12 μm filter membranes for experiments with endothelial cells, but 10 μm and 8 μm filters are similarly used. Make sure to use PVP-free membranes as PVP prevents adhesion and coating.

It is recommended to coat filter membranes for chemotaxis experiments with anchorage dependent cells such as endothelial cells. This is accomplished by placing the filter overnight at room temperature in a sterile hood in a 1:100 dilution of ECM gel in PBS (Cat.-No. E1270, Sigma). The filter is washed in assay medium prior to use.

Set-up of test compounds
It is important that the lower compartment of the chemotaxis chamber containing the test compound is loaded rapidly (within 5 min) to avoid evaporation of medium. Therefore, prepare a pipetting scheme and arrange the cups (usually Eppendorf cups) in a rack with the different test reagents according to the design of the pipetting scheme. The 48-well chemotaxis chamber allows the analysis of 12 (quadruplicate) or 16 (triplicate) different experimental conditions (e.g., different compounds or often titration curve of one compound).

Prepare dilutions of the test compound(s) in assay medium. We routinely use RPMI with 0.5 % FCS and 0.1 % BSA for chemotaxis experiments with HUVEC. Other endothelial cell populations may require slight modifications of the culture medium.

Set-up of cells
1. Trypsinize one confluent dish (100 mm or 75 cm^2 flask) of endothelial cells.

2. Collect the cells by centrifugation.

3. Resuspend in 3 ml of assay medium.

4. Count the cells and adjust the density to 5×10^5 cell per ml (a confluent 100 mm dish with HUVEC will usually yield approximately 3×10^6 cells so that you should end up with 5–7 ml of cell suspension at the desired cell density.

5. Leave the cells in a Falcon tube in a water bath at 37 °C and load the test compounds to the bottom chamber.

Set-up of chamber
The test compounds are pipetted into the lower compartment, the membrane is placed on top, the silicon gasket is applied, the lid is attached and fixed, cells are pipetted into the upper compartment and the chamber is incubated for the desired period of time

1. Place bottom plate of the chamber in the hood and orient it as preferred (usually: NP sign in upper left). Pipet between 26 μl and 28 μl of the test compounds (including proper positive and negative controls) to the bottom chamber to yield a slight positive meniscus (important to prevent the trapping of air bubbles when applying the membrane). Work rapidly and concentrated. The chamber should be loaded without pipetting mistakes in less than 5 min.

2. Cut a small piece of the filter off from one of the corners for later orientation. Orient the filter with the cut corner to the NP sign. Place filter with two forceps on chamber by allowing the filter to first make contact in the middle of the chamber to spread from there to both sides. The shiny side of the filter is oriented towards the bottom of the chamber and the dull side faces upwards. This step is critical as it easily leads to the trapping of air bubbles. The filter can be positioned somewhat, but make sure not to cause contamination between wells.

3. Place the silicon gasket in position (cut corner in upper left position).

4. Put the top plate in position, press it down, and hold it until the top plate is fixed with the screws. Tighten the screws evenly.

5. Add cells to upper compartment.

6. Carefully mix the cells which may in the meantime have sedimented to ensure an even loading of the wells.

7. Pipet 50 µl cell suspension into each well of the chemotaxis chamber (the adjusted cell density of 5×10^5/ml corresponds to 25,000 cells per well). Make sure that the cells are pipetted without trapping air. This can best be accomplished by inserting the cell-filled pipet 2/3 in the opening (make sure to not accidentally touch the filter) and rapidly ejecting the cell suspension.

8. Upon completion of loading, the chamber is inspected for any trapped air bubbles (inspect reflections of overhead light and check the size of the meniscus). It is possible to rescue wells with trapped air bubbles by completely emptying the well and refilling it, but such well may similarly be discarded by marking them on the pipetting scheme to completely exclude them from the subsequent evaluation.

9. Place the chamber into an incubator and incubate at 100 % humidity with 5 % CO_2 for the desired period of time (usually 4 h).

Harvesting and processing of the filter

1. Remove the chamber from the incubator.

2. Remove the screws while pressing the top plate down.

3. Invert the chamber and place the entire chamber on a paper towel.

4. Push the top plate down and remove the lower part of the chamber. The filter is now face up with the migrated cells on top of the shiny side of the filter.

5. Lift one side of the filter and attach the wide filter clamp (be careful not to touch the filter).

6. Lift the filter and attach the small filter clamp to the free end.

7. With the cell side up, carefully wash the bottom side of the filter (i.e., non-migrated cell side) in a petri dish with PBS. Make sure not to let the PBS rinse over the upper migrated cell-containing filter surface.

8. Wipe the cells of the non-migrated side off by drawing the filter over the wiper blade. Don't let the filter dry out during the washing and wiping procedure.

9. Clean the wiper blade with a Q-tip and repeat this step 3 times (washing bottom side in PBS followed by drawing over the wiper blade).

10. Upon removal of the non-migrated cell, the filter is fixed in Methanol for 5 min, air-dried and stained with Diff-Quik or hematoxyline (5 min followed by washes in H_2O).

11. The stained filter is placed face up on a 50 × 75 mm microscopic slide and either allowed to air dry or embedded in glycerine gelatine and a cover slip. The filter is now ready for counting.

Quantitative analysis of endothelial cell chemotaxis

Quantitation of endothelial cell chemotaxis depends to a fair degree on the technical capabilities of the lab. Consequently, only some general guidelines are discussed here.

The most straightforward quantitative analysis is the direct microscopic counting of the migrated cells. This is among the most exact analytical techniques. Visual inspection also facilitates a direct control of the migrated cells and allows the detection of aberrant migration pattern that may have resulted from the trapping of air bubbles or other experimental flaws. In turn, direct counting is laborious and time consuming. Quantitation is accomplished by counting the number of cells in at least 5 random high power field of views per well. This can be done manually counting all the cells in a microscopic field or by using a microscopic grid. Many cell culture laboratories also have image analysis systems that allow for the rapid quantitation of individual cells.

An alternative to the direct counting of cells is the automated analysis using densitometric analysis of individual well staining intensities. This similarly gives a rapid and reliable readout of endothelial cell chemotaxis. However, it is not as exact and sensitive as the direct counting of migrated cells.

Modifications of the basic system

▶ A number of modifications of the above protocol have been described. These include the development of chemotaxis chambers in different formats, including 10-well, 12-well, and 96-well plates. The 96-well plates are particularly suitable for large scale automated screening purposes as they have the format of 96-well automated ELISA or fluorescence readers. Likewise, disposable vertical chemotaxis systems have been developed by a number of different suppliers (e.g., Neuroprobe, www.neuroprobe.com, or Becton Dickinson, www.bectondickinson.com).

▶ An important control experiment to differentiate between chemotactic and chemokinetic effects is a so called checkerboard analysis. In this assay, concentrations of the test compounds are titrated in the upper and lower compartment. A truly chemotactic compound will only induce stronger cell migration than in the control if it is provided with a gradient (Fig. 2).

▶ A useful modification of the system is the transformation of a simple chemotaxis gradient experiments towards a more complex chemoinvasion assay. For this application, the filter membrane is not just coated

Pipetting scheme					Results [expressed in % of control]			
Conc. upper well	0	5	25	125	100	90	72	64
Conc. lower well	0	0	0	0				
Conc. upper well	0	5	25	125	141	111	81	73
Conc. lower well	5	5	5	5				
Conc. upper well	0	5	25	125	223	198	97	91
Conc. lower well	25	25	25	25				
Conc. upper well	0	5	25	125	342	298	184	104
Conc. lower well	125	125	125	125				

Fig. 2. Example of a checkerboard analysis of a chemotactic compound. The pipetting scheme (*left*) indicates the concentrations in the upper and the lower compartment of the chamber. The percent migration over control is given on the *right*. The 0/125 gradient leads to an approximately 3-fold induction of chemotaxis. In contrast, equal concentrations of the compound in the upper and the lower compartment (0/0, 5/5, 25/25, 125/125) lead to migration rates compared to control

with diluted 1:100 ECM but rather overlaid with a thin film of gel which is allowed to solidify on the filter. The cells migrating towards the stimulus are in this assay not just tested for their chemotactic capability but rather for their ability to invade and migrate towards a stimulus (Taraboletti, 1993). Disposable chemoinvasion assays have also been developed (e.g., Chemicon, www.chemicon.com/product/productdatasheet.asp? productitem=ecm555)

Expected Results

Lateral migration assay

The lateral migration assay is a straightforward and robust cell culture assay that is useful to analyze endothelial cell migration-inducing cytokines such as VEGF or bFGF. Generally speaking, the assay does not have a high degree of sensitivity. Yet, it is very useful as it is a simple and not very laborious assay making it in fact quite suitable for screening experiments. For example, the assay can easily be scaled up to screen several hundred wells to be analyzed within a few hours (e.g., 300 wells, corresponding to 100 compounds analyzed in triplicate). This scale of screening can for example be applied towards a functional screen to identify hybridoma supernatants that either stimulate or inhibit endothelial cell migration. Similarly, the assay is quite versatile and can be modified towards a number of other applications [e.g., differential cytochemical experiments (Kozian et al., 1997) or cell–cell interactions experiments, e.g., between endothelial cells and tumor cells (Augustin-Voss and Pauli 1992a) or between endothelial cells and pericytes).

Vertical chemotaxis assay in modified Boyden chamber

The vertical chemotaxis assay is a somewhat more delicate assay requiring careful handling in order to arrive at reliable and reproducible results. It pays for many applications to spend some time to optimize the assay, e.g., by varying the incubation time in order to determine the optimal differential between specific gradient-driven chemotaxis and random chemokinetic migration.

A solid chemotaxis experiment should always include a carefully performed checkerboard analysis. This is accomplished by titrating the concentrations of the test compound in the upper and the lower compartment. Equal concentrations of test compound in the upper and the lower compartment should not lead to a stronger migration compared to control for a truly chemotactic agent. Chemokines such as SDF-1 will induce endothelial cell migration in a strictly gradient-dependent manner (Feil and Augustin 1998). Similarly, VEGF has been reported to act primarily in a chemotactic manner, whereas bFGF stimulates endothelial chemokinesis (Yoshida et al., 1996). An example of a checkerboard analysis is shown in Fig. 2. As can be seen, only a gradient of the test compound leads to an approximately 3-fold increase in cell migration, whereas the diagonal (i.e., equal concentrations in upper and lower compartment) vary around 100 %.

Troubleshooting

Lateral migration assay

The lateral migration assay is largely devoid of any major problems. Problems in the quantitative assay may arise if repeatedly used silicon templates have not been properly washed in between applications or if proteins did dry on the silicon. This may render the silicon adhesive which leads to damage of the cells at the starting front upon removal of the silicon from the culture dish. Clean, non-adhesive templates will usually give a perfectly straight starting line of the cells.

Vertical chemotaxis experiment in modified Boyden chamber

A number of problems may be associated with the modified Boyden chamber assay. Most of them are related to handling problems and should not be encountered with some experience. Among the most critical problems is the trapping of bubbles during the assembly of the chamber. The lower chambers should be filled with a slight positive meniscus and the experiment should be prepared so that the loading is done without any delay. The same may occur in the upper compartment if a bubble is trapped below the cell. In either way, bubbles will lead to empty spots in the filter which becomes obvious during manual analysis but not necessarily during automated image analysis. Care needs also to be taken in the disassembly of the chamber and the handling of the filter during washes and staining. Make sure to strictly keep the orientation of the chamber and the filter throughout the experiment in order to avoid confusion. The worst mistake is of course the wiping of the cells from the wrong side of the membrane which unavoidably happens to the inexperienced investigator sooner or later. Yet, these are all minor problems related to the handling of the chamber which can all be easily eliminated with some experience.

References

Augustin-Voss HG, Pauli BU (1992a) Migrating endothelial cells are distinctly hyperglycosylated and express specific migration-associated cell surface glycoproteins. J Cell Biol 119:483–491.

Augustin-Voss HG, Pauli BU (1992b) Quantitative analysis of autocrine-regulated, matrix-induced, and tumor cell-stimulated endothelial cell migration using a silicon template compartmentalization technique. Exp Cell Res 198:221–227.

Boyden S (1962) The chemotactic effect of mixtures of antibody and antigen on polymorphonuclear leucocytes. J Exp Med. 115:453–466.

Falk W, Goodwin, RH, Leonard EJ (1980) A 48-well micro chemotaxis assembly for rapid and accurate measurement of leukocyte migration. J Immunol Methods. 33:239–247.

Feil C, Augustin HG (1998) Endothelial cells differentially express functional CXC-chemokine receptor-4 (CXCR-4/fusin) under the control of autocrine activity and exogenous cytokines. Biochem Biophys Res Commun. 247:38–45.

Kozian DH, Ziche M, Augustin HG (1997) The activin-binding protein follistatin regulates autocrine endothelial cell activity and induces angiogenesis. Lab Invest. 76:267–276.

Sato Y, Rifkin DB (1989) Inhibition of endothelial cell movement by pericytes and smooth muscle cells: activation of a latent transforming growth factor-beta 1-like molecule by plasmin during co-culture. J Cell Biol. 109:309–315.

Yoshida A, Anand-Apte B, Zetter BR (1996) Differential endothelial migration and proliferation to basic fibroblast growth factor and vascular endothelial growth factor. Growth Factors. 13:57–64.

Chapter 15

FACS Analysis of Endothelial Cells

Yvonne Reiss and Britta Engelhardt

▓ Introduction

Fluorescence activated cell sorting (FACS) has become a standard technique to simultaneously stain, analyze, and sort living cells based on the measurement of visible and fluorescent light emission. Since the development of the first commercial instruments (Herzenberg et al. 2002, Loken et al. 1977), the range of FACS dimensions (detectable parameters) has been improved by the use of up to 3 lasers to allow the simultaneous detection of 12 colors and 2 scatter parameters. Applications in endothelial cell biology include the characterization of newly established cell lines – i.e., after gene transfer via viral transduction or transfection (Wagner and Risau 1994, Kiefer et al. 1994, Roux et al. 1994, Greenwood et al. 1996, and corresponding chapters in this book), for the expression of various surface and activation markers, molecules involved in angiogenic processes or for cytokine expression profiles. Intracellular stainings of inactive and active kinases provide powerful tools to assess signal transduction pathways by means of FACS analysis. In addition, large scale screening of newly generated antibodies directed against endothelial surface antigens from hybridoma fusions can be achieved employing FACS analysis.

Here, we describe different approaches to detect constitutive and activation-dependent cell surface markers on endothelial cells. For large scale analysis of surface markers the authors recommend the usage of hybridoma cell supernatants and appropriate secondary reagents applying single color FACS analysis (Reiss et al. 1998, Reiss et al. 1999, Hahne et al. 1993). This approach is very convenient as commercially available antibody reagents are usually more expensive. However, if multiple colors have to be used at the same time to discriminate between subpopulations (e.g., for the analysis of endothelial progenitor cells applying markers such as CD34, CD117, and VEGFR-2), directly conjugated primary antibodies are recommended.

For a detailed and general introduction into FACS analysis and cell sorting the reader is referred to the Current Protocols in Immunology (Wiley).

Springer Lab Manual
H. Augustin (Ed.)
Methods in Endothelial Cell Biology
© Springer-Verlag Berlin Heidelberg 2004

▓ Materials

Cells

- Endothelial cells (mouse or human, primary or cell lines)

Reagents

- Primary antibodies: either purified commercially available unlabeled or fluorescently labeled antibodies or hybridoma supernatants specific for cell surface molecules present on endothelial cells (e.g., VE-cadherin, endoglin, VEGFR-2, PECAM-1, ICAM-2, MECA-32 antigen)
- Secondary antibodies labeled with fluorochromes such as FITC, PE, Cychrome, or APC; generally F(ab)'$_2$ secondary antibodies are recommended – if necessary use specifically anti-IgG or anti-IgM.
- FACS buffer: phosphate buffered saline (PBS) supplemented with 2.5 % calf serum or bovine serum albumin (BSA) and 0.1 % sodium azide.
- Fixative: 1 % (w/v) formaldehyde in PBS.
- Accutase (PAA, www.paa.at) or alternatively Hanks buffered saline solution (HBSS) without divalent cations, supplemented with 5 mM EDTA, and 25 mM HEPES
- Sodiumhypochloride
- FACS flow (BD Pharmingen, www.bdbiosciences.com)

Equipment (Fig. 1)

- 96-well round bottom microtiter plates or 12 × 75 mm polystyrene test tubes (e.g., Costar, www.labpages.com)
- Ice bath
- 8- or 12-channel pipette (e.g., Eppendorf, www.eppendorf.com)
- Buffer reservoir for multichannel pipette
- Refrigerated table-top centrifuge for test tubes or microtiter plates
- Microtiter plate shaker (e.g., Dunn, www.dunnlab.de)
- Vortex
- Minitubes and minitube rack (96-well format; Costar)
- Flow cytometer (e.g., FACScan or FACS Calibur, Becton Dickinson, www.bd.com) including computer and software (BD CellQuest software)
- any other equipment may be used accordingly but the authors have no experience with other systems.
- Color printer

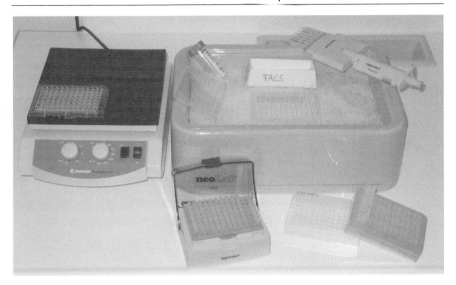

Fig. 1. Necessary equipment to perform a staining procedure: 96-well round bottom microtiter plates, microtiter plate shaker, ice bath, 8- or 12-channel pipette, buffer reservoir for multichannel pipette, minitubes, and minitube rack (96-well format)

Procedure

General considerations

▸ Control samples required: It is important to include unstained samples, isotype controls in addition to positive controls for each color which are required for the later adjustment of forward and side scatter detector settings and the determination of autofluorescence of cells.

▸ Staining in microtiter plates is suggested as lower cell numbers per sample suffice (if experienced you can stain numbers as low as 1×10^4 cells per well) and many samples can be stained simultaneously – an experienced investigator can stain up to 6 microtiter plates (= almost 600 samples) within one staining procedure!

Step-by-step procedure

1. Harvest adherent endothelial cells from tissue culture flask (i.e., T75, T25, or T12.5) by incubation with Accutase (PAA) or HBSS without divalent cations/5 mM EDTA/25 mM HEPES (5 min, 37 °C) – do **not** use trypsin because it will destroy many epitopes of cell surface molecules.

2. Wash cells in appropriate amount of FACS buffer (guideline: use 15 ml FACS buffer for 1 T25 of confluent endothelial cells).

3. Centrifuge 5 min at 250 × g and carefully aspirate supernatant without disturbing cell pellet.

4. Resuspend pellet in 1 ml FACS buffer and adjust cell number to 5×10^6 cells/ml (5×10^5 cells per 100 µl).

5. Deliver minimally 5×10^4 or ideally 5×10^5 cells in 100 µl to each well of a microtiter plate or test tube.

6. Centrifuge (microtiter plate: 4 min at 300 × g, tubes: 4 min at 250 × g), remove supernatant either by swiftly tapping microtiter plate to discard supernatants into the sink or by aspirating supernatant from tubes. Pellets need to be loosened prior to the addition of primary antibody by gently shaking microtiter plates on a microtiter plate shaker for 20 s or by vortexing tubes.

7. Optional: Pre-incubate cells with 10 µl of 10 % blocking reagent if using directly labelled primary antibodies (e.g., rat serum or rat IgG if using rat anti-mouse reagents) on ice for 10 min.

8. Dilute primary monoclonal antibodies (e.g. fluorochrome-conjugated, biotinylated, or unconjugated mAbs) to predetermined optimal concentrations (range: 1 µg to 10 µg per ml), hybridoma supernatants can be used undiluted or in a predetermined dilution. Add 100 µl antibody solution to each well and make sure cells are well suspended in the antibody solution. If not – resuspend carefully.

9. Incubate on ice for 30 min (incubation needs to be in the dark if you are using fluorescently labeled antibodies).

10. Prepare secondary reagent **now**, e.g., fluorochrome-labeled F(ab)'$_2$ secondary Ig (i.e., goat anti-rat) diluted as recommended by the distributor (usually 10 µg/ml) in FACS buffer and 10 % blocking serum (i.e., mouse serum if staining murine endothelial cells, cheaper) or purified IgG (expensive) to adsorb unspecific binding of the secondary reagent to murine or human endothelial cells – put tube on ice in the dark.

11. Wash microtiter plate twice with 200 µl FACS buffer/well (or 2 ml/tube). Remove supernatant after each centrifugation (300 × g for 4 min) and gently shake microtiter plate to loosen pellet prior to the next washing step.

12. Add secondary antibody in 40 µl (i.e., fluorochrome labeled F(ab)'$_2$ Ig or fluorochrome-conjugated avidin, streptavidin) and make sure cells are well suspended in the secondary reagent.

13. Incubate on ice for 30 min in the dark.

14. Wash 2× with 200 µl FACS buffer as described in Step 11.

15. Fix cells by adding 200 µl of cold 1 % formaldehyde in FACS buffer and transfer cells into microtubes.

16. Alternatively transfer cells into tubes appropriate for flow cytometer (12 × 75 mm Polystyrene, Falcon).

17. Aquisition and analysis of samples with flow cytometer (FACScan or FACS Calibur) using Cell Quest software.

Example

Staining of the endothelioma cell line bEndI2.3 derived from ICAM-2 mutant mice (Gerwin et al. 1999) is illustrated in Figs. 2 and 3. In this example, we describe the staining of immortalized endothelial cells using hybridoma supernatants and PE-conjugated rat secondary reagents:

1. Harvest unstimulated and cytokine-activated (TNF-α, stimulation for 16 h and 4 h) endothelioma cells as described above.

2. Apply 5×10^5 cells in 100 µl to each well of a microtiter plate.

3. Centrifuge microtiter plate (4 min at 300xg), remove supernatant by swiftly tapping microtiter plate. Gently loosen pellet using the microtiter plate shaker for 20 s.

4. Add 100 µl hybridoma supernatants (VE-Cadherin, Endoglin, MECA-32, PECAM-1, ICAM-2, ICAM-1, VCAM-1, E-selectin, P-selectin – corresponding clones see Table 1) including isotype control to each well and make sure cells are well suspended in the antibody solution.

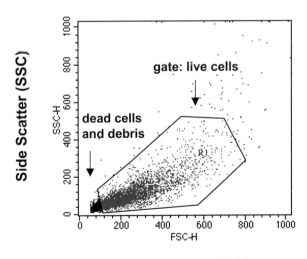

Fig. 2. Scatter analysis of ICAM-2 deficient endothelioma cells. The scatter gate is drawn so that dead cells and debris are excluded

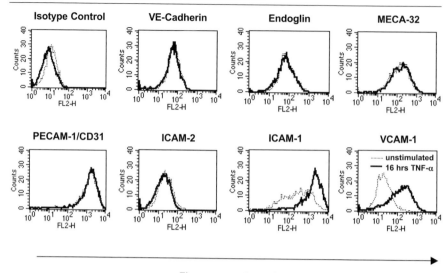

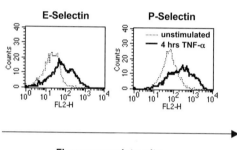

Fig. 3. Fluorescence intensity of several endothelial expressed surface and activation markers using brain-derived ICAM-2 deficient endothelioma cells (bEND I2.3) as an example. Overlays compare the surface expression of unstimulated (dotted line) and TNF-α stimulated cells (black line)

5. Incubate on ice for 30 min

6. Prepare secondary reagent: PE-conjugated F(ab)'$_2$-goat-anti rat IgG (e.g., Jackson, http://jacksonimmuno.com, or Biosource, www.biosource.com) diluted as recommended (10 µl/ml) in FACS buffer and 10% normal mouse serum (to adsorb unspecific binding of the secondary reagent to mouse endothelial cells) – put tube on ice.

7. Wash microtiter plate twice with 200 µl FACS buffer/well. Remove supernatant after each centrifugation (300 × g for 4 min) and loosen pellet prior to the next washing step.

Table 1. Cell surface molecules suitable for the characterization of mouse endothelial cells by FACS analysis

constitutively expressed EC surface molecules	antibody source
VEGFR-2 (Flk-1)	R&D – clone AF644
Tie-2	R&D – clone AF762
CD34	BD Pharmingen – clone Mec14.7
PECAM-1	BD Pharmingen – clone Mec13.3
ICAM-2	BD Pharmingen – clone 3C4
VE-Cadherin	BD Pharmingen – clone 11D4
Endoglin	BD Pharmingen or ATCC – clone MJ7/18
MECA-32 antigen	BD Pharmingen – clone MECA-32
αvβ3-integrin	BD Pharmingen – H92B8 and 2C9.G2

pro-inflammatory cytokine-induced EC surface molecules	antibody source
ICAM-1	ATCC – clone 29G1
VCAM-1	ATCC – clone MK2.7 or MK1.9
E-selectin	BD Pharmingen – clone 10E9.6
P-selectin	BD Pharmingen – clone RB 40.34

8. Add secondary antibody in 40 µl and resuspend cells well.

9. Incubate on ice for 30 min in the dark.

10. Wash 2× with 200 µl FACS buffer as described in Step 7.

11. Fix cells by adding 200 µl of cold 1 % formaldehyde in FACS buffer and transfer cells into microtubes.

12. Analyze samples using flow cytometer (FACScan or FACS Calibur) and Cell Quest software.

FACS aquisition of cell samples

Scatter gate unstained endothelial cells using forward scatter (FSC) versus side scatter (SSC) detector settings in a way that cells are displayed as a well visible population. Endothelial cells are much larger than leukocytes. It may therefore be easier to scatter cells using the FSC E0-1 rather than using the FSC E00 which is typically used for peripheral blood leukocytes. While gating endothelial cells, adjust autofluorescence to the first quadrant of the fluorescence(s) to be measured. Make sure your population is on scale. First, check fluorescence with your positive control, and compensate if necessary. Start aquisition of your samples – 10,000 events per sample are optimal.

Analysis

Draw a scatter gate on the single cell population of the endothelial cells (Fig. 2). Analyze histograms (single-color analysis) or dot blots (multi-color analysis) of gated endothelial cells. Include statistics.

Expected Results

We routinely apply FACS analysis to verify the endothelial phenotype of immortalized mouse microvascular endothelioma cells (Reiss et al. 1998, Wagner and Risau 1994, Kiefer et al. 1994, Hahne et al. 1993, see also Chap. 6). As an example, we show the surface phenotype of an ICAM-2-mutant endothelioma cell line (bEndI2.3). ICAM-2 deficient (Fig. 3) and wild type endothelioma cells (not shown) express similar levels of VE-Cadherin, endoglin, PECAM-1 and the endothelial cell-specific MECA-32 antigen. Therefore, it can be concluded that ICAM-2 deficiency does not lead to changes in the expression level of additional endothelial adhesion receptors besides ICAM-2. Upon stimulation with TNF-α for 16 h, ICAM-2 mutant endothelioma cells show upregulation of ICAM-1 and VCAM-1 on their surface to similar levels normally observed in wild type endothelium, whereas the constitutive expression of VE-Cadherin, and PECAM-1 did not change, as expected. Stimulation with TNF-α (4 h) induces comparable surface expression of E- and P-selectin. E-selectin can still be detected after 16 h of TNF-α stimulation, whereas the surface levels of P-selectin decline rapidly (not shown). Brain derived ICAM-2-mutant endothelioma cells do not establish surface expression of mucosal vascular adressin (MAdCAM-1) or peripheral lymph node addressin (PNAd, not shown).

Troubleshooting and General Considerations

▶ Determine optimal concentrations of each primary and secondary antibody by titrating in advance.

▶ In general, 10 µg/ml is a saturating concentration for most antibodies. High affinity antibodies can usually be diluted to 1 µg/ml. Normally, between 1 µg and 0.1 µg antibody per 10^6 cells in 100 µl buffer volume are optimal.

▶ When performing multi-color labeling, directly-conjugated mAbs are preferably used and can be added simultaneously.

▶ It is important to include negative and isotype controls in every experiment. Negative controls are unstained cells that are necessary for the correct adjustment of the FACS settings for the fluorescence detectors. Isotype controls are cells exposed to a subclass specific isotype (same isotype as primary antibody plus the secondary step reagent). For multi-color staining, single-color stained controls need to be included. To identify novel markers on endothelial cells, positive controls (i.e., cells which are known to express the antigen of interest) should be included in each experiment.

▶ If you notice subpopulations with respect to markers check FSC versus fluorescence and determine whether high fluorescence signals might be due to dead cells included in your preparation (appearance in FSC low) or cell aggregates (appearance in FSC high).

▶ High background staining achieved with your isotype control antibody means that pre-adsorption of your antibodies might be necessary or alternatively, additional blocking to minimize non-specific adherence of primary antibodies to your endothelial cells needs to be performed – see Steps 7 and 10 in staining procedure.

References

Coligan JE, Kruisbeek AM, Margulies DH, Shevach EM, Strober W (2001) Current Protocols in Immunology. Wiley

Gerwin N, Gonzalo JA, Lloyd C, Coyle AJ, Reiss Y, Banu N, Wang B, Xu H, Avraham H, Engelhardt B, Springer TA, Gutierrez-Ramos JC (1999) Prolonged eosinophil accumulation in allergic lung interstitium of ICAM-2 deficient mice results in extended hyperresponsiveness. Immunity 10:9–19

Greenwood J, Pryce G, Devine L, Male DK, dos Santos WL, Calder VL, Adamson P (1996) SV40 large T immortalised cell lines of the rat blood-brain and blood-retinal barriers retain their phenotypic and immunological characteristics. Neuroimmunol 71:51–63

Hahne M, Jager U, Isemann S, Hallmann R, Vestweber D (1993) Five tumor necrosis factor-inducilble cell adhesion mechanisms on the surface of mouse endothelioma cells mediate the binding of leukocytes. J Cell Biol 121:655–64

Herzenberg LA, Parks D, Sahaf B, Perez O, Roederer M, Herzenberg LA (2002) The history and future of the fluorescence activated cell sorter and flow cytometry. A view from Stanford. Clin Chem 48:1819–27

Kiefer F, Courtneidge SA, Wagner EF (1994) Oncogenic properties of the middle T antigens of polymaviruses. Adv Cancer Res 64:125–57

Loken MR, Parks DR, Herzenberg LA (1977) Two-color immunofluorescence using a fluorescence-activated cell sorter. Histochem Cytochem 25:899–907

Reiss Y, Hoch G, Deutsch U, Engelhardt B (1998) T cell interaction with ICAM-1-deficient endothelium in vitro: essential role for ICAM-1 and ICAM-2 in transendothelial migration of T cells. Eur J Immunol 28:3086–99

Reiss Y, Engelhardt B (1999) T cell interaction with ICAM-1-deficient endothelium in vitro: transendothelial migration of different T cell populations is mediated by endothelial ICAM-1 and ICAM-2. Int Immunol 11:1527–39

Roux F, Durieu-Trautmann O, Chaverot N, Claire M, Mailly P, Bourre JM, Strosberg AD, Couraud PO (1994) Regulation of gamma-glutamyl transpeptidase and alkaline phosphatase activities in immortalized rat brain microvessel endothelial cells. Cell Physiol 159:101–13

Wagner EF, Risau W (1994) Oncogenes in the study of endothelial cell growth and differentiation. Semin Cancer Biol 5:137–45

Chapter 16

Application of the RNA Interference (RNAi) Technology to Angiogenesis Research

SUNG-SUK CHAE, JI-HYE PAIK, JONATHAN SHUBERT-COLEMAN,
HENRY FURNEAUX and TIMOTHY HLA

Introduction

What is RNAi?

Important discoveries within the past decade, which demonstrated the ability of small duplex RNAs to profoundly regulate gene expression, have opened the door to a new era of gene regulation research while providing a powerful tool to specifically down-regulate genes. Originally discovered in the nematode *C. elegans*, RNA interference, or RNAi, is defined as the sequence specific silencing of gene expression brought about by the introduction of exogenous double stranded RNA (Fire et al. 1998). In the initial experiments, researchers found that the injection of long (>500bp) dsRNA into nematodes interfered with the expression of specific genes that were homologous in sequence to the injected dsRNAs. Since this initial discovery, it has been shown that long dsRNA is not the direct effector of RNAi. It is cleaved into 21–23 nt duplex RNAs, termed small interfering RNAs, or siRNAs, which are the bona fide effectors of RNAi (Elbashir et al. 2001a). Prior to this discovery, it was not possible to induce RNAi in mammalian cells since the introduction of long dsRNAs induces the interferon response, leading to cell cycle alterations and apoptosis (Elbashir et al. 2001a). However, duplex RNA less than 30 nucleotides in length does not trigger the interferon response (Elbashir et al. 2001a, Elbashir et al. 2001b), and it has become possible to silence specific genes in mammalian cells with the use of siRNA.

How does RNAi work?

RNAi occurs at the post-transcriptional level where siRNAs silence gene expression by directing the cleavage and, thus, degradation of specific mRNAs (Tuschl et al. 1999). First, siRNAs are incorporated into the ribonucleoprotein complex known as the RNA-induced silencing Complex (RISC) (Elbashir et al. 2001a, Nykanen et al. 2001). Once incorporated into the RISC, siRNA duplexes are unwound in an ATP-dependent manner. One strand of

Springer Lab Manual
H. Augustin (Ed.)
Methods in Endothelial Cell Biology
© Springer-Verlag Berlin Heidelberg 2004

the siRNA duplex, known as the antisense strand (or guide strand), is then annealed to a target mRNA to which it is complementary, and directs the cleavage of the mRNA within the complementary region (Martinez et al. 2002). The specificity of the RNAi reaction is provided by complementary base pairing between the siRNA antisense strand and the target mRNA strand and is absolutely essential for efficient degradation by the RNAi mechanism. Experiments have shown that the presence of a single base pair mismatch within the siRNA/mRNA duplex region inhibits the ability of the siRNA to direct mRNA cleavage and, thus, the ability to silence the expression of the target gene (Elbashir et al. 2001 c). The implication of these findings is that the siRNA technique is very specific and, thus, single genes can be silenced without affecting the expression of any other genes. In fact, evidence supporting the strict specificity of siRNA-induced gene silencing has recently been published. Treatment of mammalian cells with siRNAs results in the decrease of target gene mRNA levels only while no changes were observed in 36,000 other genes by cDNA microarray analysis (Chi et al. 2003).

The use of appropriate controls is the key to demonstrate the specificity of the siRNA effect. The use of more than one siRNA species that targets the given transcript which results in down regulation of the cognate mRNA and the desired phenotype is an important initial result. Dose-dependent effects of specific siRNA are generally seen. A rigorous control is to conduct a rescue experiment whereby reintroduction of the gene of interest with a silent mutation in the siRNA binding site and concomitant rescue of the desired phenotype, thus demonstrating the specificity of the silencing effect (Harborth et al. 2001, Lassus et al. 2002).

Advantages of siRNAs

Until the discovery of RNAi, the principle method of studying the effects of the loss of specific gene function in animal models involved the genetic manipulation and development of animals in which the gene of interest had been deleted. This method is comparatively time consuming and does not allow an investigator to view the effects of the loss of a particular gene product in an already developed 'normal' environment. Although antisense DNAs have been used, problems associated with non-sequence-specific effects have limited their use (Benimetskaya et al. 1995). The siRNA technology offers a promising alternative to these approaches. siRNAs can be introduced by a variety of methods into cultured cells or living organisms providing researchers with a relatively inexpensive and fast means of knocking down a particular gene of interest (Harborth et al. 2001). The use of RNAi offers the ability to investigate the effects of the loss of a specific gene product at a precise point in time in an already developed system. RNAi can be effected by introducing synthetic siRNAs as well as by the introduction of expression vectors created to produce siRNA like stem loop RNA duplex. The ability to invoke RNAi through the use of expression vectors provides further advantages to researchers. A wide variety of expression vectors, including both tissue specific and inducible expression vectors, allow investigators to further control the temporal and spatial silencing of target genes.

Use of the RNAi technology in angiogenesis research

The angiogenic response is a complex, orchestrated response of the host to various physiologic or pathologic stimuli (Carmeliet 2003). Numerous factors either acting directly on endothelial cells or indirectly through other cells can induce angiogenesis. The process of angiogenesis itself includes well orchestrated steps of endothelial cell migration, proliferation, morphogenesis, and vascular maturation. Several gene families have been implicated in these steps. Most research on angiogenesis has been focused on determination of precise regulatory steps and the hierarchy of their control. The advent of the RNAi technology provides a potentially powerful methodology to further deconvolute the angiogenic regulatory networks in both in vitro as well as in vivo systems. High-throughput RNAi screens have been performed at the genomic level in *C. elegans* (Kamath et al. 2003). Such studies have revealed that loss of function experiments in whole organisms can be readily achieved by RNAi technology. Similarly, critical metabolic routes have been discovered by comprehensive silencing of metabolic pathways in *Drosophila* (Dobrosotskaya et al. 2002). In this study, the novel finding that the sphingolipid signaling pathway regulates phospholipid (phosphotidylethanolamine) levels was revealed by comprehensive loss of function experiments in which specific enzymes were down regulated by RNAi in *Drosophila* S2 cells. Similarly, this approach may have utility in angiogenesis research, in which hierarchical signal transduction of angiogenic growth factors and receptors regulates new vascular growth. In theory, this approach may be used to deconvolute any complex phenomenon in vascular system, such as atherosclerosis, diabetes, hypertension, provided an appropriate model system is available which is amenable to siRNA delivery and suppression.

▣ Materials

For transfection with siRNA

Cells

- Endothelial cells

Reagents

- Oligofectamine (Invitrogen, www.invitrogen.com)
- Opti-MEM (Invitrogen)
- SiRNA (e.g., www.dharmacon.com or www.ambion.com)
- M199 supplemented with 30 % FBS and 3 × endothelial cell growth factors (antibiotics-free)
- SiRNA annealing buffer (30 mM HEPES-KOH, pH 7.4, 100 mM potassium acetate, 2 mM magnesium acetate)

Equipment

- CO_2 incubator

Supplies

- Tissue culture dishes

For in vivo Matrigel angiogenesis assay

Animals

Nude mice (female athymic Ncr-nu/nu, 4–6 weeks of age)

Reagents

- Anesthetics, e.g., tribromoethanol (i.e., Avertin)
- Matrigel (BD Biosciences, www.bdbiosciences.com)
- DMEM (Dulbecco's modified Eagle's medium)
- SiRNA
- Recombinant human FGF-2
- Heparin (Pharmacia and Upjohn Company)

Supplies

- 1 ml pipette tips (sterile), store at 4 °C until use
- 1 ml syringes with removable 27G needles, store at –20 °C and keep on ice
- 5 ml snap cap polystyrene tubes

- Dissecting tools
- PBS (phosphate buffered saline)
- 70 % ethanol
- 0.3 M sucrose in PBS
- 10 % formalin in PBS

Procedure

Preparation of siRNA duplex (optional)

Deprotection
Deprotect the synthetic siRNAs as described by the manufacturer.

Desalting
1. Add 1/10 (v/v) of 10 M ammonium acetate and 3 vol of ethanol. Mix well and store at –20 °C overnight or –80 °C for 2 h.

2. Centrifuge at 14,000 × g for 30 min at 4 °C.

3. Remove supernatant and rinse pellet with 1 ml of cold 75 % ethanol twice.

4. Dry the pellet and dissolve in RNAase-free water or 20 mM Tris-HCl, 1 mM EDTA, pH 7.4, 50 mM KCl.

5. Determine the concentration. (1 unit at A_{260} ~ 32 µg/ml as a rule of thumb (Elbashir et al. 2001b) but it is more accurate to consult the specific extinction co-efficient for every duplex RNA and hypochromicity of the duplex should be considered).

Annealing
1. If sense and antisense RNAs are supplied in separate tubes, hybridize them to get duplex siRNAs.

2. Incubate an equal molar ratio of single-stranded RNAs in siRNA annealing buffer for 1 min at 95 °C, then 1 h at 37 °C.

3. If necessary, check the duplex formation by 12 % nondenaturing polyacrylamide gel electrophoresis.

Transfection of endothelial cells (Ancellin et al. 2002)

1. Plate the cells a day before transfection (30–40 % confluence on the day of transfection). Use antibiotic-free growth medium.

2. Mix 50 µl of 20 µM siRNA duplex with 900 µl of Opti-MEM.

3. Mix 50 μl of oligofectamine reagent with 200 μl of Opti-MEM. Incubate 10 min at RT.

4. Mix the above mixtures – siRNA duplex to diluted oligofectamine (total vol. = 1.2 ml). Mix gently and incubate 20 min at RT.

5. Add 3.8 ml of Opti-MEM.

6. Wash cells twice with serum-free medium.

7. For 100 mm dish transfection, overlay 5 ml of siRNA-oligofectamine complexes onto the cells.

8. Incubate cells for 4 h in a CO_2 incubator.

9. Add 2.5 ml of M199 supplemented with 30 % FBS and 3× endothelial cell growth factor/heparin to the cells. Do not remove the transfection solution.

10. Analyze the cells for gene expression after 24–48 h by Western blot or Northern blot analysis.

Matrigel angiogenesis assay (Passaniti et al. 1992, for details see Chap. 20)

First day: preparation for Matrigel mixture
1. Thaw the frozen Matrigel overnight at 4 °C.

2. Keep 1 ml pipette tips (sterile) at 4 °C until use.

3. Keep 1 ml syringes with removable 27G needles at –20 °C.

Second day: subcutaneous Matrigel injection
Thoroughly and gently mix Matrigel with various compounds in a 5 ml tube on ice until there is no color interface. **Note:** Be cautious not to create bubbles. Final concentration of Matrigel matrix in the mixture must be higher than 10 mg/ml. 2 ml of mixture is required for 5 injections due to the sample loss during the procedure (Table 1).

1. Keep tubes on ice until injection.

2. Anesthesize animals and load cold 1 ml syringe by removing needles and avoiding bubbles.

3. Insert 27G needle and inject subcutaneously, 200–300 μl per injection.

Eighth day: Matrigel plug harvest
1. Dissect skin around the injected site to reveal Matrigel plugs.

2. Wash harvested Matrigel plugs in PBS and fix in 10 % formalin in PBS for 2 h.

Table 1. Protocol for s.c. Matrigel injection

Group	Matrigel (15 mg/ml)	DMEM	bFGF (0.5 mg/ml)	Heparin (10,000 U/ml)	siRNA (0.2 mM)	Control RNA (0.2 mM)	Total Volume
A	1.5 ml	490 µl	0	10 µl	0	0	2 ml
B	1.5 ml	473 µl	17 µl	10 µl	0	0	2 ml
C	1.5 ml	273 µl	17 µl	10 µl	0.2 ml	0	2 ml
D	1.5 ml	273 µl	17 µl	10 µl	0	0.2 ml	2 ml

3. Wash the fixed plugs in PBS for 30 min.

4. Store plugs in 70 % ethanol at 4 °C until processing for paraffin-embedding.

5. Cut 5 µm thick paraffin sections.

6. Perform H&E and/or CD31/PECAM staining for the neovessel formation.

Results

We tested RNAi-mediated gene silencing on human umbilical vascular endothelial cells (HUVEC) and immortalized mouse embryonic endothelial cells (MEEC). For in vivo applications, Matrigel angiogenesis system was used. Generally, duplex siRNA efficiently knocked down specific genes in vitro as well as in vivo.

When we used siRNA targeting of the $S1P_1$/EDG-1 receptor in endothelial cells, the reduction of mRNA level was about 90–95 % (Fig. 1A). Western blot analysis was not conducted because a specific antibody against the mouse $S1P_1$/EDG-1 receptor was not available. For the functional analysis, migration assay toward sphingosine-1-phosphate (S1P), a ligand for $S1P_1$/EDG-1 was also performed. siRNA for $S1P_1$/EDG-1 completely inhibited S1P induced endothelial cell migration (Fig. 1B). We were also able to silence other genes in these cells with 80–95 % efficacy.

The design of the siRNAs is largely empirical. Silencing efficacy is variable, depending on the target sequence, cell types, the confluence and passage number of the cell. In addition, the expression level and half-life of the protein affect the efficiency of RNAi.

In most cases, we observed the depletion of protein and mRNA on day 1 or 2 after transfection. Effective gene silencing of many genes was achieved with 200 nM concentration of siRNA. Therefore, 200 nM siRNA would be a reasonable starting point.

A

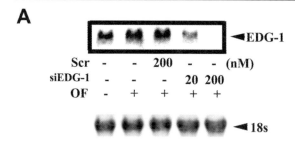

Fig. 1A,B. siRNA-mediated knockdown of expression of S1P$_1$/EDG-1 receptor and S1P-induced endothelial cell migration. A HUVEC were transfected with 20–200 nM of siRNA for S1P1/EDG-1. After 24 h, total RNA was isolated and subjected to Northern blot analysis. siRNA induced specific S1P$_1$/EDG-1 silencing but scrambled duplex RNA did not. B S1P$_1$/EDG-1 siRNA but not siRNA for β-galactosidase completely blocked cell migration toward S1P

B

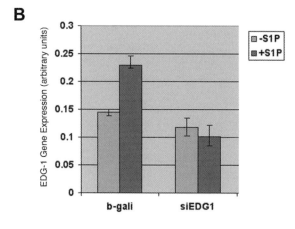

For in vivo Matrigel analysis (Passaniti et al. 1992), siRNA was used without liposomes. Although 20 μM of siRNA was efficacious in gene silencing in this system, much lower concentrations (0.5 μM) of siRNA were also equally effective. FGF-2 is a potent angiogenic factor and is used to induce angiogenesis. siRNA-mediated S1P$_1$/EDG-1 silencing completely blocked vessel formation induced by FGF-2 (Fig. 2). Blood vessel formation was determined by H&E staining by counting vessel-like structures in which red blood cells exist (data not shown).

However, CD31/PECAM immunostaining gives more definite vessel structure and allows easier quantification. It is important to point out that requirement for lipid-mediated transfection of siRNA is empirically determined and its obligatory requirement in every system is not established. There may be specific molecular mechanisms involved in siRNA uptake into cells (Fares and Greenwald 2001) and this needs to be further defined in various systems in which gene silencing is attempted.

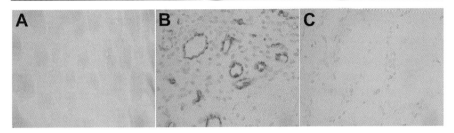

Fig. 2A-C. In vivo Matrigel angiogenesis assay. Matrigel was mixed with FGF-2 and/or siRNA and subcutaneously injected on the back of the nude mice. After 7 days, the Matrigel plug was harvested, paraffin-embedded, sectioned and immunostained with anti-CD31/PECAM antibody. FGF-2 strongly induced vessel formation. S1P$_1$/EDG-1 siRNA ablated FGF-2 induced vessel formation. Scrambled siRNA had no effect on angiogenesis. **A** Matrigel alone- no FGF-2, **B** FGF-2 + scrambled RNA, **C** FGF-2 + S1P$_1$/EDG-1 siRNA

Troubleshooting

The effectiveness of RNAi can be affected by various factors. Although a possible target sequence can be chosen according to the simple rules (http://www.rockefeller.edu/labheads/tuschl/sirna.html or http://www.bioinfo.rpi.edu/applications/sfold/index.pl), it is empirical. The procedure is generally simple but should be optimized for each gene and siRNA (Table 2).

Table 2. Troubleshooting

Problem	Presumed cause	Solution
No or weak silencing	– Non optimal target sequence	– Design new siRNA for other target sequence or use mixed siRNAs targeting multiple sites
	– Target protein with long half-life	– Check half-life of target protein – Repeat transfection e.g., every other day
	– Abundant protein	– Longer incubation – Increase the concentration of siRNA
	– Poor annealing of siRNA	– Denature siRNAs in boiling water and reanneal them
Cell death	– Excessive transfection reagents	– Reduce the amount of liposomes
	– Prolonged transfection time	– Decrease the transfection time
No vessel formation	– Inactive FGF-2	– Use fresh FGF-2
Hemorrage within Matrigel plug	– Bubble caused during mixing	– Avoid air bubbles during preparation
	– Ruptured blood vessel	– Do not inject Matrigel intradermally or intramuscularly
Poor Matrigel morphology	– Torn Matrigel plug	– Dissect Matrigel intact

References

Ancellin N, Colmont C, Su J, Li Q, Mittereder N, Chae SS, Stefansson S, Liau G, Hla T (2002) Extracellular export of sphingosine kinase-1 enzyme. Sphingosine 1-phosphate generation and the induction of angiogenic vascular maturation. J Biol Chem 277:6667–6675

Benimetskaya L, Tonkinson JL, Koziolkiewicz M, Karwowski B, Guga P, Zeltser R, Stec W, Stein CA (1995) Binding of phosphorothioate oligodeoxynucleotides to basic fibroblast growth factor, recombinant soluble CD4, laminin and fibronectin is P-chirality independent. Nucleic Acids Res 23:4239–4245

Carmeliet P (2003) Angiogenesis in health and disease. Nat Med 9:653–660

Chi JT, Chang HY, Wang NN, Chang DS, Dunphy N, Brown PO (2003) Genomewide view of gene silencing by small interfering RNAs. Proc Natl Acad Sci USA 100:6343–6346

Dobrosotskaya IY, Seegmiller AC, Brown MS, Goldstein JL, Rawson RB (2002) Regulation of SREBP processing and membrane lipid production by phospholipids in Drosophila. Science 296:879–883

Elbashir SM, Harborth J, Lendeckel W, Yalcin A, Weber K, Tuschl T (2001a) Duplexes of 21-nucleotide RNAs mediate RNA interference in cultured mammalian cells. Nature 411:494–498

Elbashir SM, Lendeckel W, Tuschl T (2001b) RNA interference is mediated by 21- and 22-nucleotide RNAs. Genes Dev 15:188–200

Elbashir SM, Martinez J, Patkaniowska A, Lendeckel W, Tuschl T (2001c) Functional anatomy of siRNAs for mediating efficient RNAi in Drosophila melanogaster embryo lysate. Embo J 20:6877–6888

Fares H, Greenwald I (2001) Genetic analysis of endocytosis in Caenorhabditis elegans: coelomocyte uptake defective mutants. Genetics 159:133–145

Fire A, Xu S, Montgomery MK, Kostas SA, Driver SE, Mello CC (1998) Potent and specific genetic interference by double-stranded RNA in Caenorhabditis elegans. Nature 391:806–811

Harborth J, Elbashir SM, Bechert K, Tuschl T, Weber K (2001) Identification of essential genes in cultured mammalian cells using small interfering RNAs. J Cell Sci 114:4557–4565.

Kamath RS, Fraser AG, Dong Y, Poulin G, Durbin R, Gotta M, Kanapin A, Le Bot N, Moreno S, Sohrmann M, Welchman DP, Zipperlen P, Ahringer J (2003) Systematic functional analysis of the Caenorhabditis elegans genome using RNAi. Nature 421:231–237

Lassus P, Rodriguez J, Lazebnik Y (2002) Confirming specificity of RNAi in mammalian cells. Sci STKE 2002, PL13

Martinez J, Patkaniowska A, Urlaub H, Luhrmann R, Tuschl T (2002) Single-stranded antisense siRNAs guide target RNA cleavage in RNAi. Cell 110:563–574

Nykanen A, Haley B, Zamore PD (2001) ATP requirements and small interfering RNA structure in the RNA interference pathway. Cell 107:309–321

Passaniti A, Taylor RM, Pili R, Guo Y, Long PV, Haney JA, Pauly RR, Grant DS, Martin GR (1992) A simple, quantitative method for assessing angiogenesis and antiangiogenic agents using reconstituted basement membrane, heparin, and fibroblast growth factor. Lab Invest 67:519–528

Tuschl T, Zamore PD, Lehmann R, Bartel DP, Sharp PA (1999) Targeted mRNA degradation by double-stranded RNA in vitro. Genes Dev 13:3191–3197

Part III:
Functional Assays in Vivo

Murine Bone Marrow Transplantation Models that Enable the Study of EPC Recruitment

Toshinori Murayama, Oren M. Tepper, and Takayuki Asahara

Introduction

It has recently been established that postnatal neovascularization is not restricted to angiogenesis, but also includes vasculogenesis (Asahara et al. 1997, Shi et al. 1998, Isner and Ashara 1999, Gunsilius et al. 2000). During adult vasculogenesis, bone marrow (BM)-derived endothelial progenitor cells (EPCs) are mobilized to the systemic circulation in response to certain cytokines or pharmacologic agents, and are recruited to ischemic tissue where they differentiate in situ. As there is currently no marker that can distinguish between BM-derived (vasculogenesis) versus resident endothelial lineage cells (angiogenesis), murine BM transplantation (BMT) models offer a powerful technique for studying the process of postnatal vasculogenesis. BM transplantation (BMT) models that enable one to distinguish BM-derived EPCs are designed for donor cells from either (1) transgenic mice in which the expression of marker gene is driven by endothelial-specific promoter (Asahara et al. 1999a,b, Takahashi et al. 1999, Llevadot et al. 2001, Murayama et al. 2002, Zhang et al. 2002), or (2) transgenic mice with a ubiquitously expressed marker gene or retroviral-infected wild-type BM, followed by endothelial staining (Crosby et al. 2000, Jackson et al. 2001, Lyden 2001 et al., Edelberg 2002 et al., Grant 2002 et al., Hess et al. 2002, Sata et al. 2002, Werner et al. 2002). Such models allow for:

a) Identification of BM-derived EPCs in situ
b) Quantification of BM-derived EPCs contribution in postnatal neovascularization
c) Assessment of recruitment of BM-derived EPCs in response to various stimuli

Springer Lab Manual
H. Augustin (Ed.)
Methods in Endothelial Cell Biology
© Springer-Verlag Berlin Heidelberg 2004

■ Materials

Donor mice

Four to eight weeks of age (male, if Y chromosome detection planned), one donor mouse typically yields enough BM cells to transplant to 5 recipient mice

- B6.129-Kdr[tm1Jrt]; Flk1/LacZ mice (Shalaby et al. 1995)
- FVB/N-Tg(TIE2-lacZ)182Sato; Tie2/LacZ mice (Schlaeger et al. 1997)
- FVB/N-Tg(TIE2GFP)287Sato; Tie2/GFP mice (Motoike et al. 2000)
- B6.129S7-Gt(ROSA)26Sor; ROSA mice (Friedrich et al. 1991)
- C57BL/6-Tg(ACTbEGFP)1Osb; eGFP mice (Okabe et al. 1997)

Recipient mice

Background of mice should correspond to donors (syngeneic); 4–8 weeks of age (female, if Y chromosome detection planned)

- FVB
- C57BL/6

   All mice are available from the Jackson Laboratory, www.jaxmice.jax.org

Reagents

- HISTOPAQUE-1083 (Sigma, www.sigmaaldrich.com)
- Phosphate-buffered saline with 5 mM of EDTA (PBS-E)
- Appropriate anesthesia, e.g., 2,2,2-Tribromoethanol (Avertin®) solution (20 mg/ml)

Equipment

- Low-speed centrifuge
- Irradiator, e.g., Gammacell 40 Exactor (MDS Nordion, www.nordion.com)
- 15-ml conical centrifuge tubes
- 1-ml tuberculin syringe with 25-G needle
- 3-ml syringe with 18-G needle
- 1-ml insulin syringe with 28-G needle
- Prefilter or cell strainer (∅ 70 µm)
- Scissors, forceps, and a scalpel
- Pasteur pipette
- Optional: mouse pie cage (Braintree Scientific, Inc., www.braintreesci.com)

Procedure

1. Add 3 ml of HISTOPAQUE-1083 to each 15-ml conical centrifuge tube, and bring to room temperature.

2. First irradiation of recipient mice (in a mouse pie cage, or an equivalent device) with 4.5 Gy for FVB, or 6 Gy for C57BL/6.

3. Sacrifice donor mice and harvest the femurs and tibias into cold PBS-E.

4. (Working under hood is recommendable).. Remove the muscle from the bones, and cut off the ends of the bones with a scalpel. Flush out BM cells from the remaining central portion of the bones with a tuberculin syringe and PBS-E. Each bone should be flushed extensively until red BM cells disappear inside the bones.

5. Carefully lay the BM cell suspension (i.e., drop by drop) onto HISTOPAQUE in a 15-ml conical tubes.

6. Centrifuge at $900 \times g$ for 25 min at RT **without** brake.

7. Discard the upper layer to within 0.5 cm (1/4 inch) of the opaque interface (mononuclear cells; MNCs) with a Pasteur pipette.

8. Collect MNCs-rich interface carefully by 18-G needle on a 3-ml syringe into a new conical tube.

9. Suspend them into 15 ml of PBS-E for washing.

10. Centrifuge at $1200 \times g$ for 5 min at 4 °C.

11. Discard supernatant and resuspend an MNC pellet in 2 ml of PBS-E.

12. Filter MNC suspension through 70-μm cell strainer.

13. Count MNCs and prepare them at $2-3 \times 10^7$ cells / ml in PBS (A total of 2×10^7 MNCs are expected from both lower extremities/mouse).

14. Second irradiation of recipients with the same dose as the first one.

15. Anesthetize them with Avertin solution (10–15 μl/g body weight).

16. Systemic administration (through lateral tail vein or intracardiac) of $2-3 \times 10^6$ of BM-MNCs/mouse with an insulin syringe.

17. Keep mice warm on a hotplate for recovery from anesthesia.

18. Wait for 4–6 weeks until BM reconstitution.

Troubleshooting

Confirmation of BM reconstitution

eGFP mice: direct flow-cytometry analysis (Manfra et al. 2001).

ROSA mice: methylcellulose semisolid colony assay with X-gal staining (Asahara et al. 1999a) or fluorescein-di-α-D-galactopyranoside staining followed by flow-cytometry analysis (Castro et al. 2002).

Tie2/LacZ or Tie2/GFP mice: real-time (quantitative) PCR analysis of genomic DNA from peripheral blood or BM cells.

The above techniques should result in more than 90 % BM reconstitution at 4–6 weeks following BMT.

Irradiation dose

Since the relationship between the irradiation dose and BM reconstitution ratio is known to depend on dose rate and fractioning (Down 1991 et al., Van Os et al. 1993), researchers should consider not only a total dose but also a dosing schedule. Our procedure here is based on the condition of 0.55 Gy / min and 3-h interval of two equal fractions by Gammacell 40 Exactor (^{137}Cs). We have also found that C57BL/6 mice are more radioresistant than FVB, so that different doses have been applied to each strain. Since ^{137}Cs has a high energy, an at least 1-inch-thick lead shield is required to attenuate the radiation. A clinical lead protector gown (for researcher) or thin lead shield (for mice) is not helpful.

Difficulties in detecting marker signal of donor cells

An important issue to be noted is organ/tissue-specific expression level of Tie2 and Tie2-promoter-driven LacZ gene (Murayama et al. unpublished data), e.g., almost all endothelial cells (ECs) express Tie2/LacZ gene in the limb muscle, but only a small percentage of ECs do it in the heart from adult Tie2/LacZ mice. Although the varied expression of the Tie2/LacZ gene found in mature ECs cannot necessarily be applied to differentiating EPCs in situ, researchers should be aware of this potential limitation and should therefore not underestimate the contribution of BM-derived EPCs. Therefore, we recommend approaching an initial study with multiple donor models outlined above. In addition, weak GFP signal from Tie2/GFP or eGFP BMT mice can be amplified by anti-GFP antibody and/or confocal microscopy. Also, the expression level of "ubiquitous" marker gene in ROSA or eGFP mice may vary in each tissue/organ. Researcher should confirm the expression of genes in the interested cells in transgenic mice at first before BMT.

Sudden death after donor cell injection

Since aggregated cells may cause pulmonary embolization and sudden death, we recommend gentle tapping of the syringe prior to injection.

BMT using nude rats

While syngeneic mice are the best choice of recipients for transplantation of transgenic mouse BM, certain experimental models are often technically difficult to perform in mice and better suited for larger animals such as rats. Although we have demonstrated successful BMT from Tie2/LacZ mice to nude rats (Walter et al. 2002), the optimal condition of this procedure has not been determined yet. It should be noted that rat effector natural killer cells are radioresistant and capable of rejecting allograft even in nude rats (Engh et al. 1998). The BMT failure of nude rats may result from an inadequate number of injected donor cells (Mohiuddin et al. 2000) or a graft versus host effect.

Acknowledgements. This chapter is supported in part by Establishment of International COE for Integration of Transplantation Therapy and Regenerative Medicine (COE program of the Ministry of Education, Culture, Sports, Science and Technology, Japan).

References

Asahara T, Murohara T, Sullivan A, Silver M, Van der Zee R, Li T, Witzenbichler B, Schatteman G, Isner JM (1997) Isolation of putative progenitor endothelial cells for angiogenesis. Science 275:964–967

Asahara T, Masuda H, Takahashi T, Kalka C, Pastore C, Silver M, Kearne M, Magner M, Isner JM (1999a) Bone marrow origin of endothelial progenitor cells responsible for postnatal vasculogenesis in physiological and pathological neovascularization. Circ Res 85:221–220

Asahara T, Takahashi T, Masuda H, Kalka C, Chen D, Iwaguro H, Inai Y, Silver M, Isner JM (1999b) VEGF contributes to postnatal neovascularization by mobilizing bone marrow-derived endothelial progenitor cells. Embo J 18:3964–3972

Castro RF, Jackson KA, Goodell MA, Robertson CS, Liu H, Shine HD (2002) Failure of bone marrow cells to transdifferentiate into neural cells in vivo. Science 297:1299

Crosby JR, Kaminski WE, Schatteman G, Martin PJ, Raines EW, Seifert RA, Bowen-Pope DF (2000) Endothelial cells of hematopoietic origin make a significant contribution to adult blood vessel formation. Circ Res 87:728–730

Down JD, Tarbell NJ, Thames HD, Mauch PM (1991) Syngeneic and allogeneic bone marrow engraftment after total body irradiation: dependence on dose, dose rate, and fractionation. Blood 77:661–669

Edelberg JM, Tang L, Hattori K, Lyden D, Rafii S (2002) Young adult bone marrow-derived endothelial precursor cells restore aging-impaired cardiac angiogenic function. Circ Res 90:E89–93

Engh E, Benestad HB, Strom-Gundersen I, Vaage JT, Bell EB, Rolstad B (1998) Role of classical (RT1.A) and nonclassical (RT1.C) MHC class I regions in natural killer cell-mediated
bone marrow allograft rejection in rats. Transplantation 65:319–324

Friedrich G, Soriano P (1991) Promoter traps in embryonic stem cells: a genetic screen to
identify and mutate developmental genes in mice. Genes Dev 5:1513–1523

Grant MB, May WS, Caballero S, Brown GA, Guthrie SM, Mames RN, Byrne BJ, Vaught T, Spo-
erri PE, Peck AB, Scott EW (2002) Adult hematopoietic stem cells provide functional
hemangioblast activity during retinal neovascularization. Nat Med 8:607–612

Gunsilius E, Duba HC, Petzer AL, Kahler CM, Grunewald K, Stockhammer G, Gabl C, Dirn-
hofer S, Clausen J, Gastl G (2000) Evidence from a leukaemia model for maintenance of
vascular endothelium by bone-marrow-derived endothelial cells. Lancet 355: 1688–1691

Hess DC, Hill WD, Martin-Studdard A, Carroll J, Brailer J, Carothers J (2002) Bone marrow
as a source of endothelial cells and NeuN-expressing cells after stroke. Stroke 33:1362–
1368

Isner JM, Asahara T (1999) Angiogenesis and vasculogenesis as therapeutic strategies for
postnatal neovascularization. J Clin Invest 103:1231–1236

Jackson KA, Majka SM, Wang H, Pocius J, Hartley CJ, Majesky MW, Entman ML, Michael LH,
Hirschi KK, Goodell MA (2001) Regeneration of ischemic cardiac muscle and vascular
endothelium by adult stem cells. J Clin Invest 107:1395–1402

Llevadot J, Murasawa S, Kureishi Y, Uchida S, Masuda H, Kawamoto A, Walsh K, Isner JM,
Asahara T (2001) HMG-CoA reductase inhibitor mobilizes bone marrow-derived
endothelial progenitor cells. J Clin Invest 108:399–405

Lyden D, Hattori K, Dias S, Costa C, Blaikie P, Butros L, Chadburn A, Heissig B, Marks W, Witte
L, Wu Y, Hicklin D, Zhu Z, Hackett NR, Crystal RG, Moore MA, Hajjar KA, Manova K,
Benezra R, Rafii S (2001) Impaired recruitment of bone-marrow-derived endothelial and
hematopoietic precursor cells blocks tumor angiogenesis and growth. Nat Med 7:
1194–1201

Manfra DJ, Chen SC, Yang TY, Sullivan L, Wiekowski MT, Abbondanzo S, Vassileva G,
Zalamea P, Cook DN, Lira SA (2001) Leukocytes expressing green fluorescent protein as
novel reagents for adoptive cell transfer and bone marrow transplantation studies. Am J
Pathol 158:41–47

Mohiuddin MM, Ildstad ST, DiSesa VJ (2000) Establishment of fully xenogeneic (mouse—
>rat) bone marrow chimeras: evidence for normal development and clonal deletion of
mouse T cells. Transplantation 69:731–736

Motoike T, Loughna S, Perens E, Roman BL, Liao W, Chau TC, Richardson CD, Kawate T, Kuno
J, Weinstein BM, Stainier DY, Sato TN (2000) Universal GFP reporter for the study of vascular development. Genesis 28:75–81

Murayama T, Tepper OM, Silver M, Ma H, Losordo DW, Isner JM, Asahara T, Kalka C (2002)
Determination of bone marrow-derived endothelial progenitor cell significance in
angiogenic growth factor-induced neovascularization in vivo. Exp Hematol 30: 967–972

Okabe M, Ikawa M, Kominami K, Nakanishi T, Nishimune Y (1997) 'Green mice' as a source
of ubiquitous green cells. FEBS Lett 407:313–319

Sata M, Saiura A, Kunisato A, Tojo A, Okada S, Tokuhisa T, Hirai H, Makuuchi M, Hirata Y,
Nagai R (2002) Hematopoietic stem cells differentiate into vascular cells that participate
in the pathogenesis of atherosclerosis. Nat Med 8:403–409

Schlaeger TM, Bartunkova S, Lawitts JA, Teichmann G, Risau W, Deutsch U, Sato TN (1997)
Uniform vascular-endothelial-cell-specific gene expression in both embryonic and adult
transgenic mice. Proc Natl Acad Sci U S A 94:3058–3063

Shalaby F, Rossant J, Yamaguchi TP, Gertsenstein M, Wu XF, Breitman ML, Schuh AC (1995) Failure of blood-island formation and vasculogenesis in Flk–1-deficient mice. Nature 376:62–66

Shi Q, Rafii S, Wu MH, Wijelath ES, Yu C, Ishida A, Fujita Y, Kothari S, Mohle R, Sauvage LR, Moore MA, Storb RF, Hammond WP (1998) Evidence for circulating bone marrow-derived endothelial cells. Blood 92:362–367

Takahashi T, Kalka C, Masuda H, Chen D, Silver M, Kearney M, Magner M, Isner JM, Asahara T (1999) Ischemia- and cytokine-induced mobilization of bone marrow-derived endothelial progenitor cells for neovascularization. Nat Med 5:434–438

Van Os R, Thames HD, Konings AW, Down JD (1993) Radiation dose-fractionation and dose-rate relationships for long-term repopulating hemopoietic stem cells in a murine bone marrow transplant model. Radiat Res 136:118–125

Walter DH, Rittig K, Bahlmann FH, Kirchmair R, Silver M, Murayama T, Nishimura H, Losordo DW, Asahara T, Isner JM (2002) Statin therapy accelerates reendothelialization: a novel effect involving mobilization and incorporation of bone marrow-derived endothelial progenitor cells. Circulation 105:3017–3024

Werner N, Priller J, Laufs U, Endres M, Bohm M, Dirnagl U, Nickenig G (2002) Bone marrow-derived progenitor cells modulate vascular reendothelialization and neointimal formation: effect of 3-hydroxy–3-methylglutaryl coenzyme a reductase inhibition. Arterioscler Thromb Vasc Biol 22:1567–1572

Zhang ZG, Zhang L, Jiang Q, Chopp M (2002) Bone marrow-derived endothelial progenitor cells participate in cerebral neovascularization after focal cerebral ischemia in the adult mouse. Circ Res 90:284–288

Chapter 18

Balloon Denudation of Blood Vessels

VOLKHARD LINDNER

Introduction

Balloon catheter denudation of arteries is a widely used model to study the vascular response to injury. This technique was originally described in a rabbit model by Baumgartner and Studer (Baumgartner and Studer 1966) and much of the characterization of this model in the rat has been performed by Clowes, Schwartz and Reidy (Clowes et al. 1983b, Clowes et al. 1983a, Reidy et al. 1983, Clowes et al. 1986). This model mimics a variety of aspects associated with angioplasty procedures including proliferation of smooth muscle cells (SMC) and development of intimal hyperplasia, adventitial cell proliferation with accompanying fibrosis and constrictive remodeling (Smith et al. 1999, Wilcox et al. 1996), as well as endothelial cell regeneration. This chapter is focused on the aspects of endothelial cell regeneration associated with this model. The technical details of this procedure are described here for the common carotid artery and aorta of the rat. However, with bigger catheters this procedure can be adopted for larger species. The technical details of endothelial denudation of the mouse carotid artery with a guide wire from an angioplasty catheter are described elsewhere (Lindner et al. 1993). A skilled person with extensive experience can perform the rat balloon denudation procedure within five minutes from start to finish. However, it will take many surgeries to develop consistent results especially when intimal hyperplasia is studied with this model.

Materials

Animals

The technique described here is suitable for rats with a body weight of more than 250 g. If necessary, denudation of the aorta can be performed in smaller animals by inserting the catheter via the common carotid artery as opposed to the external carotid artery. Obviously, these procedures require the

Springer Lab Manual
H. Augustin (Ed.)
Methods in Endothelial Cell Biology
© Springer-Verlag Berlin Heidelberg 2004

approval of the Institutional Animal Care and Use Committee and all local and federal or state guidelines need to be followed. In addition, prior to performing this procedure, the principal investigator and the person conducting the surgery should consult with the institutional veterinarian.

Anesthesia

Although a number of anesthetics, analgesics, and sedatives are approved for the use in rats, we routinely use a combination of ketamine and xylazine for intraperitoneal injection. In addition, your institutional policies may require the inclusion of a small amount of atropin in this mixture. For preparation of the anesthetic mixture, we use stock concentrations of 50 mg/ml for ketamine and 20 mg/ml for xylazine. Ketamine solutions of higher concentrations (i.e., 100 mg/ml) should be diluted 1:1 with sterile saline prior to use. For practical purposes, we add 1.7 ml of xylazine (20 mg/ml) to one 10 ml vial of ketamine (50 mg/ml). After combining both drugs, the cocktail should be used within in a few weeks as mixing reduces the shelf life of the individual components. For a rat weighing 350 g, 0.75 ml of cocktail drawn into a 1 ml syringe with a suitable needle attached (e.g., 25 gauge) will be injected intraperitoneally. Handling of the animal is facilitated if the injection is given after a brief inhalation of halothane. This can be done by placing the animal in a glass bell jar to which 2–3 ml of halothane have been added. It is crucial that the tip of the injection needle has entered the peritoneum without penetrating any abdominal organs. It is common experience that people new to this procedure often fail to completely penetrate the abdominal muscle layer with the needle and this will result not only in insufficient depth of anesthesia, but it may also give rise to serious ulcerations of the abdominal wall.

Surgical instruments (sterile)

- Scalpel
- Kelly hemostat, straight, 14 cm (2 ×)
- Hartmann hemostat, curved, 9 cm
- Extra delicate Vannas scissors, straight, 7.5 cm
- Stevens scissors, straight blunt tips, 10.8 cm
- Micro dissecting forceps, serrated, half-curved, 10 cm
- Micro dissecting forceps, curved, 12 cm, with suture platform
- Jewelers forceps, straight, 10.8 cm
- Michel wound clip applying forceps, 12.5 cm
- Michel wound clips, 7.5 mm

Equipment and supplies

- Platform for rat
- Rubber bands

- Light source (fiber optic/goose neck)
- Hair clipper
- One 10 ml syringe
- 1 ml syringes
- 25 gauge needles (0.5 mm)
- One three-way stopcock
- One 2 French Fogarty embolectomy catheter
- polyethylene tubing (inner diameter 0.8 mm)
- Gauze
- 3/0 silk suture, black braided
- Cotton applicators
- 70 % ethanol and betadine
- Surgical gloves
- Evan's blue dye (5 % solution in saline, sterile filtered)

Procedure

Denudation of the common carotid artery

Within a few minutes of intraperitoneal injection of the anesthetic, the anesthetized animal will be placed with its back on the surgery platform and the legs will be gently strapped to the platform with rubber bands. One additional rubber band holding the head by the upper incisors is sufficient to prevent the head from moving during the procedure.

The balloon catheter is attached to a three-way stopcock with one 10 ml syringe and a 1 ml syringe filled with saline attached to the other ports. The air in the catheter is replaced with saline by alternatingly pulling a vacuum with the 10 ml syringe followed by injection of saline from the 1 ml syringe. This process is repeated 3 to 4 times so that all air is removed from the Latex balloon and inflation of the balloon is achieved with saline alone. The three-way stopcock can then be removed and the 1 ml syringe alone attached to the catheter. A 5 cm long piece of polyethylene tubing with an inner diameter just wide enough to slide over the catheter (0.8 mm) can be used to facilitate the insertion of the catheter into the artery. For that purpose, the distal side of the tubing is cut at a sharp angle before sliding it over the catheter (Fig. 1). The tip of the catheter can then be placed in a tube with sterile saline until the denudation procedure.

Sufficient depth of anesthesia is verified by toe pinch (review your institutional animal procedure manual). The hair along the ventral side of the neck is removed with the hair clipper and disinfectants are applied. A 3 cm midline incision from the lower jaw towards the sternum is performed with the scalpel. The exposed subcutaneous tissue is dissected with dissecting scissors and the neck muscles underlying the left salivary gland are exposed.

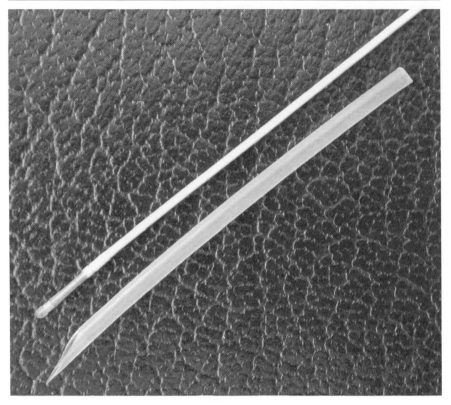

Fig. 1. Two French Fogarty embolectomy with beveled polyethylene tubing serving as trochar for insertion into arteriotomy

The Kelly hemostat is used to grab a small piece of fascia at the dorsal base of the gland and resting it on the platform to expose the neck muscles overlying the carotid bifurcation. The sternomastoid muscle is separated from the sternohyoid muscle by blunt dissection. Pushing the straight sternohyoid muscle towards the trachea, the carotid bifurcation and the external carotid artery are identified and the vessel is exposed by gentle dissection of the external carotid with the micro dissecting forceps. Using the curved micro dissecting forceps, a 20 cm long piece of 3/0 silk suture is placed around the proximal external carotid artery just above the carotid bifurcation. This will serve as the proximal ligature. The Kelly hemostat is used to hold the ends of the ligature (not to be tied off at this point) and it is placed on the animal's abdomen with gentle traction in caudal direction. Dissection of the external carotid artery can now continue in the distal direction. Great care should be taken not to damage small side branches leaving the external carotid. A second ligature is then placed around the external carotid artery as far downstream from the proximal ligature as possible and it is tied off. Gentle tension in cranial direction is applied to this ligature by attaching the Hartmann hemostat. In order to prevent blood loss during the arteriotomy, the proximal ligature is now pulled towards the right by the hemostat's weight and lack of pulsation between the 2 ligatures will indicate sufficient tension to stop blood flow. A small arteriotomy is now made close to the distal ligature with Vannas scissors. The tip of the tubing around the catheter can now be inserted into the artery with the bevel facing down. Sometimes a slight rotation of the tubing is necessary to facilitate entry of the catheter into the arteriotomy hole. The catheter tip is advanced to the proximal ligature. At this point, the proximal ligature has to be loosened around the vessel so that the catheter can be advanced past the ligature. Gentle tension in cranial direction on the distal ligature can help straighten the vessel and facilitate insertion of the catheter. After successful insertion of the catheter, blood loss from the arteriotomy can be controlled by applying tension to the proximal ligature. For denudation of the common carotid artery, the catheter will be advanced up to the first marking (approx. 5 cm). The catheter tip will then be located in the aortic arch and the balloon will now be inflated with saline (approx. 20–50 µl, depending on the size of the vessel). The inflated catheter will now be withdrawn, and resistance indicates that the balloon is inflated too much to enter the common carotid. While gently pulling in the catheter, the inflation is reduced until the catheter can be felt to enter the common carotid. Catheter inflation should be maintained at this position while the catheter is drawn towards the arteriotomy under constant rotation. Some resistance should be felt while withdrawing the catheter. Lack of resistance indicates insufficient inflation. When the inflated balloon is near the proximal ligature, it will be deflated and advanced into the arch and the procedure repeated 2 more times. After this, the tip of the catheter is withdrawn past the proximal ligature and the proximal ligature is tied off while the catheter is still inside the external carotid artery. The catheter can now be removed. It is important to examine the arteriotomy for bleeding as there often is retrograd blood flow originating from the small side branches located between the two ligatures. If bleeding occurs, it can be easily stopped by placing an additional ligature just proximal of the arteriotomy. After aligning the wound edges of the skin incision, it is closed with Michel wound clips (usually 3 are sufficient) and the animal is allowed to recover.

Denudation of the thoracic and abdominal aorta

The procedure is readily modified for denudation of the aorta by advancing the catheter to the third marking (approx. 15 cm). At that position the tip of the catheter will be in the abdominal aorta. If the catheter cannot be advanced significantly beyond the first marking, it is likely that it entered the heart. Pulling the catheter back beyond the first marking and repositioning the head and sometimes elevating the head will facilitate advancement of the catheter into the thoracic aorta. It is very important not to over-inflate the abdominal segment of the aorta as this will result in aneurysm formation which tend to rupture announced a few days later. Around the second marking, the catheter can be felt to pass the diaphragm. At this point, substantially more inflation is needed to denude the thoracic aorta, often in excess of 50 μl, and inflation needs to increase gradually as the catheter reaches the aortic arch. Proper inflation can be judged by experiencing a certain amount of resistance during catheter withdrawal. This denuding procedure should also be performed 3 times and under constant back and forth rotation to assure complete denudation.

Another variant of this procedure is the denudation of the common carotid artery and aorta via the femoral artery. This procedure has the advantage that blood flow in the carotid artery is not altered as the external carotid is not ligated by this approach. This approach is more difficult to perform in smaller rats and it likely causes hindlimb ischemia as the femoral artery is ligated in the process.

Endothelial regeneration

The left common carotid artery of the rat normally has no side branches. Following removal of all endothelial cells with the balloon catheter, endothelial regeneration therefore proceeds only from the carotid bifurcation and the aortic arch. As these anatomical branch points are readily identified, they represent the most useful landmarks for the assessment of endothelial regeneration. Macroscopically, endothelial regeneration can be assessed by intravenous injection of 0.5 ml of a 5 % Evans blue solution (in saline, sterile filtered). Evans blue can be injected via the tail vein in the anesthetized rat. It should be allowed to circulate for 5 min prior to sacrifice. The anesthetized rat can then be exsanguinated, the vasculature perfused with phosphate buffered saline (PBS) or lactated Ringer's solution with a catheter (18 gauge) placed in the abdominal aorta or heart. The perfusate can be allowed to drain from a vein such as the jugular vein. When the perfusate drains as a clear liquid from the vein, one can start the perfusion fixation at physiological pressure with the appropriate fixative. As opposed to alcohol based fixatives, there is no discoloration of the tissues and Evans blue stained vessels with aldehyde based fixatives (4 % buffered formaldehyde or paraformaldehyde). The common carotid arteries and the entire aorta can be excised in one piece. For practical reasons, it is advisable to clear these vessels from adherent connective tissue and fat in situ, i.e., prior to excision of the vessels.

Endothelial regeneration in the carotid artery proceeds at a rate of no more than 1–2 mm per week and usually endothelial outgrowth stops at 6 weeks after denudation leaving the center portion of the common carotid artery permanently denuded of endothelium (Reidy et al. 1983, Lindner et al., 1990). Endothelial proliferation and migration occurs at the leading edge whereas endothelial cells further away from the leading edge are no longer undergoing proliferation. SMC on the luminal surface of the denuded segment (pseudo-endothelium) can adopt properties normally associated with endothelium (Lindner et al. 1995, Couper et al., 1997). Permeability to Evans blue in these chronically denuded vessels is usually reduced compared to acutely denuded vessels and Evans blue staining intensity is therefore reduced. The most suitable vessel for studies on endothelial regeneration, however, is the thoracic aorta where the endothelium will grow out from the intercostal arteries following complete denudation. Denuded areas will appear blue while reendothelialized areas will be white. Endothelial outgrowth from the intercostal arteries proceeds in the shape of an oval around the intercostal ostia with more growth in the longitudinal direction of the vessel (Fig. 2). Within a week of denudation, these areas of reendothelializa-

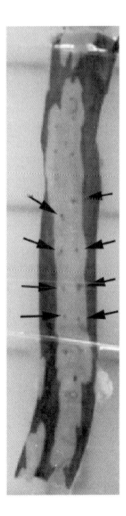

Fig. 2. En face view of a rat aorta 2 weeks after balloon catheter denudation. *Dark areas* represent still denuded areas (Evans blue stained) whereas *light areas* around intercostal arteries co-localize with regenerated endothelium. *Arrows* indicate intercostal branch points

tion begin to merge. Any area with regenerated endothelium not associated with an arterial branch point is most likely due to incomplete denudation and these areas should be excluded from quantitative measurements of endothelial regeneration. Even a single viable endothelial cell left behind after the denudation procedure will provide an additional site from where regeneration will proceed. Endothelial regeneration from intercostal arteries is extremely reproducible and predictable with approx. 50% of the luminal surface of the thoracic aorta getting reendothelialized within 2 weeks following denudation (Fig. 2) (Lindner and Reidy 1996). Unlike the carotid artery, complete reendothelialization usually occurs in the thoracic aorta within 6 weeks.

The excised carotid artery can be cut open longitudinally and mounted flat between glass slides for acquisition of images with subsequent measurements of endothelial outgrowth. The thoracic aorta is preferably cut open longitudinally along the ventral (denuded) side before mounting it temporarily between two glass slides in the presence of PBS. After obtaining a digital image of the mounted vessel, endothelial regeneration is readily quantified with image analysis software. We use the program NIH Image available for free downloading for both MacOS and Windows operating systems from http://rsb.info.nih.gov/nih-image/download.html .

The surface area of the aorta ranging from the descending thoracic aorta to the diaphragm will be quantified as well as the area covered by regenerated endothelium. The latter will be apparent as white areas around the intercostal arteries. Endothelial regeneration will then be expressed as a percentage of reendothelialized area over total vessel surface area.

Examination of endothelium with en face methods

Rat thoracic aorta segments are ideally suited for examination with en face techniques of immunohistochemistry or in situ hybridization (Lindner and Reidy, 1993). Immunohistochemistry or in situ hybridization procedures would be followed by the Haeutchen procedure (Schwartz and Benditt 1973) which allows one to isolate the luminal cell layer (endothelium) and mount it on slides for visualization under the microscope. Following balloon denudation, quiescent endothelium close to the intercostal ostium as well as proliferating and migrating endothelium at the leading edge can be examined on the same specimen. Endothelial-leukocyte interactions of aortic endothelium can also be studied with this method (Lindner and Collins 1996).

▨ References

Baumgartner HR, Studer A (1966) Effects of vascular catheterization in normo- and hyper-cholesteremic rabbits. Pathol Microbiol 29:393–405

Clowes AW, Clowes MM, Reidy MA (1986) Kinetics of cellular proliferation after arterial injury. III. Endothelial and smooth muscle growth in chronically denuded vessels. Lab Invest 54:295–303

Clowes AW, Reidy MA, Clowes MM (1983a) Kinetics of cellular proliferation after arterial injury. I. Smooth muscle growth in the absence of endothelium. Lab Invest 49:327–333

Clowes AW, Reidy MA, Clowes MM (1983b) Mechanisms of stenosis after arterial injury. Lab Invest 49:208–215

Couper LL, Bryant SR, Eldrup-Jorgensen J, Bredenberg CE, Lindner V (1997) Vascular endothelial growth factor increases the mitogenic response to fibroblast growth factor-2 in vascular smooth muscle cells in vivo via expression of fms-like tyrosine kinase-1. Circ Res 81:932–939

Lindner V, Collins T (1996) Expression of NF-κB and IκB-α by aortic endothelium in vivo. Am J Pathol 148:427–438

Lindner V, Fingerle J, Reidy MA (1993) Mouse model of arterial injury. Circ Res 73:792–796

Lindner V, Giachelli CM, Schwartz SM, Reidy MA (1995) A subpopulation of smooth muscle cells in injured rat arteries expresses platelet-derived growth factor-B chain mRNA. Circ Res 76:951–957

Lindner V, Majack RA, Reidy MA (1990) Basic fibroblast growth factor stimulates endothelial regrowth and proliferation in denuded arteries. J Clin Invest 85:2004–2008

Lindner V, Reidy MA (1993) Expression of basic fibroblast growth factor and its receptor by smooth muscle cells and endothelium in injured rat arteries. An en face study. Circ Res 73:589–595

Lindner V, Reidy MA (1996) Expression of VEGF receptors in arteries after endothelial injury and lack of increased endothelial regrowth in response to VEGF. Arterioscler Thromb Vasc Biol 16:1399–1405

Reidy MA, Clowes AW, Schwartz SM (1983) Endothelial regeneration. V. Inhibition of endothelial regrowth in arteries of rat and rabbit. Lab Invest 49:569–575

Schwartz SM, Benditt EP (1973) Cell replication in the aortic endothelium: A new method for study of the problem. Lab Invest 28:699–707

Smith JD, Bryant SR, Couper LL, Vary CP, Gotwal PJ, Koteliansky VE, Lindner V (1999) Soluble transforming growth factor-beta type II receptor inhibits negative remodeling, fibroblast transdifferentiation, and intimal lesion formation but not endothelial growth. Circ Res 84:1212–1222

Wilcox JN, Waksman R, King SB, Scott NA (1996) The role of the adventitia in the arterial response to angioplasty: the effect of intravascular radiation. Int J Radiat Oncol Biol Phys 36:789–796

Chapter 19

Peripheral Hind Limb Ischemia Models

STEPHAN SCHIRMER, IMO HOEFER, and IVO BUSCHMANN

Introduction

Three types of vessel growth are currently described in the textbooks as occuring under physiological and pathological conditions. The first and earliest in the course of life is known as vasculogenesis, describing the de-novo formation of tubular, blood-conducting structures during the early embryonic development, evolving from hemangioblasts, i.e., very little differentiated cells. Although recent publications point to a role of adult vasculogenesis, this chapter will focus on angiogenesis and arteriogenesis, two distinct mechanisms of vessel growth after birth.

Angiogenesis is referred to as capillary sprouting, i.e., formation of new capillary sprouts from pre-existing capillaries or intussusceptive growth. These newly evolved vessels consist of a single layer of endothelium, lacking surrounding smooth muscle cells, and are prone to rupture (Buschmann and Schaper 2000). Moreover, capillaries are not capable of conducting hemodynamically relevant amounts of blood flow to ischemic territories. Tissue hypoxia is known to be one of the strongest stimuli to induce angiogenesis; low pO_2 results in upregulation of the hypoxia–inducible factor-1α (HIF-1α), enhancing the expression of different growth factors, e.g., vascular endothelial growth factor (VEGF). The mitogenic effect of VEGF on endothelial cells is mainly mediated via the two receptors Flk-1 and Flt-1.

Arteriogenesis differs from angiogenesis in that it refers to pre-existing collateral pathways. Once a pressure gradient across a major artery becomes relevant, pre-existing arteriolar anastomoses (resistance arteries) are being recruited and transformed into functional collateral arteries (conductance arteries), large enough to conduct efficient amounts of blood flow to avoid detrimental effects of tissue hypoxia. This is, however, not a new observation but was described in detail more than a century ago. Arteriogenesis is independent of low oxygen levels; in fact, arteriogenesis occurs in regions which are not directly affected by the arterial stenosis. For example, in the case of femoral artery occlusion, arteriogenesis can be observed in the non-

Springer Lab Manual
H. Augustin (Ed.)
Methods in Endothelial Cell Biology
© Springer-Verlag Berlin Heidelberg 2004

ischemic thigh, whereas only the distal limb or foot becomes ischemic. In other words, collateral growth takes place in oxygen-rich tissue to provide arterial blood flow to hypoperfused territories outside the risk region.

While tissue ischemia will inevitably lead to the above-described angiogenesis, an increase in shear forces exerted by blood flow trying to bypass a functional stenosis or occlusion in an area of physiological oxygen partial pressure will result in the following steps: activation of the endothelium, increased production of NO, Intercellular Adhesion Molecules (e.g., ICAM-1), chemokines, and other inflammatory cytokines (such as Monocyte Chemoattractant Protein 1, Granulocyte-Macrophage Colony Stimulating Factor, Tumor Necrosis Factor Alpha), invasion of circulating monocytes and their precursors (Buschmann et al. 2003, Busch et al. 2003), activation of these invading cells and their further secretion of cytokines such as TNF-α or basic Fibroblast Growth Factor (FGF-2 or bFGF). After stenosis or occlusion of a large conducting artery, blood follows the intravascular pressure gradient (i.e., high-pressure proximal, low-pressure distal to the occlusion). These fundamental hemodynamic changes increase blood flow through pre-existing arteriolar bypasses, raising shear stress (up to 20-fold) as the crucial stimulus for the enlargement of the arterioles by proliferation of endothelial and smooth muscle cells.

The invasion of hematopoietic-derived endothelial precursor cells or even pluripotent stem cells might play an additional important role in both the formation of new capillaries and the modification of small arterioles into collateral arteries.

Several animal models have been established in order to mimic the human situation in various settings of PAOD. The following description of three different established in-vivo models therefore focuses on the different aspects regarding the occurrence of angiogenesis and arteriogenesis and how they can be achieved. All models are described as rabbit models; for other purposes very similar techniques can be used in different animals (mice, rats, pigs). However, both possibilities and means of parameter acquisition and perfusion measurements as well as the transfer into the clinical situation are distinctively different when using different species.

Non-ischemic hind limb model

In the non-ischemic model of femoral artery ligation, New Zealand White Rabbits (NZWR) with an average weight of about 2.5 to 4 kg are anesthetized with an intravenous injection of ketamine hydrochloride (40–80 mg/kg) and xylazine (8–9 mg/kg body weight). Supplementary dosages of anesthetic may be given as needed. The medial aspect of the upper leg is shaven from the inguinal area to the knee. After skin disinfection using preferably an alcoholic tincture, an incision of approx. 2 cm is made about 2 cm distal of the inguinal ligament into the direction of the knee. Careful preparation through the subcutaneous adipose tissue towards the adductor muscle is performed. A small incision into the muscle is needed to bare the trunk of the femoral artery, vein, and nerve. The femoral artery is now carefully dissected and exposed while the vein and nerve are to be carefully protected from damage. The artery is ligated using a 2-0 silk suture at two places 1 cm apart to prevent the proliferation of bridging collaterals which would immediately restore perfusion. The ligation is placed distal to the branches of the profound femoral artery as well as the circumflex femoral artery; both provide the origin of already existing arteriolar anastomoses which proliferate into functional arteries circumventing the occlusion. For therapeutic approaches, the femoral artery can be cannulated with a sterile polyethylene catheter, placed with the tip approximately 1 cm distal to the branches of the two above-mentioned vessels and pointing upstream. Linked to an osmotic minipump which is implanted subcutaneously into the lower lateral part of the abdominal region, the effects of a continuous intraarterial infusion of various agents directly into the collateral circulation can be examined. The pump can be inserted directly via the small skin incision above the femoral artery and carefully pushed forward after preparation of a subcutaneous path towards the lower abdomen. For the implantation procedure only the distal one of the two aforementioned sutures is performed, the second is carefully tightened around the catheter, closure of which must be avoided to ensure substance delivery. Afterwards, the incision is sutured by careful adaptation of the skin (Hoefer et al. 2001). After receiving an intramuscular injection of tetracycline or cephazolin (15 mg/kg/day) as an antibiotic prophylactic as well as an analgetic (buprenorphine, 0.04 mg/kg), the animals are housed individually with access to water and chow ad libidum. No gangrene or gross impairment of hind limb function was observed during the weeks following the operation. Neither 3 h, 24 h, 3 days nor 7 days after femoral artery ligation in this non-ischemic model were any changes in LDH A, HIF 1-α, ATP, ADP, AMP, adenosine, inosine, hypoxanthine, lactate, or creatine phosphate observed in the quadriceps, adductor, peroneus, or gastrocnemius muscle (Deindl et al. 2001). Therefore, we can conclude that the double ligation of the superficial femoral artery does not result in ischemia, and as ischemia is known to be the main stimulus for angiogenesis (Ito et al. 1997), this model is not suitable to study angiogenesis in the peripheral circulation, but is well suited to analyze collateral artery growth, i.e., arteriogenesis (Fig. 1A).

Ischemic hind limb model

In the ischemic hind limb model, animals are anesthesized with the same dosages of ketamine hydrochloride and xylazine administered intravenously. Again, the upper right leg is shaven spaciously and aseptical conditions are maintained by disinfection tincture. This time, the longitudinal

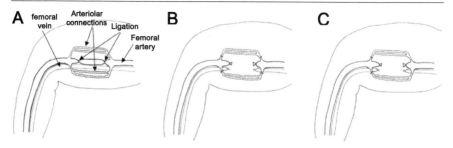

Fig. 1. A Non-ischemic hind limb model. Femoral artery and vein are left intact and blood flow to the distal limb is secured by pre-existing arteriolar connections located in the thigh. **B** Ischemic hind limb model. Femoral artery and vein are excised and the branches serving as stem vessels for pre-existing collateral arteries are occluded. **C** Semi-ischemic hind limb model. Femoral artery and vein excised. In contrast to the ischemic model, flow to the thigh is secured by leaving the stem vessels patent, while the re-entry region is occluded

skin incision is longer, reaching from distal to the inguinal ligament to proximal to the patella. Again, subcutaneous adipose tissue is removed and the adductor muscle is cautiously divided and incised. The femoral artery is carefully exposed and dissected along its entire length, including the proximal parts of the popliteal and saphenous arteries. All its major branches, the inferior epigastric (from the external iliac artery) and deep femoral artery, the lateral circumflex and superficial epigastric artery, are ligated with a 4-0 silk suture. Ischemia is induced when the proximal origin (distal external iliac artery just above the inguinal ligament) as well as the distal end (popliteal/saphenous artery) are ligated with a 2-0 or 4-0 silk suture, respectively, and the complete common and superficial femoral arteries, from the branch of the external iliac to the point distally where it bifurcates to form the saphenous and popliteal arteries, are excised (Pu et al. 1993). Excision of the femoral artery results in retrograde propagation of thrombus and occlusion of the external iliac artery. Consequently, the only vessel supplying the limb with blood flow sufficient to prevent major necrosis is the internal iliac artery. After skin adaptation, the incision site is sutured carefully and the animals receive an intramuscular injection of antibiotic and analgetic. In this model, an increase in distinct phosphates and creatine kinase as well as HIF1-α and LDH A in both the upper and lower leg (m. quadriceps and adductor, m. peroneus and gastrocnemius), indicating ischemia due to insufficient blood supply, can be observed. Not all but some of the operated animals show varying degrees of superficial tissue necrosis in their distal calves or toes and muscle atrophy. Furthermore, an increase in endogenous VEGF expression can be measured. This model of total femoral artery excision may be used as model of persistent ischemia, follow-up of up to 90 days reveals continuous ischemic conditions (Pu et al. 1994); as ischemia stimulates angiogenesis, this model is widely accepted for the study of various angio-genic agents (parameter acquisition and perfusion measurements see below) (Fig. 1B).

Semi-ischemic hind limb model

Recently, a third hind limb model has been established, presenting a means of inducing ischemia in part of the hind limb only, i.e., upholding physiolog-ical blood flow and perfusion conditions in the thigh while generating ischemia in the calf (Rissanen et al. 2003). Again, after shaving and disinfec-tion, the skin is incised and the superficial femoral artery is dissected and exposed carefully, preventing vein and nerve from damage. The complete superficial femoral artery is then excised. Additionally, to make sure collat-eral circulation in the thigh will not be sufficient to restore perfusion of the lower leg and thereby prevent ischemia, re-entry paths of pre-existing collat-eral anastomoses from the lateral circumflex and deep femoral artery are lig-ated with a 4-0 silk suture. Skin is again sutured prudently and the animals housed as described above. Necrosis, atrophy, or fibrosis are reported to occur in about half the animals operated, but solely in calf and not once in thigh muscles. Five weeks after the initial operation, MRI, phosphocrea-tine/ATP ratio measurements at rest, and perfusion determination in semi-membranous muscle underline that only the calf but not the thigh is indeed ischemic (Rissanen et al. 2003). In conclusion, a model of sustained but restricted ischemia in the rabbit hind limb is set up that combines features of both the above-mentioned models of femoral artery ligation and excision (Fig. 1C).

Data acquisition, measurements, and evaluations

The ascribed animal models mimic different pathological circumstances that allow us the examination of different mechanisms of vessel growth by various methods of measurement and quantification, depending on whether the focus is on angiogenesis or arteriogenesis.

Capillary counting

As mentioned above, excision of the femoral artery leads to severe ischemia and hypoxia in the hind limb. As a result, angiogenesis occurs throughout the whole limb depending on the degree of ischemia. The angiogenic response can be easily assessed by quantification of capillaries within the tissue. Typ-ically, the increase in the number of capillaries per fiber or in the number of capillaries per area is measured. However, if the size of the muscle fiber changes due to atrophy or hypertrophy, as can, for example, be the case dur-ing critical ischemia, capillary density per area represents only a very unre-liable parameter. It should also be taken into account that atrophy might occur to a larger extent in settings of inhibited angiogenesis and that less atrophy can be detected after stimulation of angiogenesis. In extreme cases, this means that although a compound is indeed anti-angiogenic, no difference or even an increase in capillary density per area is observed. The capillary to fiber ratio avoids this pitfall and should therefore be used instead.

Furthermore, the diversity of capillary density in different muscles may be even more critical. For example, the capillarity of the fast-twitch white gastrocnemius region (mainly type IIb fibers) is very low relative to the adjacent red region of the gastrocnemius (mainly type IIa fibers) (Prior et al. 2003). If not carefully chosen, tissue samples from the same muscle show areas with significant differences among the number of capillaries. This in fact also does not allow direct comparison of samples from different muscles. Tissue samples for capillary counting thus have to be chosen very carefully to avoid artificial differences in capillary density. Therefore, for a morphometric assessment of angiogenesis and efficacy of pro- and anti-angiogenic properties of substances, the quantification of the capillary to fiber ratio is the easiest and most reliable method.

For immunohistochemical examinations, defined tissue samples of the different regions of interest should be cryopreserved. To prevent gross damage of the tissue structure, samples are mounted on cork and embedded with a suitable medium (e.g., TissueTec). The samples are then slowly frozen in methylbutaneate –150 °C before long-term storage at –80 °C. Before cutting with a cryomicrotome, the samples should be allowed to warm up to –20 °C. Tissue sections (5–20 μm thickness) are placed onto silicone-coated slides and fixed with acetone. After careful rinsing with PBS, the sections are then incubated with a primary antibody directed against CD-31 (PECAM-1) at 4 °C over night (monoclonal mouse anti-human, diluted 1:100 in PBSA) in a moist chamber. For rabbit tissue, the clone JC/70A has been proven to cross-react. The next day, after thorough washing with PBS, the sections are incubated with a FITC-coupled anti-mouse antibody (diluted 1:200) for 4 h in the moist chamber at RT. For better contrast, an anti-actin antibody can be used as a third antibody, staining the surrounding muscle fibers (e.g., monoclonal mouse anti-human muscle actin, clone HHF35, diluted 1:100 in PBSA). Again, incubation at 4 °C overnight is necessary. Thorough washing with PBS (several times) and Tween is necessary afterwards. Finally, nuclear staining can be achieved with 7-aminoactinomycin D (diluted 1:1,000 in PBS) or TOTO-3 (diluted 1:2,000 in PBS), both of which need an incubation period of only 1–5 min at RT. Embedding of samples with Fluromount G is the last step before microscopy and evaluation. Since capillary density as well as capillary to fiber ratio differ significantly among tissue samples and between different muscles, blinded quantification is inevitable. It must also be certain that the number of fields per section chosen is large enough.

Post-mortem angiography

Capillary counting is a reliable and powerful tool to investigate the process of angiogenesis. However, since arteriogenesis is a distinct mechanism, involving the whole thigh, histological quantification of collateral arteries is made practically impossible. Post-mortem angiography using high-resolution contrast agents and techniques serves this need. Different contrast agents can be used, all having different advantages and disadvantages. Nevertheless, all contrast agents commonly used for post-mortem angiography are based on two components: a contrast giving compound (e.g., lead) and gelatine. The gelatine fraction allows the warmed contrast medium to be infused at body temperature, but hardens when put on ice. Thus, the angiograms can be performed days or weeks after the infusion of the contrast medium when stored at –20 °C. As mentioned, different contrast agents are available and commonly used: barium, bismuth, and lead. Barium has the advantage of being cheap and available in large amounts. However, its contrast properties are comparably poor, making it suitable mainly for larger animal models. Lead-based contrast medium is also easy to make, but lead is a very harmful substance and disposal is more complicated. However, its contrast properties are very good, making it suitable also for very small species. Finally, bismuth is a relatively untoxic substance with very good contrast properties, combining the advantages of the previous two. The main disadvantage is that mixing it is very time-consuming and the amounts achieved are small. It is therefore not the first choice when large amounts are needed (e.g., pig hind limb).

Generally, all contrast agents have to be infused into the vasculature under controlled conditions, to make direct comparisons possible. In the rabbit hind limb model, the abdominal aorta is dissected and cannulated downstream. To insure maximal vasodilation for diameter assessment, adenosine is injected into the aorta, and the contrast medium (body temperature) is then constantly infused into both hind limbs under pressure controlled conditions for a specified time period (e.g., 80 mmHg, 8 min). Subsequently, the contrast medium is allowed to gel by placing the hind limbs on crushed ice for at least 60 min. According to Longland's definition, collateral arteries always consist of a defined stem, mid-zone, and re-entrant. To avoid misidentification of arteries, three-dimensional assessment of vessels is needed. Therefore, angiograms of each single hind limb are taken at two different angles in a low kV Balteau radiography apparatus. Under stereoscopic view, each collateral vessel can then be identified and counted (Fig. 2).

Post-mortem angiography furthermore allows one to estimate the diameter of arteries and to compare, for example, the diameter of collateral arteries under different conditions. It should be noted that this technique is not a quantitative but a qualitative one.

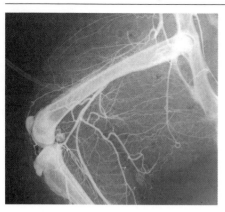

Fig. 2. Post-mortem angiogram one week after ligation of the right femoral artery in the rabbit hind limb

Morphology and hemodynamics

It has to be noted that morphological assessment of vascular growth does not necessarily translate into the functional aspects of the vessels. An increased number of capillaries, for example, leads to only a small increase in blood flow to the dependent regions, not sufficient to replace a large occluded artery such as the femoral artery. According to the law of Hagen-Poiseuille, laminar flow through tubes or vessels depends on different parameters. However, the main factor is the diameter of the vessel as it is in the fourth power. This means that for an occluded femoral artery in the rabbit hind limb (approx. 2 mm of diameter), more than a billion newly built capillaries would be needed, a mass that would not even fit in the existing dimensions. The importance of vessel size does also mean that even only small errors in estimating the vessel diameter lead to large errors when blood flow is calculated. The relative error, however, increases with decreased species size; but even if the diameter could be adequately and accurately measured, other parameters such as pressure difference and length can hardly be quantified and derived from morphological techniques (e.g., angiograms, histology).

To improve the expressiveness of morphological techniques which can also be used in clinical practice, angiographic score systems have been developed. A recent study comparing the normally used Rentrop classification (Rentrop et al. 1989) with myocardial blood flow in pig hearts as assessed by the "gold standard", the use of microspheres revealed, however, that there was no correlation between angiographic scoring and tissue perfusion (Fuchs et al. 2001). This underlines the need for functional and quantitative assessment of vascular growth. A number of different techniques have been developed and established to serve this need. The "gold standard" for tissue perfusion and blood flow measurements is constituted by fluorescent or radioactive microspheres (Unger 2001). Since a detailed step-by-step protocol of these techniques would not fit in a book chapter, the authors would like to refer to the respective literature.

▓ Expected Results

The inhibition and in particular the stimulation of vessel growth serves one aim: the modulation of tissue perfusion. As indicated above, morphometric analysis from angiographic or histological findings does not serve as a reliable parameter when tissue perfusion needs to be assessed. Nevertheless, results can contribute to a better understanding of mechanisms taking place in the processes of angiogenesis or arteriogenesis.

In an ischemic hind limb model, a significant increase in capillary density (capillaries/mm2) in the adductor muscle can be found on day 5 after occlusion (351 ± 36 vs. 216 ± 24 capillaries/mm2, $p < 0.05$) which diminishes after day 10 (260 ± 22, $p < 0,05$ vs. day 5) (Hershey et al. 2001). However, after stimulation with TGF-β_1, a factor proven to increase perfusion by vessel growth, van Royen et al. found no effect on the number of capillaries in comparison to the PBS-treated control group (170.5 ± 36.0 vs. 160 ± 43.8, p=NS) (Van Royen et al. 2002). Again, the varying place of tissue harvesting makes comparisons difficult.

The number of angiographically visible collateral arteries in a non-ischemic hind limb model increases significantly after 1 week of PBS treatment (control) (15.4 ± 2.2 vs. 6.66 ± 1.2, $p < 0.05$) and even more after treatment with MCP-1 (30.1 ± 3.3, $p < 0.05$ vs. control) under stereoscopic view. GM-CSF or TGF-β_1 also lead to an increase in the number of collateral arteries when identified by their distinct properties of being composed of stem, midzone, and reentry (GM-CSF 25.6 ± 3.6, TGF-β_1 24.6 ± 4.1 capillaries/mm^2). Focusing solely on the number of arteries, however, leads to a misconception, as in the course of arteriogenesis the process of pruning leads to a growth of few large-caliber collaterals while a number of earlier-recruited arteriolar anastomoses regress again. Thus, an inverse relationship can be observed between the number of angiographically visible arteries and the conductance index one and four weeks, respectively, after femoral artery occlusion (Hoefer et al. 2001). The above-mentioned score systems that have been developed are today not convincing when it comes to accuracy, reliability, and reproducibility.

Acknowledgements. We thank Mrs. Katja Speiser for providing figure 1.

References

Busch HJ, Buschmann I, Mies G, Bode C, Hossmann KA (2003) Arteriogenesis in hypoperfused rat brain. J Cereb Blood Flow Metab 23(5):621–628

Buschmann I, Schaper W (2000) The pathophysiology of the collateral circulation (arteriogenesis). J Pathol 190:338–342

Buschmann I, Busch HJ, Mies G, Hossmann KA (2003) Therapeutic induction of arteriogenesis in hypoperfused rat brain via granulocyte-macrophage colony-stimulating factor. Circulation 108(5):610–615

Deindl E, Buschmann I, Hoefer I, Podzuweit T, Boengler K, Vogel S, Van Royen N, Fernandez B, Schaper W (2001) Role of ischemia and of hypoxia-inducible genes in arteriogenesis after femoral artery occlusion in the rabbit. Circ Res 89(9):779–786

Fuchs S, Shou M, Baffour R, Epstein SE, Kornowski R (2001) Lack of correlation between angiographic grading of collateral and myocardial perfusion and function: implications for the assessment of angiogenic response. Coron Artery Dis 12:173–178

Hershey JC, Baskin EP, Glass JD, Hartman HA, Gilberto DB, Rogers IT, Cook JJ (2001) Revascularization in the rabbit hindlimb: dissociation between capillary sprouting and arteriogenesis. Cardiovasc Res 49:618–625

Hoefer I, van Royen N, Buschmann I, Piek J, Schaper W (2001) Time course of arteriogenesis following femoral artery occlusion in the rabbit. Cardiovasc Res 49:609–617

Ito WD, Arras M, Winkler B, Scholz D, Htun P, Schaper W (1997) Angiogenesis but not collateral growth is associated with ischemia after femoral artery occlusion. Am J Physiol 273:H1255-H1265

Prior BM, Lloyd PG, Yang HT, Terjung RL (2003) Exercise-induced vascular remodeling. Exerc Sport Sci Rev 31:26–33

Pu LQ, Sniderman AD, Brassard R, Lachapelle KJ, Graham AM, Lisbona R, Symes JF (1993) Enhanced revascularization of the ischemic limb by angiogenic therapy. Circulation 88:208–215

Pu LQ, Jackson S, Lachapelle KJ, Arekat Z, Graham AM, Lisbona R, Brassard R, Carpenter S, Symes JF (1994) A persistent hindlimb ischemia model in the rabbit. J Invest Surg 7:49–60

Rentrop KP, Feit F, Sherman W, Thornton JC (1989) Serial angiographic assessment of coronary artery obstruction and collateral flow in acute myocardial infarction. Report from the second Mount Sinai-New York University Reperfusion Trial. Circulation 80:1166–1175

Rissanen TT, Markkanen JE, Arve K, Rutanen J, Kettunen MI, Vajanto I, Jauhiainen S, Cashion L, Gruchala M, Narvanen O, Taipale P, Kauppinen RA, Rubanyi GM, Yla-Herttuala S (2003) Fibroblast growth factor 4 induces vascular permeability, angiogenesis and arteriogenesis in a rabbit hindlimb ischemia model. FASEB J 17:100–102

Unger E (2001) Experimental evaluation of coronary collateral development. Cardiovasc Res 49:497–506

Van Royen N, Hoefer I, Buschmann I, Heil M, Kostin S, Deindl E, Vogel S, Korff T, Augustin H, Bode C, Piek JJ, Schaper W (2002) Exogenous application of transforming growth factor beta 1 stimulates arteriogenesis in the peripheral circulation. FASEB J 16:432–434

Chapter 20

In Vivo Matrigel Angiogenesis Assays

Tony Passaniti and Michele I. Vitolo

Introduction

The in vivo Matrigel angiogenesis assay, or Matrigel plug assay, was developed to examine neovascularization in rodents in response to specific angiogenic factors, independently of the presence of tumor cells (Kibbey et al. 1992, Passaniti et al. 1992, Grant et al. 1993, Capogrossi and Passaniti 1999). Since its inception, a variety of agents have been used to deliver an angiogenic stimulus, including recombinant cytokines, adenoviral vectors, low dose tumor cells, and even telomerase-transfected endothelial cells (Muhlhauser et al. 1995, Klement et al. 2000, Akhtar et al. 2001, Yang et al. 2001). The applications of the assay include (1) the study of the basic mechanisms underlying angiogenesis at the molecular or cellular level and (2) the in vivo evaluation of the activities of angiogenic and angiostatic compounds or cytokines (Capogrossi and Passaniti 1999).

Initial studies using this method showed that cellular cytokines, such as fibroblast growth factors (basic, acidic), vascular endothelial growth factor, hepatocyte growth factor, platelet activating factor-4, and tumor necrosis factor-alpha were potent angiogenic cytokines in vivo while transforming growth factor-β, interleukin-alpha, and interferon-inducible protein-10 were potent inhibitors of angiogenesis (Grant et al. 1993, Passaniti et al. 1992). These factors were delivered within the Matrigel as soluble proteins, as the products of adenoviral vectors dissolved in Matrigel, or as the products of tumor cells that were infected with specific adenoviral vectors encoding the angiogenic fibroblast or vascular endothelial cell growth factors (Muhlhauser et al. 1995). More recently, since the demonstration that circulating endothelial precursor cells can contribute to angiogenesis de novo, the Matrigel assay has also been used to induce angiogenesis using mature endothelial cells.

Other in vivo angiogenesis assays have been developed, some of which are separate topics in this manual, including the corneal angiogenesis assay, the chick chorionalloantoic assay, and ischemic or injury models to measure

Springer Lab Manual
H. Augustin (Ed.)
Methods in Endothelial Cell Biology
© Springer-Verlag Berlin Heidelberg 2004

vascular response. One of the advantages of the in vivo Matrigel assay is its ease of use without the need for surgical intervention. Since the assay can be performed in rodents, both inhibitors and activators of angiogenesis can be tested by direct inclusion of the compounds within the injected gel or by systemic delivery. Recent modifications of the assay described below also allow for direct placement of activators and/or inhibitors within the Matrigel plug after solidification.

A variety of approaches have been taken to quantitate the in vivo angiogenesis response. These include direct counting of vessels within tissue sections, image analysis of vessels or cellular response within the gel, measurement of fluorescence intensity within vessels after intravenous injection of labeled molecules, and measurement of total hemoglobin recovered from the implanted gels (Vukanovic et al. 1992). These approaches will be described in the following protocol, along with techniques that have been shown to enhance the angiogenic response or to optimize the ability to detect specific inhibition of the response using angiogenesis inhibitors.

▨ Outline

Summary of Matrigel plug assay (see text for details)

Before the assay: prepare reagents and animals
- Order mice and Matrigel
- Thaw Matrigel at 4 °C overnight
- Prepare syringes and needles
- Re-suspend angiogenic factors or inhibitors
- Prepare cells or viruses for infection
Optional:
Two to three days before the assay: Protocol "Treatment with viral expression vectors"

First day: Matrigel mixtures and injections
- Prepare Matrigel mixtures with angiogenic or anti-angiogenic factors
- Set up animal cages and randomize animals for treatment groups
- Inject Matrigel with or without angiogenic factors
- Protocol "Treatment with viral expression vectors"
Optional:
Day 2: optional sponge implants
- Prepare mice for anesthesia and implantation of sponges

Days 6–14: dissection and analysis
- Prepare reagents and mice for euthanasia and dissection
- Organize tissue for specific histological or blood analysis
- Analyze immunohistochemical or direct vessel visualization data

Optional:
- Protocol "Blood flow visualization"
- Protocol "Hemoglobin analysis"

Materials

For first day: Matrigel mixtures and injections

Reagents

- Matrigel (BD Biosciences, www.bdbiosciences.com or Trevigen Inc., www.trevigen.com)
- Cytokines (e.g., Fibroblast growth factor-2, R&D Systems, www.rndsystems.com)
- Heparin (Gibco/BRL, www.gibcobrl.com)
- DMEM media (Biofluids, www.biofluids.com or Invitrogen, www.invitrogen.com)
- Adenoviral stock (optional)

Supplies

- 3 ml syringes, 19 gauge and 25 gauge needles
- Alzet mini-pumps (optional ; Alzet, www.alzet.com)

Equipment

- Animal cages and water bottles

Cells

- 3T3 fibroblasts can be used to introduce adenoviral vectors expressing angiogenic factors.
- Endothelial cells can be used to elicit an angiogenic response.

Animals

- Female C57BL6/J mice (Jackson Laboratory, www.jaxmice.jax.org or Harlan Sprague-Dawley, www.harlan.com), 6–8 weeks of age
- Athymic nu/nu or SCID mice (optional)

For day 6–14: dissection and analysis

Reagents

- Formalin (37.7 % stock from Mallinkrodt Chemicals, www.mallchem.com) diluted 1:10 (10 % formalin) with phosphate-buffered saline (PBS) to a final concentration 3.77 %
- Glutaraldehyde (25 % stock from Sigma-Aldrich, www.sigmaaldrich.com) diluted to 1 % final concentration
- 70 % ethanol
- Paraffin, xylene, Trichrome stain, hematoxylin, eosin
- Antibodies

Supplies

- Sterile scissors
- Tweezers
- Glass slides

Equipment

- Microscope
- Computer
- Image analysis software

For optional or alternative analyses

Reagents

- Adenoviral vectors (e.g., AdCMV.NLSâGal; AdCMV.FGF; AdCMV.VEGF; AdCMV.p53)
- Avertin anesthetic (Sigma-Aldrich)
- Drabkin hemoglobin analysis kit (Sigma-Aldrich)
- X-gal substrates and buffers for β-galactosidase detection
- FITC-Dextran for i.v. injections

Supplies

- Polyvinyl sponges (Hydron pellets, Elvax®)
- Scalpel blades
- Tweezers
- Surgical sutures

Equipment

- Electric clipper to shave hair

Procedure

Before the assay: prepare reagents and animals

1. Order 6–8 week old female mice from suppliers (Jackson Laboratory or Harlan Sprague-Dawley).

2. Order 10 ml of Matrigel in DMEM (Dulbecco's Modified MEM with 4.5 g glucose/L and glutamine) or in phenol-free DMEM from suppliers (BD Biosciences or Trevigen).

3. Thaw Matrigel by placing at 4 °C the day before the assay.

4. Reconstitute 25 µg basic FGF (FGF-2) in physiological saline containing carrier BSA (1 mg/ml) to 50 µg/ml. Store one aliquot at 4 °C and the rest at −80 °C.
   Note: As with most cytokines and antibodies, avoid repeated cycles of freeze-thawing, which may reduce activity.

5. Prepare heparin stock solution by reconstituting vial of 100,000 units to 10,000 units/ml with DMEM.

6. Prepare 3 ml syringes and 19 gauge and 25 gauge needles for injections. Store syringes at 4 °C.

7. Prepare angiogenic inhibitors, cells, or viruses for infection.
   Note: If preferred, order and prepare 6–8 week old scid or nude mice (Jackson Laboratory or Harlan Sprague-Dawley) in isolator cages with filters.

First day: Matrigel mixtures and injections

1. Remove thawed liquid Matrigel from refrigerator and place in ice bucket at 4 °C.

2. Transfer bFGF (FGF-2) to 4 °C. This factor will be used to induce a positive response to test inhibitors.

3. Prepare 3 ml of Matrigel/FGF angiogenic mixture for injection of at least 5 mice per data point:
   a) Dilute bFGF (50 µg/ml) to final 150–300 ng/ml by adding 9–18 µl bFGF to 150 µl of DMEM. Use DMEM alone for a negative control.
   b) Place 3 ml Matrigel in 15 ml sterile conical tube.
   c) Add 50 µl of diluted bFGF to Matrigel tube in three aliquots, inverting the tube after each addition of bFGF.
   Note: Thorough mixing of bFGF or other angiogenic factors with the Matrigel is essential for reproducible responses in the assay.

4. Keep mixtures on ice until the time point of injection into the mice to avoid gel formation.

5. Aspirate 3 ml of Matrigel mixture into 3 ml syringe using a 19 gauge needle, both previously stored at 4 °C. Avoid bubbles. Replace needle with 25 gauge needle.

6. Place syringes at RT before injecting mice, but avoid matrigel polymerization. This prevents bubble formation and maintains a more intact gel after injection.

7. Pick up mice from their cages by the dorsal skin flap, lay the mouse in the palm of the hand and wrap the tail around the small finger, exposing the abdomen.

8. Inject 0.5 ml of Matrigel mixture subcutaneously in each mouse along the abdominal midline. Correct placement of gel results in a noticeable subcutaneous bump.

9. If testing angiogenic agents, include agents within gel mixtures as described above.

10. If testing angiostatic agents, include agents within the gel mixtures or begin treating animals systemically throughout the angiogenic response depending on the pharmacokinetics, solubility, and clearance rates:
a) intraperitoneal injections with a 25 gauge needle
b) subcutaneous injections with a 25 gauge needle
c) slow-release mini-pumps (e.g., Alzet) implanted in the nuchal area
Note: No heparin is necessary to elicit blood vessel formation when bFGF is used. Heparin may be included within the Matrigel/FGF mixture if hemoglobin analysis is used for quantitation (see below) and must be titrated between 1 and 10 U/ml to minimize non-specific vessel leakage.

For treatment with viral vectors, see "Treatment with viral expression vectors", for "Optional sponge implants" on day 2, see the according protocol.

Day 6–14: dissection and analysis

1. Sacrifice mice after 6–14 days.

2. Use 70 % ethanol to wipe the abdominal midline at site of injection. This will highlight the area of the plug.

3. Excise the perimeter of the plug including skin and peritoneal muscle with sterile scissors.

4. Place the intact skin, gel, and muscle in tubes containing PBS-buffered 10 % formalin (containing 1.0 % glutaraldehyde if desired).

5. After overnight incubation in fixative, prepare the gel plugs for histological stains or antibodies by progressive dehydration in increasing ethanol concentrations and embed in paraffin under vacuum using established protocols.

6. Deparaffinize 5 μm sections with Xylene, rehydrate with decreasing ethanol concentrations, and stain with Masson's Trichrome stain, hematoxylin/eosin (H&E), or specific antibodies. This will reveal the edges of the plug, which include skin and muscle borders.

7. Antibody staining: Factor VIII-related antigen (vWF) is routinely used for detection of endothelial cells (DakoCytomation, www.dakocytomation.com) although this often requires signal amplification (e.g., Tyramine system, Dupont, www.dupont.com). Briefly, peroxide and Pronase-treated slides are reacted with vWF antibody (1:1,000 dilution) at 4 °C (16 h) in PBS containing a blocking agent such as non-fat dry milk or bovine serum albumin. The Vector ABC kit is then used (biotin/avidin/HRP) to visualize the reaction product (Vector Laboratories, www.vectorlabs.com). Alternatively, antibodies to PECAM-1 (e.g., CD-31; Chemicon International, www.chemicon.com) or collagen IV (Sigma-Aldrich) can be used with FITC-conjugated secondary antibodies. Certain lectins, such as Ulex europeus and Griffonia simplificans (Sigma-Aldrich), although not specific for endothelial cells, can also be used to visualize blood vessels.

8. Use a standard imaging software [Universal Imaging Corp. (Image-1), www.image1.com or NIH Image, http://rsb.info.nih.gov/nih-image/] to capture TIF images and quantitate vessel area.

9. Vessels or total cellular infiltrates are counted adjacent to the skin or muscle. Calculate the mean area per field ($\times 10^5$ μm^2) from 10–20 fields (200×) and express with standard errors from the mean. Calculate p values.

10. Alternative analysis: To visualize intact vessels in the gel plug (without fixative), the overlying skin can be resected and the gel can be placed in a petri plate under a dissection scope at low power (6 ×–10 ×). To improve visibility, squashes of the tissue can be prepared by placing a glass slide over the gel plug and photographing the surface vessels. With direct illumination the small, tortuous vessels within the gel plug can be observed.

11. Alternative analysis: adenoviral β-galactosidase detection. Fix gel plugs for 2–3 h in fresh 10 % formalin buffered with PBS. Wash samples 3 times at 30 °C for 30 min each in PBS prior to staining in X-Gal buffer (35 mM potassium ferricyanide, 1 mM MgCl$_2$, 0.02 % NP-40, and 0.01 % deoxycholate) to which 1 mg/ml (final) X-Gal (5-bromo-4-chloro-3-indolyl β-

D-galactopyranoside) is added fresh from a 50 mg/ml stock solution in DMSO:DEPC water (50:50, v:v). After 3 days, wash tissue in PBS until the rinse solution is colorless and process for paraffin-embedding and sectioning as described above.

Note: Each experiment described above should be performed with at least 5 mice per treatment group, but should be repeated two more times to yield a total of 15 mice for statistical purposes.

Optional or Alternative Analyses

Treatment with viral expression vectors

In some applications where an angiogenic or anti-angiogenic stimulus is necessary, mice may be pre-injected with adenovirus 3 days prior to implantation of the Matrigel or adenovirus may be added within the Matrigel at day 0. The LacZ gene encoding β-galactosidase can be used as a control.

1. Prepare adenovirus stocks at 1×10^8 pfu/ml.

2. Inject 0.5 ml adenovirus (5×10^7 pfu) in PBS either
   a) 3 days prior to Matrigel implant at the same midline abdominal sites that will be later injected with Matrigel or,
   b) included in the Matrigel plug.

3. If virus is injected prior to Matrigel implant, expression of adenoviral cDNA is allowed to proceed in vivo for 2–3 days. The animals are then injected subcutaneously with 0.5 ml liquid Matrigel containing FGF or other components, if needed.

Optional sponge implant, second day (application of factors to the gel)

The optional sponge implant allows a slow, local release of angiogenic or anti-angiogenic factors.

1. On the first day, 500 μl Matrigel was injected subcutaneously on the ventral abdominal midline as described above.

2. Allow Matrigel to solidify for 24 h.

3. Anesthetize mice using Avertin (2,2,2-tribromoethanol; Sigma-Aldrich). Follow Institutional Review Board (IRB) approved protocols.

4. Shave a portion of the abdomen where Matrigel plug is located.

5. Make a small nick (0.5 cm) in the skin with the aid of a #15 surgical blade (center of Matrigel plug).

6. With the aid of the same surgical blade, make a smaller nick in the Matrigel plug under the skin.

7. Insert a small piece of Polyvinyl sponge (pre-irradiated with 2,000 Gy of gamma irradiation and containing 1 µl of solution) in the nick in the middle of the Matrigel plug using tweezers. Sponge may contain angiogenic factors and/or inhibitors.

8. Close the wound with a suture and allow mice to recover from anesthesia.

9. Observe mice the next day to determine the condition of the wound.

10. For systemic delivery of inhibitors of angiogenesis, implant Alzet minipumps containing 200 µl of endostatin (8 mg/ml) or other appropriate control subcutaneously into some mice.

Blood flow visualization (FITC Dextran injection for in vivo Matrigel assay)

1. One to two weeks after injection of the gel plug, inject 200 µl FITC-Dextran (MW 2×10^6 Daltons; Sigma-Aldrich) through the tail vein.

2. After 3–5 min, sacrifice mice and remove Matrigel plug with a #15 surgical blade and a pair of tweezers.

3. Place Matrigel plug in 10 % formalin.

4. Use phase contrast microscopy and fluorescence microscopy to visualize the plug and the presence of perfused blood vessels, respectively.

5. Capture TIF images with a video camera as described above and process with Canvas 5.0 Image analysis software (Deneba [ACD Systems], www.deneba.com).

6. To quantitate the fluorescent vessels, extract the gel plugs by crushing the gel in PBS. Centrifuge to remove insoluble material. Save the supernatants.

7. Collect total plasma (approx. 1 ml) in heparinized tubes after cardiac puncture from mice that were injected with FITC-Dextran. Centrifuge to remove insoluble material.

8. Determine fluorescence units using a fluorescent microplate reader. Create a standard curve using the FITC-Dextran used for injection. Express blood vessel fluorescence as a ratio of Matrigel plug fluorescence: plasma fluorescence.

Hemoglobin analysis

1. For hemoglobin analysis, use heparin (1–10 U/ml) within the injected Matrigel plug as described above.

2. Separate the gel plugs from the surrounding skin with a #15 surgical blade while retaining the peritoneal muscle. Weigh each plug before proceeding.

3. Transfer the tissue to an Eppendorf tube containing 0.5 ml of de-ionized water.

4. After overnight incubation at 37 °C, RBC will lyse and release hemoglobin.

5. Crush the tissue with the rubber end of a syringe barrel and centrifuge samples at $200 \times g$ for 10 min to remove tissue pieces and gel.

6. Collect the supernatant which contains hemoglobin for use in the Drabkin assay (Sigma-Aldrich) to estimate hemoglobin concentration.
   Note: A detailed protocol of the Drabkin assay is described in the Drabkin kit (Sigma-Aldrich) with the following modifications. Stock solutions of hemoglobin are used to generate a standard curve. Samples (25–50 µl) of tissue supernatant are added to 1 ml Drabkin reagent. After color development and quantitation of hemoglobin protein, results are expressed relative to total protein in the supernatant or to the weight of the original gel plug.

Expected Results

The in vivo Matrigel assay was used in both mice and rats to elicit neovascularization in response to a variety of angiogenic factors or to test the ability of specific compounds to inhibit angiogenesis. Numerous applications and methods of analysis have been described. In this review, results from three specific applications will be discussed: (1) histological response after FGF injection, (2) inhibition analysis using endostatin in polyvinyl sponges, and (3) angiogenic response from endothelial cell co-injection.

Time course and histological studies with FGF revealed an initial (2–6 h) neutrophil response, invasion of cells into the gel within 24 h, and detection of capillary-sized vessels inside the gel within 72 h. Persistent vessels with clear lumen containing red blood cells (RBCs) surrounded by fenestrated endothelial cells were readily apparent up to 3 weeks following injection of the mice (Fig. 1). Antibody staining of tissue sections also revealed increases in vWF staining within 24 h of implantation. Hemoglobin content increased from 2 to 6 days after injection. Most studies have employed mice between 2 to 3 months of age because the angiogenic responses of older mice (24–30 months of age) are often compromised (Pili et al. 1994). This "inhibitor" effect of age appears to involve a slower and less intense response to FGF since the angiogenic response does improve in a two-week assay, although not to the extent as in young mice.

Studies with the in vivo Matrigel angiogenesis assay have utilized both angiogenic and anti-angiogenic factors. Cytokines (FGF, TNF, HGF), adenoviral vectors (FGF, VEGF), tumor cells (LNCaP, 3T3 infected with Ad.FGF-

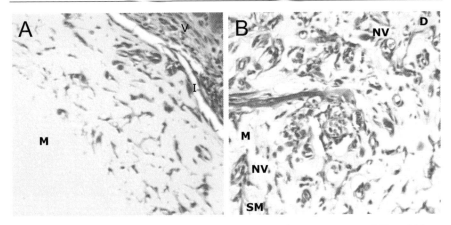

Fig. 1A,B. In vivo angiogenesis in the Matrigel plug assay, basic response to FGF, and histological analysis. Young (6–8 week) female C57BL mice were injected subcutaneously at the abdominal midline with 0.5 ml of Matrigel (**A**), or with 0.5 ml of Matrigel containing 200 ng/ml bovine FGF-2 (**B**). One week later, gel plugs were dissected and fixed in buffered formalin prior to paraffin embedding and sectioning. Tissue was stained with Masson's Trichrome stain to visualize existing vessels in the skin (*V*), interstitial collagens (*I*), Matrigel (*M*), and neovessels (*NV*) arising from the dermal (*D*) or skeletal muscle on the opposite side of the plug (*SM*)

1), and other molecules (IL-15, NO, thrombin) have all been used to stimulate blood vessel formation. Delivery of VEGF with recombinant adenoviral vectors has resulted in potent angiogenic responses in nude or scid mice. Inhibitors of angiogenesis used in this assay have included anti-proliferative cytokines (TGFβ, IL-1, IP-10), protease inhibitors (Batimastat), glycosidase inhibitors (CST, castanospermine), and adenoviral vectors coding for tumor suppressor genes (Ad.p53). Expression of p53 was induced by prior inoculation of the angiogenic implant site, resulting in inhibition of FGF-mediated angiogenesis within Matrigel plugs dissected from the same site. In cell culture studies, over-expression of p53 in adenovirus-infected endothelial cells was found to inhibit cell differentiation without affecting cell proliferation.

Matrigel plugs can be implanted with polyvinyl sponges impregnated with angiogenic (FGF) factors or with inhibitors such as endostatin (Fig. 2). These approaches deliver a localized angiogenic or angiostatic stimulus that directs or inhibits vessel formation, respectively. In these studies, the neovessels were visualized by i.v. injection of FITC-Dextran several minutes before dissection of the plug. Light level and fluorescence microscopy then revealed the presence of neovessels directed toward the angiogenic factor. The presence of endostatin inhibitor within the pellet clearly inhibited neovessel formation. The advantage of this mode of visualization is that functionality of the vessels can be verified.

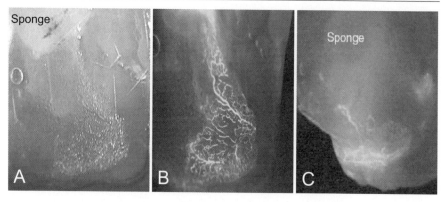

Fig. 2A–C. Endostatin inhibits in vivo angiogenesis in the Matrigel plug assay: implantation of polyvinyl sponges within matrigel. Matrigel (0.5 ml) was injected subcutaneously and 24 h later mice were anesthetized, and each plug was implanted with polyvinyl sponges (5 × 5 × 1 mm) containing an unknown angiogenic factor derived from Matrigel. Seven days later, mice were injected intravenously with FITC-Dextran and sacrificed 5 min later. Gel plugs were re-sected and placed under a dissecting microscope equipped with visible and UV light. **A** Angiogenic factor applied to sponge induces vessels (2.5× magnification); visible light showing location of sponge. **B** UV light showing vessels (2.5× magnification) in the same Matrigel plug. **C** Angiogenic factor + 1.6 mg Endostatin (within Alzet mini-pump, s.c.) results in a weaker angiogenic response in a separate Matrigel plug [originally published by Akhtar et al. (2001) Endothelium 8:221–234; reprinted with permission from Taylor and Francis, Inc.]

In an alternative application of the assay (Fig. 3), the ability of human endothelial cells, instead of angiogenic cytokines, was used to mediate an angiogenic response and the formation of vessels in SCID mice was tested. Mice were implanted with bFGF (FGF-2), early passage (young) primary human endothelial cells, or late passage (old) human endothelial cells that were engineered to express the human telomerase gene (hTERT) (Yang et al. 2001). Gel implants were harvested two weeks later and stained with either H&E or with a human-specific antibody that reacts with collagen IV, an extracellular matrix product of endothelial cells. Host vessel invasion of Matrigel implants was stimulated in the presence of bFGF (upper left panel) or endothelial cells (middle and right panels). However, H&E staining did not differentiate human from host vessels (middle and right upper panels). Human basement membrane collagen antibody reproducibly reacted with human microvessels in Matrigel (middle and lower right panels). Intravenous injection of red microspheres via the tail vein confirmed the appearance of red tracer within these vessels, suggesting that mouse neovessels contained functional human endothelial cells. Interestingly, the injection of late passage (old) primary endothelial cells did not stimulate angiogenesis, indicating that only telomerized or young primary endothelial cells are angiogenically competent in this xenobiotic model.

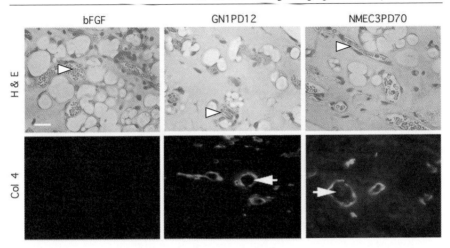

Fig. 3. Telomerized endothelial cells mediate in vivo angiogenesis in the Matrigel plug assay. Two to three week old male SCID mice (Taconic, www.taconic.com) were used as hosts for these implants. Primary human dermal microvascular endothelial cells were harvested, washed twice, and re-suspended in serum-free EGM2 basal medium (1×10^5 cells/μl). Ten μl (1×10^6 cells) were mixed with 0.5 ml of Matrigel on ice and the mixture was implanted into the ventral midline thoracic tissue of each mouse by subcutaneous injection using a 25 gauge needle. Red blood cells (*arrowheads*), stained by H&E (human and mouse cells), are visible within human-specific type IV collagen immunoreactive lumenal structures (*arrows*) (human cells only) derived from both early passage primary endothelial cells (GN1PD12) and late passage telomerized (primary cells over expressing telomerase) endothelial cells (NMEC3PD70). *Bar* = 10 μm [originally published by Yang et al. (2001) Nat. Biotechnol. 19: 219–224; reprinted with permission from Nature Publishing Group]

Troubleshooting

Before the assay: prepare reagents and animals

Solidification of Matrigel before use:
Matrigel is thawed slowly at 4 °C overnight (not in a 37 °C water bath) to prevent gelling prior to preparation of cytokine solutions. If needed on the day of the assay, the frozen Matrigel can be thawed until ice crystals melt by rotating the tube in the palms of the hands. It is then immediately placed on ice for the addition of cytokines.

First day: Matrigel mixtures and injections

▶ Excessive bleeding:
   It is important to maintain a constant source and preparation lot of heparin from a single manufacturer because quality varies from lot to lot. Frozen aliquots of concentrated heparin can be stored at –70 °C

almost indefinitely. Titration of dose is necessary to avoid non-specific vascular leakage.

▶ Use of larger rodents:
If rats are used for angiogenic assays, up to 2 ml of Matrigel can be injected per rat and it is often advantageous to pre-gel the Matrigel/FGF mixtures within the syringe prior to injection.

▶ Variability in the response:
The site of injection should be kept constant when comparing positive and negative responses because of variable responses at other sites. Usually only one injection per mouse is used.

▶ Loss of gel plug:
Failure to see a distinct 'bump' during s.c. injection is an indication of an intraperitoneal implant which cannot be recovered. Injection at an intra-dermal site occurs rarely, but is indicated by high back pressure on the syringe which leads to injection of only 50 µl of solution. Injections near the inguinal area tend to diffuse into the subcutaneous fat pad and do not form a distinct, recoverable 'bump' on the animal.

▶ Weak response to angiogenic agents:
Combinations of angiogenic cytokines can also be used in the assay to achieve synergistic effects. Some cytokines have not been shown to be active in the assay. Although bovine FGF-2 is a potent angiogenic factor, recombinant human FGF-2 has been a poor angiogenic factor in these assays.

Day 6–14: Dissection and analysis

▶ Inadequate response time:
The time of dissection depends on the specific angiogenic factor used. For FGF, potent responses are observed after 7 days, whereas for HGF, implants are maintained for 14 days.

▶ Failure to recover gel:
See notes about site of injection and subcutaneous location.

▶ Detachment of Matrigel from skin and/or muscle layers:
Avoid prolonged fixation. Usually, overnight incubation in formalin at RT is adequate for complete fixation.

▶ Lack of vessel infiltration into Matrigel:
Gaps may have formed between the skin and/or muscle layers and the cold Matrigel upon injection of the plug. To minimize this, bring Matrigel to RT after loading syringe and then inject subcutaneously with a firm continuous push of the syringe barrel.

▶ Weak vWF antibody staining:
Neovascular endothelial cells contain fewer storage granules of vWF. Signal amplification of the antibody binding and antigen retrieval methods

(pronase pre-treatment or microwave application) are recommended prior to reaction with antibodies.

Blood flow visualization (FITC-Dextran injections for in vivo Matrigel assay)

▶ Normalization of fluorescence:
Collect mouse plasma to measure circulating levels of Dextran. This will control for variability of i.v. injections.

▶ Large statistical variability in the fluorescence data:
Rinse the gel plugs in PBS immediately after dissection to remove excess Dextran from the surfaces of the plug. Increase number of mice used for each test compound.

Hemoglobin analysis

▶ Low levels of hemoglobin:
The hemoglobin determination requires sufficient extraction of hemo-globin from the gel plugs. Low levels of heparin allow some neovessel leakage into the gel in vivo which can then be extracted from the gel plug in vitro. It is important to use gels lacking FGF as controls.

▶ Excessive bleeding:
Maintain consistent lots of heparin and carry out a dose response to choose appropriate amount as above.

▶ Low hemoglobin values:
Increase the volume of extract and/or decrease volume of Drabkin reagent.

References

Akhtar N, Carlson S, Pesarini A, Ambulos N, Passaniti A (2001) Extracellular matrix-derived angiogenic factor(s) inhibit endothelial cell proliferation, enhance differentiation, and stimulate angiogenesis in vivo. Endothelium 8:221–234

Capogrossi M, Passaniti A (1999) An in vivo angiogenesis assay to study positive and nega-tive regulators of neovascularization; Chap. 30: in Methods in molecular medicine, vas-cular disease: molecular biology and gene therapy protocols, ed. by Baker AH, Humana Press Inc., Totowa, NJ. pp. 367–384

Grant DS, Kleinman HK, Goldberg ID, Bhargava MM, Nickoloff BJ, Kinsella JL, Polverini P, Rosen EM (1993) Scatter factor induces blood vessel formation in vivo. Proc Natl Acad Sci U S A 90:1937–1941

Kibbey MC, Grant DS, Kleinman HK (1992) Role of the SIKVAV site of laminin in promotion of angiogenesis and tumor growth: an in vivo Matrigel model. J Natl Cancer Inst 84:1633–1638

Klement G, Baruchel S, Rak J, Man S, Clark K, Hicklin DJ, Bohlen P, Kerbel RS (2000) Con-tinuous low-dose therapy with vinblastine and VEGF receptor-2 antibody induces sus-tained tumor regression without overt toxicity. J Clin Invest 105:R15-24

Muhlhauser J, Pili R, Merrill MJ, Maeda H, Passaniti A, Crystal RG, Capogrossi MC (1995) In vivo angiogenesis induced by recombinant adenovirus vectors coding either for secreted or non-secreted forms of acidic fibroblast growth factor. Hum Gene Ther 6:1457–1465

Passaniti A, Taylor RM, Pili R, Guo Y, Long PV, Haney JA, Pauly RR, Grant DS, Martin GR (1992) A simple, quantitative method for assessing angiogenesis and anti-angiogenic agents using reconstituted basement membrane, heparin, and FGF. Lab Invest 67:519–528

Pili R, Guo Y, Chang J, Nakanishi H, Martin GR, Passaniti A (1994) Altered angiogenesis underlying age-dependent changes in tumor growth. J Natl Cancer Inst 86:1303–1314

Vukanovic J, Passaniti A, Hirato T, Traysman RJ, Isaacs JT (1992) Anti-angiogenic effects of the Quinoline-3-carboxamide, linomide. Cancer Res 53:1833–1837

Yang J, Nagavarapu U, Relloma K, Sjaastad M, Moss WC, Passaniti A, Herron GS (2001) Telomerized human microvasculature is functional in vivo. Nat Biotech 19:219–224

Chapter 21

Analysis of Vascular Permeability in Vivo

Katja Issbrücker, Hugo H. Marti, and Matthias Clauss

Introduction

Edema formation in ischemic and hypoxic tissues is mediated by the production of hypoxia-inducible factors such as the vascular endothelial growth factor (VEGF) (Marti et al. 2000 and Schoch et al. 2002). In this context, VEGF, although better known as the major angiogenesis-inducing factor, was first described as a vascular permeability-increasing factor (Senger et al. 1983). In addition, leakiness of blood vessels is a striking observation in many tumor vessels believed to be formed by ischemia-induced angiogenesis. This has led to the concept that vascular permeability is required for tumor angiogenesis (Dvorak et al. 1999). The possible link between angiogenesis and vascular permeability/edema formation is also of major clinical interest in respect to diseases other than tumors. Edema formation accompanies angiogenesis in many ischemic disorders such as myocardial infarction or stroke. Furthermore, therapeutic attempts to overcome ischemic conditions by application of angiogenic factors such as VEGF are hampered by severe edema formation (van Bruggen et al. 1999). It is a still unresolved question whether vascular permeability is an unavoidable feature of angiogenesis or only a side effect of it, which by mere coincidence occurs in parallel and, therefore, could be counteracted.

If vascular permeability deserves our attention in respect to studies on angiogenesis or on angiogenic factors, how can it be best assessed? The preferable in vivo assay should be easy, quick, and reliable. One commonly used and comparably simple method to assess vascular leakiness is the use of fluorescent or chromogenic compounds which are injected intravenously. The extravasation of the dye into the parenchyma can be measured and compared to standard conditions. In this chapter, two examples of this technique are presented. First, we analyze vascular permeability changes in the brain studied in an example of exposure to hypoxia. Furthermore, a therapeutic intervention, namely treatment with a p38 MAPK inhibitor, is shown. In this model, systemic cardiac perfusion prior to analysis of vascular leakage is

Springer Lab Manual
H. Augustin (Ed.)
Methods in Endothelial Cell Biology
© Springer-Verlag Berlin Heidelberg 2004

needed in order to increase the signal to noise ratio for detection of the fluorescent marker leakage. A second example describes modifications of the well known Miles assay which simply estimates the concentration of extravasated dye in the dermis of animals treated with single injections of individual permeability factors (Miles and Miles 1952). In contrast to the original protocol, this chapter describes a modification in which the permeability factor is injected in the skin of the ear and fluorescent dye is used instead of Evan's Blue. One advantage of this modification is a better quantification by analyzing tissue extracts from the whole ear.

Materials

Animals

- Mice (C57BL/6 and others), other rodents

Reagents & solutions

- 0.5 M borate buffer (pH=10)
- 70 % and 100 % ethanol
- Sodium-fluorescein (Sigma; www.sigmaaldrich.com)
- FITC-labeled BSA (Sigma)
- PBS

Supplies

- Liquid nitrogen
- Anesthetics (e.g., Forene®)
- 1 ml syringes
- 30G (Gauge) × 1/2 needles (0.3 × 13 mm)
- 1.5, 2, and 15 ml polypropylene tubes
- 96-well microtiter plate (e.g., Dynatech MicroFluor, white)

Equipment

- 37 °C water bath
- Mouse restraining chamber for tail vein injection (e.g., Lab Animal Science)
- Tissue mixer (e.g., Ultra Turrax T8, IKA Laboratory Technology, www.ika.de)
- Cooled centrifuge (e.g., Biofuge fresco, Heraeus, www.heraeus-instruments.de)
- Analytical balance
- Microplate fluorescence reader (e.g., MWG Biotech Lambda Fluoro 320, www.mwg-biotech.com)

Procedure

21.1 Protocol for measurement of vascular permeability in the brain (Fig. 1)

Injection with sodium-fluorescein

1. Warm mouse tail in 37 °C water bath for 3–10 min.
   Note: Intravenous injection should only be started when tail veins are clearly visible.

2. Fill 300 µl of sodium-fluorescein (prepared at 6 mg/ml in PBS, and stored in darkness at –20 °C) into 1 ml syringe.

3. Fix mice in a mouse restraining chamber for tail vein injection to get easy access to the tail.

4. Inject 200 µl of sodium-fluorescein into a tail vein using a 30G needle.

5. Allow dye to circulate for 30 min.

Perfusion of the left heart ventricle

1. Anesthetize mouse (e.g., with Forene®).

2. Immobilize mouse on the back on an operation table.

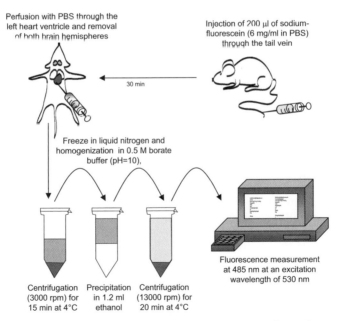

Fig. 1. Flow diagram of the individual steps for the determination of vascular permeability in the brain

3. Open thorax and remove ribs to get access to the heart.

4. Fill a 20G needle (or a Butterfly) connected to a rubber tube with heparin.

5. Puncture the left ventricle with the 20G needle or the Butterfly connected to the rubber tube.

6. Make a small incision in the right atrium or ventricle.

7. Perfuse with 10 ml PBS, preferentially using a peristaltic pump to keep perfusion pressure constant.

8. Move head of the mouse from time to time from left to right and vice-versa to avoid blood clotting in carotid arteries and to ensure brain perfusion.
   Note: Perfusion was successful when the color of the liver gets pale.

Isolation of the brain tissue and fluorescence measurement

1. Decapitate the animal and remove the skin from the scull.

2. Open the scull using a scissor starting at the foramen ovale and gently break away the bone of the scull using a fine but stable forceps.

3. Cut the brain nerves including the chiasma opticum and gently mobilize the brain with a small spoon.

4. Remove the brain onto a glass slide and cut both brain hemispheres and the cerebellum.

5. Put each brain tissue specimen in a pre-weighted 2 or 15 ml tube.

6. Measure weight of tube and brain using an analytical balance and calculate brain weight.

7. Snap freeze tubes in liquid nitrogen and store at –80 °C in darkness.
   Note: Tissue can be stored at –80 °C for several weeks.

8. Thaw samples and add 500 µl 0.5 M borate buffer (pH=10).

9. Homogenize samples with a tissue mixer for 45 s on ice at reduced light.
   Note: Homogenizer should be cleaned in water, 70 % ethanol, and distilled water after each sample.

10. Centrifuge (3,000 rpm) for 15 min at 4 °C

11. Add 300 µl of supernatant I to 1.2 ml of 100 % ethanol to precipitate proteins, vortex.

12. Store rest of supernatant I for determination of protein concentration.

13. Centrifuge (13,000 rpm) for 20 min at 4 °C to obtain supernatant II.

14. Collect supernatant II in fresh 1.5 ml tubes and keep in darkness at –4 °C until further use.

15. Measure empty microtiter plate in the Fluorescence Reader to obtain background levels.

16. Pipette 100 µl of each sample in triplicate to 96-well plate.

17. Measure fluorescence of the supernatant II at 530 nm and at an excitation wavelength of 485 nm in the Fluorescence Reader.

18. Determine protein content of samples (supernatant I) using standard methods and calculate results as relative fluorescence units (rfu) per ng total protein

Optional:

19. Calculate results as relative fluorescence units (rfu) per mg of brain tissue.

21.2 Protocol for measurement of vascular permeability in the ear (modified Miles assay)

Injection with Evans Blue Dye

1. Warm up mouse tail in 37 °C water bath for 3–10 min
   Note: Intravenous injection should only be started when tail veins are clearly visible

2. Fill 0.5 ml Evans Blue Dye (prepared at 12 mg/ml in PBS and stored at 4 °C) into 1 ml syringe

3. Fix mice in a mouse restraining chamber for tail vein injection to get easy access to the tail

4. Inject 200 µl of Evans Blue Dye into a tail vein using a 30G needle

5. Allow dye to circulate for 5 min

Injection of vascular permeability modifying factors into the ear

1. Anesthetize mouse (e.g., with Forene®).

2. Inject intradermal vascular permeability factors (such as 0.2 µg VEGF in 20 µl PBS) into the left and solvent (saline) into right ear with a 20G needle and a 1 ml syringe.

3. Sacrifice animals 4 to 8 min later and take photographs

Modification for quantification of vascular permeability in the ear (modified Miles assay)

1. Instead of Evans Blue Dye, inject i.v. with 2.5 mg FITC-labeled BSA (Sigma) in 100 μl PBS

2. Omit taking photographs in Step 8

3. Remove ears by using scissors

4. Dissect tissue with scalpel blades into 1 mm³ blocks

5. Add 0.5 ml protease K (1 mg/ml) per ear in PBS buffer and incubate overnight at RT

6. Centrifuge (3,000 rpm) for 15 min at 4 °C

7. Measure fluorescence of the supernatant at 530 nm and at an excitation wavelength of 485 nm in the Fluorescence Reader

8. Present results as ratio between left and right ear, a ratio greater than one indicates vascular permeability increasing activity

Expected Results

We measured vascular permeability under various conditions and applications. In one model, vascular permeability changes were studied in the brain under hypoxic conditions. Hypoxia, which induces the expression of VEGF, was achieved by substituting oxygen with nitrogen after a gradual adaptation

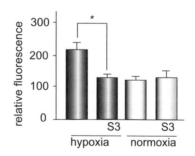

Fig. 2. p38 MAPK inhibition abrogates vascular permeability in the brain. Adult mice were intraperitoneally injected with 20 μg SB203580 (S3) or vehicle and were exposed to normobaric hypoxia at 8 % oxygen for 24 h or were kept at room air. To quantify vascular permeability of brain vessels 200 μl sodium-fluorescein (6 mg/ml in PBS) was injected through the tail vein and the fluorescence of the brain parenchyma determined as described in the methods section. p38 MAPK inhibition completely blocks the hypoxia-induced increase in vascular permeability, whereas it has no effect under normoxic conditions. Values are mean ± SEM. (n = 6). *P < 0.05. (Figure modified from the original Figure 2B on page 263, as published by Issbrucker et al. in FASEB J 17: 262–264, 2003; reprinted with permission)

time of 1 h. Mice exposed to normobaric hypoxia at 8 % oxygen for 24 h showed a two times higher vascular permeability in the brain (Fig. 2). Furthermore, intraperitoneal application of a small chemical inhibitor of the p38 MAP kinase (SAPK-2a) completely abolished the hypoxia-induced vascular permeability in the brain (Issbrucker et al. 2003). In conclusion, the described method of vascular permeability measurement can be used to assess drugs for therapeutic interventions in the setting of increased vascular permeability/edema formation under hypoxic conditions.

In a second model, we describe the use of a modified Miles (Fig. 3) assay suitable for quantitative evaluation. Instead of using systemic hypoxia for

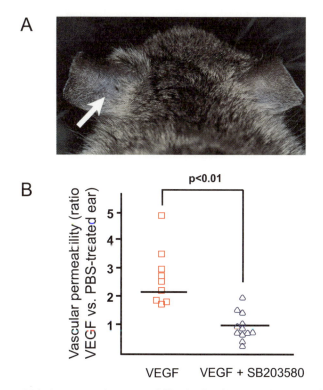

Fig. 3. A VEGF induces vascular permeability in the skin. Mice (8–10 weeks old) were i.v. injected with 200 μl Evan's Blue dye. After 5 min, VEGF was injected into the left ear and saline into the right ear. VEGF-induced increase in vascular permeability is observed by extravasation of the blue dye in the left ear (*arrow*). **B** p38 MAPK inhibition abrogates vascular permeability in the skin. Mice (8–10 weeks old) were intraperitoneally injected with 10 μg SB203580 or vehicle. After 30 min, 200 μl sodium-fluorescein was injected through the tail vein and the fluorescence in the VEGF-treated versus untreated ears was determined as described in the methods section. p38 MAPK inhibition completely blocked the VEGF-induced increase in vascular permeability. Values are mean ± SEM. (n = 8). *P < 0.01. (Figures were modified from the original figures 6A and 7B on pages 1325 and 1326, published first by Clauss et al. in Blood 97: 1321–1329, 2001 Copyright American Society of Hematology, reprinted with permission)

(local) production of the hypoxia inducible VEGF, in this model the vascular permeability factor was applied by local injection. As observed in the model of cerebral edema, also in this assay, i.p. injection of the p38 MAPK inhibitors prior (30 min) to the treatment with VEGF abolished VEGF induced vascular permeability (Clauss et al. 2001). In consequence, p38 MAP kinase activity appears to be essentially involved in VEGF-induced vascular permeability.

In conclusion, the proposed techniques of testing conditions or factors inducing vascular permeability are relatively simple and give quick results. They are suitable for testing small molecular weight chemical and other inhibitors of vascular permeability. Increase in vascular permeability induced mostly both by gap formation and by vesiculo-vacuolar organelles (VVOs) can be addressed with this assay system (Feng et al. 2002). As a considerable limitation, this method cannot differentiate from increase in vascular permeability caused by changes in blood flow and/or dilatation of blood vessels. For instance, VEGF is known to increase both trans- and paracellular transport in vivo but also to cause vessel dilation via generation of NO. It can not be excluded that the combination of these three parameters contributes to the VEGF-induced vascular permeability. In order to exclude influences from additional parameters such as blood flow, measurements using defined and single perfused vessels such as described for isolated frog mesenteric vasculature would be more useful (Bates et al. 1999). However, this assay is established only in the frog and is considerably more difficult to perform. Taken together, the use of dye labeled biological macromolecules such as fluorescent labeled dyes is a suitable method to analyze principles and factors of vascular permeability changes despite of the limitations described above.

Troubleshooting

Vascular permeability in the brain

▶ Low fluorescence signal
  – Fluorescent dye not injected i.v.. Needle not properly feeding the vein, do not use animals which show liquid accumulation in the tail after injection
  – Not exact dye volume injected
  – Tissue homogenization did not work. Make sure that tissue appears mashed after homogenization, check that supernatant is clear after ethanol precipitation and contains fluorescence
  – Avoid exposure to light at all steps to minimize degradation of fluorochrom
▶ High fluorescent background
  – Cardiac perfusion with PBS incomplete. Liver should turn white during perfusion

- Blood clotting in carotid arteries. Check whether brain has turned white after perfusion
▶ High standard variation
- i.v. injection not reproducibly done
- Imperfect perfusion

Vascular permeability in the ear

▶ Pitfalls through i. v. injection of dye. See above first point under "Vascular permeability in the brain"
▶ Too low or too high values (ratios between left and right ear)
- Hemorrhage formation caused by injection. Watch out for bleeding that may occur due to needle caused damage of vessels, omit these values
- Protease digest insufficient: protease should be kept aliquoted at $-20\,°C$
▶ High standard variation
- Different incubation times. Keep all incubations times as exact as possible, use stopwatch
- Keep animals warm by using a lamp to avoid differences in tissue perfusion.
▶ Further pitfalls
- Keep the time after dye injection as short as possible (optimal at 5 min) in order to maintain a high gradient of the dye concentration between blood and tissue
- Avoid too long incubations of mice after injection of permeability factors. After 10 min, an increase of background can be observed and the difference between permeability factor treated ear to control ear is quenched

▦ References

Baba M, Oishi R, Saeki K (1988) Enhancement of blood-brain barrier permeability to sodium fluorescein by stimulation of mu opioid receptors in mice. Naunyn-Schmiedeberg's Arch. Pharmacol. 337:423–428

Bates DO, Lodwick D, Williams B (1999) Vascular endothelial growth factor and microvascular permeability. Microcirculation 6:83–96

Clauss M, Sunderkotter C, Sveinbjornsson B, Hippenstiel S, Willuweit A, Marino M, Haas E, Seljelid R, Scheurich P, Suttorp N, Grell M, Risau W (2001) A permissive role for tumor necrosis factor in vascular endothelial growth factor-induced vascular permeability. Blood 97:1321–1329

Dvorak HF, Nagy JA, Feng D, Brown LF, Dvorak AM (1999) Vascular permeability factor/vascular endothelial growth factor and the significance of microvascular hyperpermeability in angiogenesis. Curr Top Microbiol Immunol 237:97–132

Feng, D, J. A. Nagy, Dvorak HF, Dvorak AM (2002). Ultrastructural studies define soluble macromolecular, particulate, and cellular transendothelial cell pathways in venules, lymphatic vessels, and tumor-associated microvessels in man and animals. Microsc Res Tech 57:289–326

Issbrucker K, Marti HH, Hippenstiel S, Springmann G, Voswinckel R, Gaumann A, Breier G, Drexler HC, Suttorp N, Clauss M (2003) p38 MAP kinase--a molecular switch between VEGF-induced angiogenesis and vascular hyperpermeability. FASEB J 17:262–264

Marti HJH, Bernaudin M, Bellail A, Schoch H, Euler M, Petit E, Risau W (2000) Hypoxia-induced vascular endothelial growth factor expression precedes neovascularization after cerebral ischemia. Am J Pathol 156:965–976

Miles A and Miles E (1952) Vascular reactions to histamine histamine-liberator and leukotaxin in the skin of guinea pigs. J. Physiol. 118:228–257

Schoch HJ, Fischer S, Marti HH (2002) Hypoxia-induced vascular endothelial growth factor expression causes vascular leakage in the brain. Brain 125:2549–2557

Senger DR, Galli SJ, Dvorak AM, Perruzzi CA, Harvey VS, Dvorak HF (1983) Tumor cells secrete a vascular permeability factor that promotes accumulation of ascites fluid. Science 219:983–985

van Bruggen NH, Thibodeaux H, Palmer JT, Lee WP, Fu L, Cairns B, Tumas D, Gerlai R, Williams SP, van Lookeren Campagne M, Ferrara N (1999) VEGF antagonism reduces edema formation and tissue damage after ischemia/reperfusion injury in the mouse brain. J Clin Invest 104:1613–1620

Chapter 22

Validating the Manipulation of Endothelial Cell Signaling in Vivo

JOHN S. RUDGE and GAVIN THURSTON

Introduction

The ultimate goal of much current research in angiogenesis and vascular biology is to manipulate the cellular and intracellular signaling of endothelial cells, and thereby alter the course of diseases such as cancer, chronic inflammation, and diabetes. Toward this end, a variety of powerful potential therapeutic agents have been developed. These reagents target either the extracellular events involved in signaling (e.g., growth factor binding to cellular receptor) or the intracellular events downstream of ligand-receptor binding (e.g., tyrosine phosphorylation of receptor). Reagents that either activate or block signaling have been developed.

The final utility of a particular therapeutic agent depends, of course, on whether it can alter the course of a disease. However, an important intermediate endpoint is whether the agent affects the specific signaling pathway to which it is targeted. Typically, potential therapeutics are tested for their effects in cell culture or biochemical systems and, later, these agents are tested using in vivo models. Ideally, a strong mechanistic link can be established between the in vitro and in vivo actions of an agent. For example, if an agent has been shown to block tyrosine kinase activity of a particular receptor in a cell-free system, then it should also reduce receptor kinase activity in a disease model. Establishing such a link addresses several crucial questions. First, is the agent having the anticipated effect on cellular signaling in vivo? Second, is the dosing optimal or adequate? And third, does the signaling system that is being targeted really have a critical role in the disease process?

This chapter describes methods to begin to assess the in vivo effects of potential therapeutic agents that modify angiogenic receptor signaling. In particular, methods are described to determine the dose of a potential protein-based therapeutic, and whether the specific receptor signaling system is being affected. As an example, we will describe protocols with the angiopoietin/Tie2 ligand receptor system (Davis et al. 1996, Yancopoulos et al. 2000). We describe an ELISA method to measure the amount of a delivered

Springer Lab Manual
H. Augustin (Ed.)
Methods in Endothelial Cell Biology
© Springer-Verlag Berlin Heidelberg 2004

angiopoietin reagent in serum of mice, and we describe an immunoprecipitation-Western blotting method to determine whether the levels of Tie2 receptor phosphorylation in the lung are affected.

Because we draw our examples from the angiopoietin/Tie2 system, some of the reagents are very specific to that system. However, it should be emphasized that much of the procedures and the overall approaches that are described are applicable to other growth factor systems involved in angiogenesis such as the VEGF and PDGF systems. The main point to emphasize is that the information derived from such assays is crucial to establish a role of a particular signaling pathway in normal or pathologic angiogenic processes.

ELISA for circulating reagents, using Ang1-F1-Fc-F1 as an example

The protein Ang1-F1-Fc-F1 is a recombinant form of Angiopoietin-1 that contains the fibrinogen-like domain of human Ang1, a human Fc domain of human IgG1, and another fibrinogen-like domain of human Ang1 (Davis et al. 2003). By virtue of its Fc domain, this molecule dimerizes to form a homodimer containing four Ang1 fibrinogen-like domains. This molecule has also been called BowAng1 and Ang1TFD (Ang1 tetra fibrinogen domain).

Principle

A critical component in assessing the efficacy of a reagent in any animal model system is to know the levels of active reagent circulating in the blood stream. The most accurate and reliable method is a sensitive ELISA to track the reagent over time in serum or plasma. The example we will use is measurement of a recombinant form of angiopoietin-1, called $Ang1-F_1-Fc-F_1$, that contains the human Fc portion of IgG (Davis et al. 2003). The method described will work in serum or plasma.

The approach, generally applicable to other Fc-tagged proteins, is to capture the reagent by coating the wells of a 96-well microplate with a ligand or receptor, and then to report using an anti-Fc antibody linked to an enzyme. This approach serves as a functional ELISA, because the reagent being tested – either ligand or circulating soluble receptor – must bind to its cognate receptor/ligand in order to be detected.

In this particular example, the assay involves coating 96-well microplates with the angiopoietin receptor Tie2 which will capture $Ang1-F_1-Fc-F_1$. Standards and samples are pipetted into the wells, and functionally active $Ang1-F_1-Fc-F_1$ binds and becomes immobilized. After washing away unbound substances, an enzyme-linked polyclonal antibody specific for the Fc domain of $Ang1-F_1-Fc-F_1$ is added to the wells. Following a wash to remove any unbound antibody-enzyme reagent, a substrate solution is added and allowed to develop, so that the color is proportionate to the amount of Ang1-F_1-Fc-F_1 captured in the first step. The color development is stopped, the intensity of the color is measured colorimetrically, and this intensity is compared to diluted standards.

Materials

- Chinese hamster ovary cell-produced Ang1-F_1-Fc-F_1 (Regeneron, www.regeneron.com)
- 0.05 M Carbonate-Bicarbonate Buffer, pH 9.6 (Sigma, Cat. No. C-3041, www.sigmaaldrich.com)
- Human Tie2 receptor mouse Fc (human Tie2 R-mu Fc lot 970811, 0.228 mg/ml, Regeneron)
- BSA concentrate (Kirkegaard & Perry, Cat. No. 50-61-01, www.kpl.com)
- Milk Blocking Solution (Kirkegaard & Perry, Cat. No. 50-82-00)
- Wash buffer concentrate (Kirkegaard & Perry, Cat. No. 50-63-01)
- ImmunoPure normal mouse serum (VWR, Cat. No. 31881, www.vwrsp.com)
- Goat Anti human Fc. IgG HRP (Sigma, Cat. No. I-0170)
- TMB substrate (Sigma, Cat. No. T-8665)
- Stop solution, 2N sulphuric Acid (Mallinckrodt, Cat. No. H381, www.mallchem.com)

Procedure

1. Coat Immulon 96-well plate with 100 µl/well of human Tie2 mouse Fc at 5 µg/ml in Carbonate-Bicarbonate Buffer and incubate overnight at 4 °C.

2. Wash 2 × 300 µl/well with 1× wash buffer.

3. Block plate with 300 µl/well of blocking solution. Incubate 1 h at RT.

4. Serially dilute CHO-produced Ang1-F_1-Fc-F_1 standard from 100 ng/ml in 3-fold dilutions down to 0.14 ng/ml. Make these dilutions in diluent (1:15 dilution of BSA concentrate in water). For spiking normal mouse serum, reserve some of the BSA diluent and add normal mouse serum at 1:100 or the lowest dilution of sample used and prepare the standard curve as above. Plate samples and standards at 100 µl/well in duplicate. Incubate for 2 h at RT.

5. Wash 4 × 300 µl/well with 1× wash buffer.

6. Dilute the goat anti-human Fc IgG HRP (1:20,000) in diluent. Add 100 μl/well to assay plate and incubate for 1 h at RT.

7. Wash 4 × 300 μl/well with 1× wash buffer, and then drain on absorbent paper towels.

8. Add 100 μl/well of TMB Substrate and develop at RT for 15 min.

9. Add 100 μl of stop solution, 2N sulphuric acid.

10. Read plate at 450–570 nm.

▨ Expected Results

Figure 1 shows a typical standard curve achieved by titrating Ang1-F_1-Fc-F_1 from 100 ng/ml to 0.14 ng/ml with or without normal mouse serum spiked at a dilution of 1/100. It is important to read unknown values off the linear portion of the standard curve (Fig. 1).

Once the standard curve is shown to be reproducible, it is possible to measure circulating Ang1-F_1-Fc-F_1 in mice after a single bolus injection of the protein. In the following experiment, a single bolus of 25 mg/kg of BowAng1 was injected subcutaneously into C57BL/6 mice (n=3) and tail bleeds taken at 1 h, 2 h, 4 h, 6 h, 24 h, 3 days, and 7 days. In addition, prior to the injection, a tail bleed should be taken to ensure that there is no interference in the ELISA from the serum of the animals used in the experiment. Serum should be prepared from the blood and frozen at –80 °C until run in the ELISA. On the day of the ELISA, unknown samples should be thawed on

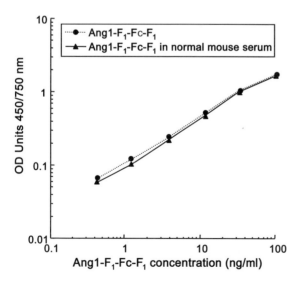

Fig. 1. Measured OD of test ELISA using dilutions of Ang1-F_1-Fc-F_1. Measurements were done with or without addition of normal mouse serum

Fig. 2. Measurement of Ang1-F_1-Fc-F_1 in mouse serum after a single subcutaneous injection (25 mg/kg)

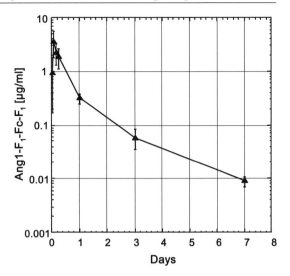

ice and then introduced to the ELISA plate at several dilutions in order to ensure a reading in the linear portion of the standard curve. Values obtained from the standard curve can then be represented graphically (Fig. 2).

As shown in Figure 2, after a 25 mg/kg bolus the maximal serum concentration (C_{max}) of Ang1-F_1-Fc-F_1 is 3.7 µg/ml and the time (T_{max}) at which Ang1-F_1-Fc-F_1 achieves this concentration is 2 h.

Troubleshooting

▶ In this assay, which uses Tie2 linked to mouse Fc, it is important to make sure that the reporting anti-human Fc antibody does not cross-react with mouse Fc on the plate.

▶ We have found that using the extracellular domain of Tie2 (ExTEK) as the capture reagent in this assay reduces the sensitivity.

▶ Do not use dilutions of less than 1:5 for your test sample as this can interfere with the assay.

▶ The time constraints of the ELISA should be strictly adhered to in order to ensure reproducibility.

▶ Samples should be stored –80 °C prior to assay and should not be freeze-thawed more than twice.

▶ When developing new assays based on this principle, it is important to compare incubating the test reagent for 2 h or overnight to see if a longer incubation improves the sensitivity of the assay.

▶ When developing new assays based on this principle, it is best to compare standard curves with and without a serum spike.

22.2 Immunoprecipitation of Tie2 and Western blotting for phosphotyrosine

Once the pharmacokinetic profile of a reagent, in this case Ang1-F_1-Fc-F_1, has been determined, it is important to obtain a functional readout to ensure that the molecule is active in vivo. Such an assay can then be used to determine whether the agent is capable of producing a functional effect in a disease model. A direct way of determining functional activity, in the case of Ang1-F_1-Fc-F_1, is to look at phosphorylation of the angiopoietin receptor Tie2. Such a test involves administering Ang1-F_1-Fc-F_1 to mice, sacrificing them at specific time points, homogenizing select tissues, and immunoprecipitating Tie2 with a specific antibody. The immunoprecipitated Tie2 is then run on a Western blot, and its level of activation determined by probing with an antibody to phosphotyrosine. The amount of Tie2 protein can then be determined by reprobing the blot with a Tie2-specific antibody.

▨ Materials

- Protease inhibitor tablets (Boehringer-Ingelheim, Cat. No. 1697-498, www.boehringer-ingelheim.com)
- Protein G Sepharose® beads (Pharmacia [Pfizer], Cat. No. 17-0618-01, www.pharmacia.com [www.pfizer.com])
- Immobilon-P membrane (PVDF membrane, Integrated Separation Systems, Cat. No. SE151104, 0.45 μm pore size, e.g., Selectscience, www.selectscience.net)
- Anti-Tie2 monoclonal antibody (Upstate Biotechnologies, mAb33, Cat. No. 05–584, www.upstatebiotech.com).
- Anti-Phosphotyrosine antibody (monoclonal IgG2b, Upstate Biotechnologies, Cat. No. 05-321)
- HRP-conjugated goat anti-mIgG (Promega, Cat. No. W4021, www.promega.com)
- ECL 1+2 Detection Reagent (Amersham, Cat. No. RPN 2105, www5.amershambiosciences.com)
- β-mercaptoethanol (14.3 M solution from Sigma, Cat. No. M-6250)
- Tris-Glycine SDS running buffer (10× solution from Novex, e.g., Helixx Technologies, www.helixxtec.com)
- 4–12 % Tris-Glycine gel (Novex)
- 10× Tris-Glycine solution (SEPRABUFF, Enprotech, Cat. No. SA100034, www.enprotech.com)
- BCA Assay (Pierce Biotechnology, Cat. No. 23225, www.piercenet.com)

Procedure

Tissue Homogenization

1. Make up incomplete lysis buffer:
   - 20 mM Tris, pH 7.6 (from 1 M stock)
   - 150 mM NaCl (FW 58.44)
   - 1 mM EDTA (from 0.5 M stock)
   - 5 mM benzamidine (FW 156.6)
   - 50 mM NaF (FW 41.99)
   - 1 mM sodium orthovanadate (from 200 mM stock)

2. Make 50 ml up complete lysis buffer, combining the following:
   - Incomplete lysis buffer (42.25 ml)
   - 1 % NP40 (5 ml of 10 % NP40 stock)
   - 1 Complete™ tablet (containing protease inhibitors)
   - 0.5 % Deoxycholic acid [2.5 ml of 10 % stock (make 10 % w/v stock in 10 mM Tris, 1 mM EDTA, pH 8.0)]
   - 0.1 % SDS (250 µl of 20 % SDS stock)

3. Homogenize frozen samples in 10 % weight/volume (e. g., for a 0.13 g heart you would require 1.3 ml complete lysis buffer) ice-cold complete lysis buffer with Polytron, highest speed, for 30–60 s. Keep homogenization times consistent between animals and organs, if possible. While homogenizing, to prevent homogenate from getting warm, keep tube with tissue in a beaker filled with ice.

4. Centrifuge homogenates for 10 min at $1,000 \times g$ at 4 °C.

5. Remove supernatant and assay protein levels using a BCA Assay.

6. Store homogenate at –80 °C prior to running the gel.

Immunoprecipitation

1. Wash beads
   - Determine total amount of beads required for the experiment. Plan to use 100 µl Protein G Sepharose® beads per sample. Wash 10 % more than you need.
   - Vortex beads in original container to thoroughly mix them. Using wide-orifice tips, pipet as much as you will need for the experiment into a 2 ml microcentrifuge tube.
   - Mark volume level on tube.
   - Spin tube at 10,000 RPM in microcentrifuge for 20 s
   - Aspirate off liquid, being careful to not disturb beads.
   - Add 1 ml complete lysis buffer, invert tube several times, and spin tube again at 10,000 RPM for 20 s.

- Repeat wash 2× for a total of 3 × 1 ml washes.
- After the third wash, remove all the liquid and bring contents up to the level of the mark with complete lysis buffer.

2. Incubate 1–5 mg protein sample in a volume of 500 μl complete lysis buffer with 50 μl washed beads. Incubate for 1 h at 4 °C with rotation.

3. After incubating for 1 h, spin tubes at 10,000 RPM on the microcentrifuge for 5 min. Transfer supernatant to fresh tubes. Save bead pellet at –20 °C.

4. Add 2–5 μg of Anti-Tie2 (mAb 33). Incubate for 2 h to overnight at 4 °C with rotation.

5. After one-hour incubation, add 50 μl of washed protein G beads to each sample and incubate at 4 °C with rotation for at least 2 h or overnight.

6. Spin samples at 10,000 RPM on microcentrifuge at 4 °C for 5 min.

7. Remove supernatant and store it at –80 °C. Wash beads 3 × 1 ml complete lysis buffer.

8. After aspirating the last wash, add 60 μl of 2× sample buffer to each tube of beads:
   - 125 mM Tris-HCl, pH 6.8 at 25 °C
   - 4 % w/v SDS
   - 20 % glycerol
   - 0.2 % w/v bromphenol blue
   - 0.715 M β-mercaptoethanol (5 % of a 14.3 M solution from Sigma, Cat. No. M-6250; add BME to 2× sample buffer in fume hood immediately before use. Keep all tips, etc. used with BME in a closed container. Store any unused sample buffer with BME at –20 °C)
   If the gel will not be run immediately, store beads in sample buffer at –20 °C.

Running and Transferring a Gel

1. Run 4–12 % Tris-Glycine gel at 125 V for ~2 h until the 44 kD marker is at the bottom of the gel.
2. Approximately 30 min before gel will be finished running, prepare for transfer:
   - Prepare 1.2 l Tris-Glycine running buffer with 20 % methanol. (Make up from 10× Tris-Glycine solution, 1 bottle makes up 1 l 10× solution - do not add methanol directly to 10× Tris-glycine solution.)
   - Cut Immobilon-P PVDF membrane to blot size. Soak membrane in methanol for 20 s, in Milli-Q water for 2 min, then in used transfer buffer.

- Soak sponges in fresh transfer buffer under a loose cassette in the transfer apparatus.
- Soak two pieces of thick blotting paper in used transfer buffer.

3. When the gel has finished running, disassemble apparatus, and pry open gel cassette. Cut foot off gel, and, holding by the bottom, carefully transfer the gel to a container with fresh transfer buffer.

4. Keeping all materials wet with transfer buffer, assemble cassette. Roll any air bubbles out of the stack using a pipet after second piece of blocking paper is placed. Place assembled cassette into transfer apparatus.

5. Run the transfer at 4 °C with the transfer apparatus on a stir plate, stirring. Run overnight at 33 V.

6. Take apart the transfer cassette, mark the membrane, and rinse the membrane in ST (200 mM NaCl, 15 mM Tris, pH 7.5) for 2 min. Stain the gel in Coomassie blue, shake it overnight at RT.

PTyr Probing

1. After ST wash, transfer the membrane to 20 ml 1 % BSA in TBS (20 mM Tris-HCl, pH 7.5, 150 mM NaCl). Incubate shaking overnight at 4 °C or for 1 h at RT.

2. Make up 1st and 2nd antibody incubation buffer: 1 % BSA + TBS + 0.05 % Tween-20.

3. Use 20 ml buffer to dilute 1st antibody, anti-phosphotyrosine monoclonal antibody, 1:7500. Incubate membrane in primary incubation buffer for 1 h at RT with moderate shaking.

4. Wash: 3 × 10 min, 30 ml each wash, 0.1 % TBST with shaking (0.1 % TBST– 20 mM Tris-HCl, pH 7.5, 150 mM NaCl, 0.1 % Tween-20).

5. Dilute 2nd antibody, HRP-conjugated goat anti-mouse IgG 1:50,000 in 20 ml incubation buffer. Incubate membrane in secondary incubation buffer for 1 h at RT with moderate shaking.

6. Wash:
   - 3 × 10 min, 30 ml each wash, 0.1 % TBST with shaking
   - 3 × 5 min, 30 ml each wash, 0.3 % TBST with shaking
   - 2 × 5 min, 30 ml each wash, TBS with shaking

7. Blot membrane between 2 pieces of 3MM Whatman paper to remove excess TBS. **Note:** Do not let membrane dry out completely.

8. Soak blot in 1:1 mixture of ECL 1+2 Detection Reagent, mixed just before use, for 2 min. If placing two membranes in one dish, flip half-way through incubation.

9. Blot membrane between the 3MM Whatman paper again, place between transparencies, put into a cassette, and expose to DuPont Film (Reflection NEF-496, without intensifying screen) as soon as possible. Expose for 1 min, 5 min, 30 min, and 1 h.

Stripping protocol

There are two acceptable protocols for stripping.

2-mercaptoethanol method:
    Submerge the membrane in stripping buffer (100 mM 2-mercaptoethanol, 2 % SDS, 62.5 mM Tris-HCl, pH 6.7) in a well-sealed container and incubate at 50 °C for 30 min with occasional agitation.
    Wash the membrane for 3 × 10 min in TBST at RT using large volumes of wash buffer.

Glycine method:
    Submerge the blot in 0.2 M Glycine (pH2.8) at RT overnight with mild agitation.
    Wash the membrane for 3 × 10 min in TBST at RT using large volumes of wash buffer.

Reprobing for Tie2

Block the membrane
1. After the final wash, transfer the membrane to 20 ml 2.5 % BSA in TBST (20 mM Tris-HCl, pH 7.5, 150 mM NaCl, 0.05 % Tween-20). Incubate shaking overnight at 4 °C or for 1 h at RT.

2. Make up 1st and 2nd antibody incubation buffer: 2.5 % BSA + TBS + 0.05 % Tween-20.

3. Use 20 ml buffer to dilute 1st antibody, mAb33 (1:10,000 of 2 mg/ml stock). Incubate membrane in primary incubation buffer for 1 h at RT with moderate shaking.

4. Wash: 3 × 10 min., 30 ml each wash, 0.1 % TBST with shaking, (0.1 % TBST= 20 mM Tris-HCl, pH 7.5, 150 mM NaCl, 0.1 % Tween-20).

5. Dilute 2nd antibody, HRP-conjugated goat anti-mIgG (Promega, Cat. No. W4021) 1:50,000 in 20 ml incubation buffer. Incubate membrane in secondary incubation buffer for 1 h at RT with moderate shaking.

6. Wash:
   - 3 × 10 min, 30 ml each wash, 0.1 % TBST with shaking
   - 3 × 5 min, 30 ml each wash, 0.3 % TBST with shaking
   - 2 × 5 min, 30 ml each wash, TBS with shaking

7. Blot membrane between 2 pieces of 3MM Whatman paper to remove excess TBS. **Note:** Do not let membrane dry out completely.

8. Soak blot in 1:1 mixture of ECL 1+2 Detection Reagent (Amersham, Cat. No. RPN 2105, www5.amershambiosciences.com), mixed just before use, for 2 min. If placing two membranes in one dish, flip half-way through incubation.

9. Blot membrane between the 3MM Whatman paper again, place between transparencies, put into a cassette, and expose to DuPont Film (Reflection NEF-496, without intensifying screen) as soon as possible. Expose for 1 min, 5 min, 30 min, and 1 h.

Expected Results

Figure 3 shows representative blots where 100 µg of Ang1-F_1-Fc-F_1 or Ang1-F_2-Fc-F_2, was injected intravenously and the animals sacrificed at 15 min or 24 h (Fig. 3A,B). Lungs were removed and homogenized in lysis buffer, and

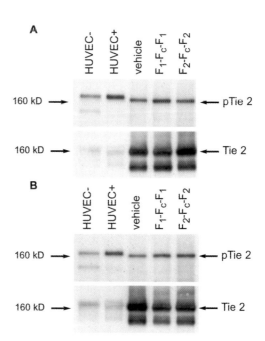

Fig. 3A,B. Blots showing levels of phospho-Tie2 and Tie2 receptor after IV administration of Ang1-F_1-Fc-F_1 and Ang1-F_2-Fc-F_2 for 15 min and 24 h followed by immunoprecipitation and Western blotting

Tie2 was immunoprecipitated. The upper panels in Figure 3 show the result of the phosphotyrosine blot, and the lower panels show the result of reprobing for Tie2. In most tissue blots, a second band is present in the Tie2 blot, which may represent the extracellular domain of Tie2 that is not phosphorylated. It is also noteworthy to mention that Ang1-F_2-Fc-F_2[1] is capable of phosphorylating Tie2 in lung at 24 h, consistent with previous findings in vitro (Teichert-Kuliszewska et al. 2001). The lanes labeled HUVEC – and HUVEC + show results from human umbilical vein endothelial cells treated without or with Ang1-F_1-Fc-F_1 for 10 min in vitro and processed in the same way as the lung tissue (Fig. 3).

Troubleshooting

▶ All treatment conditions should be run in triplicate

▶ It is imperative that a positive control is run on every blot. We use human umbilical vein endothelial cells (HUVECs) that have been treated with Ang1-F_1-Fc-F_1 for 10 min, lysed and stored at –80 °C in aliquots. For each experiment, 200–250 μg of the HUVEC lysate is immunoprecipitated and run on the gel. This controls for the molecular weight of Tie2 (approx. 160 kD, although human Tie2 is slightly larger than mouse Tie2) and the probing antibodies.

▶ A pilot experiment should be run when probing different tissues, because of differing amounts of Tie2 in different tissues (e.g., 1 mg is sufficient for lung, but 4 mg is necessary for liver).

▶ If you are going to use non-fat milk instead of BSA for blocking prior to PTyr probing, make sure that it is low in phosphatases.

▶ We include a pre-clearance step in the protocol where the protein sample is incubated with protein G beads prior to the actual immunoprecipitation. This is not always necessary but helps if you are getting high background on your blots.

Sodium orthovanadate should be activated for maximal inhibition of protein phosphotyrosyl-phosphatases (see below).

▶ Prepare a 200 mM solution of sodium orthovanadate.

▶ Adjust the pH to 10.0 using either 1 N NaOH or 1 N HCl. The starting pH of the sodium orthovanadate solution may vary with lots of the chemical. At pH 10.0 the solution will be yellow.

---

[1] Ang1-F_2-Fc-F_2 contains the fibrinogen-like domain of human Ang2, a human Fc domain of human IgG1, and another fibrinogen-like domain of human Ang2. By virtue of its Fc domain, this molecule dimerizes to form a homo-dimer containing four Ang2 fibrinogen-like domains. This molecule has also been called BowAng2 and Ang2TFD, Ang2 tetra fibrinogen domain

- ▶ Boil the solution until it turns colorless (approximately 10 min).
- ▶ Cool to RT.
- ▶ Readjust the pH to 10.0 and repeat Steps 3 and 4 until the solution remains colorless and the pH stabilizes at 10.0.
- ▶ Store the activated sodium orthovanadate as aliquots at −20 °C.

Acknowledgements. We thank Nick Papadopoulos for helpful discussions, Donna Hylton, Liz Pasnikowski, Sherry Xu, and Shelly Jiang for technical expertise and assay development, and Scott Staton for help with the plates.

References

Davis S, Aldrich TH, Jones PF, Acheson A, Compton DL, Jain V, Ryan TE, Bruno J, Radziejewski C, Maisonpierre PC, Yancopoulos GD (1996) Isolation of angiopoietin-1, a ligand for the TIE2 receptor, by secretion-trap expression cloning. Cell 87:1161–1169

Davis S, Papadopoulos N, Aldrich TH, Maisonpierre PC, Huang T, Kovac L, Xu A, Leidich R, Radziejewska E, Rafique A, Goldberg J, Jain V, Bailey K, Karow M, Fandl J, Samuelsson SJ, Joffe E, Rudge JS, Daly TJ, Radziejewski C, Yancopoulos GD (2003) Angiopoietins have distinct modular domains essential for receptor binding, dimerization and superclustering. Nat Struct Biol 10:38–44

Teichert-Kuliszewska K, Maisonpierre PC, Jones N, Campbell AI, Master Z, Bendeck MP, Alitalo K, Dumont DJ, Yancopoulos GD, Stewart DJ (2001) Biological action of angiopoietin-2 in a fibrin matrix model of angiogenesis is associated with activation of Tie2. Cardiovasc Res 49:659–670

Yancopoulos GD, Davis S, Gale NW, Rudge JS, Wiegand SJ, Holash J (2000) Vascular-specific growth factors and blood vessel formation. Nature 407:242–248

Chick Chorioallantoic Membrane Assay: Growth Factor and Tumor-induced Angiogenesis and Lymphangiogenesis

Martin Hagedorn and Jörg Wilting

Introduction

The chick chorioallantoic membrane (CAM) is a transient, densely vascularized organ, which is located underneath the shell membrane and eggshell. Because of this anatomical location, it is easily accessible for experimental studies. It serves as a respiratory organ, supplies the embryo with calcium dissolved from the shell, and stores and concentrates urea. Regarding its nutritive functions, the CAM has been compared to the placenta. In contrast to the placenta, it contains lymphatic vessels accompanying all mayor blood vessels (Oh et al. 1997).

The CAM develops by fusion of the avascular chorion with the strongly vascularized allantois. The chorion consists of an ectodermal epithelium and a single layer of mesodermal cells. The allantois is made up of an endodermal epithelium and several layers of mesoderm containing blood islands and lymphangioblasts (Papoutsi et al. 2001). The primary vascular plexus of the CAM differentiates into a vascular tree with arteries and veins. This process is completed around day 12 (DeFouw et al. 1989). The proliferative index (thymidine-labeling) of endothelial cells in the chick CAM is 23 % during days 8–10, and drops to 3 % on day 11 (Ausprunk et al. 1974). To avoid interference with the endogenous angiogenesis of the CAM, we recommend to perform angiogenesis assays after day 12 (Wilting et al. 1991). Involution of the chick CAM starts around days 18–19, and hatching takes place at days 20–21.

From day 4 of development on, the allantois becomes densely vascularized. During differentiation of the blood vessels, the capillary plexus is shifted into the chorionic epithelium (Leeson and Leeson 1963). Thereby, the blood-air-barrier becomes extremely thin. The arterial and venous conduit vessels remain within the mesodermal stroma and are accompanied by lymphatic vessels. Prox1-positive lymphangioblasts are found in the mesoderm of the allantois at day 4 already.

Springer Lab Manual
H. Augustin (Ed.)
Methods in Endothelial Cell Biology
© Springer-Verlag Berlin Heidelberg 2004

It is known for some time that the CAM is a suitable site for the cultivation of tissues (Willer 1924), microorganisms (Woodruff and Goodpasture 1931), and tumors (Murphy and Rous 1912). The CAM has then also been used to study the activities of angiogenic growth factors (Folkman 1974) and lymphangiogenic growth factors (Oh et al. 1997). It has been recommended to apply angiogenic growth factors on the CAM at day 9 or 10 (Folkman 1974), but a considerable number of carrier effects (so-called false positive responses) have been reported with this method (Jakob et al. 1978, Wilting et al. 1991). Most likely, this is due to interference with the endogenous angiogenic activity of the CAM, which is at its peak during that period. To circumvent these problems, we have applied growth factors on the differentiated CAM (day 13) and evaluated the results after three days (Wilting et al. 1991). This method is suited to discriminate between different growth factor effects and is also applicable for lymphangiogenic growth factors (Oh et al. 1997). An angiogenic agent, like vascular endothelial growth factor A (VEGF-A), increases the amount of capillaries in the application area, thereby inducing growth of capillaries into the stroma of the CAM. This effect is well visible under a stereomicroscope. Inhibitors can be tested in combination with VEGF-A and their effect can be compared with controls (Hagedorn et al. 2002). The "spoke wheel" vessel pattern, often regarded as a positive sign of angiogenesis in the CAM, is difficult to interpret. In fact, the capillaries in such regions have vanished, and one could also interpret this result as an anti-angiogenic event. However, one should keep in mind that some growth factors may induce "arteriogenesis" or "venogenesis", but criteria to identify such effects in the CAM assay have not been reported yet.

In case of lymphangiogenic growth factors, such as VEGF-C and -D, inspection of the application area under a stereomicroscope is not sufficient. Lymphatics are hardly visible, and therefore it is necessary to identify them with specific antibodies (anti-Prox1) or by intralymphatic injection of corrosion cast media, such as Merkox-blue (Oh et al. 1997).

Grafting of tumors and tumor cell lines on the chick CAM has been performed to study their invasiveness and their interaction with the blood and lymphatic vessels of the CAM (Armstrong et al. 1982, Brooks et al. 1994, Murphy and Rous 1912, Papoutsi et al. 2000). The method gives insight into the behaviour of tumor cells, and helps to decide whether or not to use mammalian animal models for further studies. Since the chorion is an epithelial barrier, only tumor cells with sufficient invasive potential are able to cross this barrier and get access to the vasculature. Tumor angiogenesis and lymphangiogenesis can be measured and can be used to estimate the potency of tumors to form hematogenic or lymphogenic metastases in vivo.

In summary, the CAM assay performed with the differentiated CAM is highly suited to study growth factor effects. All types of growth factors can be used that affect either of the following cell types: epithelial cells, fibrocytes, vascular smooth muscle cells, pericytes, blood vascular endothelium, lymphatic endothelium. The CAM/tumor assay is a powerful tool to study behaviour of tumor cells in vivo, including their interaction with epithelial cells, endothelial cells, pericytes/vascular smooth muscle cells, and the lymphatic endothelium.

Materials

Animals

- Fertilized White Leghorn chicken eggs (Gallus gallus)

Cells

- Tumor cell lines: U87 human glioma, C6 rat glioma, 10AS or ARIP pancreatic carcinoma, B16 or A375 melanoma cells (and others; e.g., from ATCC, www.atcc.org)

Growth factors

- Human recombinant vascular endothelial growth factors (VEGFs): VEGF-A, -B, -C, -D (establish your own baculovirus production or use commercial source, e.g., Reliatech, www.reliatech.de)

Equipment

- Egg incubator (e.g., BSS160 without rolls, Ehret Laboratory Technolgy, http://manuf.labworld-online.com/ehret/)
- Stereomicroscope with connected digital camera and flexible cold light source (e.g., Nikon SMZ800/Coolpix 950, www.nikon-instruments.com)
- Storage card reader (e.g., Compact Flash Cards) for transfer of digital images to a computer
- Cryostat (e.g., Leica CM 3050S, www.leica.com)
- Ultracut S microtom (Leica)
- Confocal laser scanning microscope (Leica or Olympus, www.olympus.com)
- Leica DMR microscope
- Magnetic stirrer

Small material

- Egg lamp for trans-illumination (Ehret, can be self-made)
- Metal saw, ca. 1–2 mm thick blade (see Fig. 1)
- Fine forceps and iris scissors (e.g., ophthalmologic surgical equipment)
- Sterile Pasteur pipettes

- Sterile scalpels
- Leukosilk or 3M Durapore tape 2.5 cm (no plastic tape; 3M, www.mmm.com)
- Graphite pen
- Cotton swaps
- Nunc Thermanox™ tissue culture cover slips (13 mm diameter; Nalge Nunc International, www.nalgenunc.com)
- Slide-A-Lyzer™ MINI dialyse unit (0.1–0.5 ml, 10,000 MWCO, Pierce Biotechnology, www.piercenet.com)

Solutions

- Locke-solution (physiological solution for chick embryos)
- Stock solutions
  Solution 1   94.27 g NaCl   1 l $_{dd}H_2O$
  Solution 2   12.0 g KCl   1 l $_{dd}H_2O$
  Solution 3   15.80 g $CaCl_2\cdot2H_2O$ 1 l $_{dd}H_2O$

  Mix 100 ml Solution 1 + 37 ml Solution 2 + 21 ml Solution 3, add $_{dd}H_2O$ to 1 l, add penicillin-streptomycin as for routine cell culture

- Hanks-solution
  Stock solution A:
  40 g NaCl
  2 g KCl
  0.7 g $CaCl-2H_2O$
  Dissolve in 250 ml $_{dd}H_2O$, add 1 g $MgSO_4-7H_2O$

  Stock solution B:
  0.6 g $Na_2HPO_4-1H_2O$
  0.3 g KH_2PO_4
  Dissolve in 250 ml $_{dd}H_2O$

  Stock solution C:
  1.75 g $NaHCO_3$
  Dissolve in 250 ml $_{dd}H_2O$

  Take 25 ml of each stock solution A, B and C; add 425 ml $_{dd}H_2O$

- Dalton's solution
  4% $K_2Cr_2O_7$ (potassium bichromate), adjust to pH 7.4 with 5N KOH (Solution A)
  3.4% NaCl solution (Solution B)
  Take 25 ml of Solution A and B; add two clean vials of 500 mg osmi-

umtetroxide (toxic!); add 50 ml $_{dd}$H$_2$0; shake over night; store at 4 °C. Can be used for several months.

For sterilisation of instruments, cleaning of eggs
- 70 % ethanol

For intralymphatic injection of vessels
- Merkox-blue (Okenshoji) (work under a hood!)

Fixative for immunohistochemistry and confocal imaging
- 4 % paraformaldehyde in PBS, pH 7.4

Fixatives for semi- and ultrathin sectioning
- 3 % glutaraldehyde and 2 % formaldehyde in 0.12 M sodium cacodylate buffer
- Dalton's solution

For staining of semithin sections
- 1 % methylene blue and 1 % azure II (Sigma-Aldrich, www.sigmaaldrich.com)

For permeabilization
- PBS, pH 7.4 + 0.1 % Triton X-100 (PBS-T)

For embedding
- Prolong Antifading Kit (Molecular Probes)
- Epon resin (resin for semi- and ultra-thin sectioning; Serva, www.serva.de)

Antibodies and lectins
- QH-1, for quails only (Developmental Studies Hybridoma Bank, www.uiowa.edu/~dshbwww)
- Anti-von Willebrand Factor (antigen retrieval often needed; DAKO, www.dakocytomation.com)
- Anti-Prox1 antibody (Reliatech)
- Appropriate secondary antibodies (e.g., Molecular Probes, www.probes.com)
- SNA-1 lectin coupled to fluorescein (Vector Laboratories, www.vectorlabs.com)

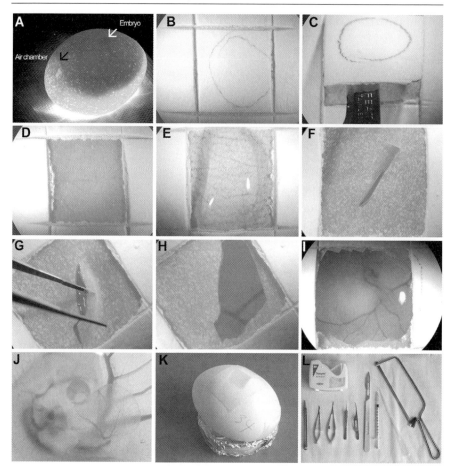

Fig. 1. A Trans-illumination (mark the air chamber and embryo position). **B** Saw the incisions above the embryo, and in the middle of the air-chamber. **C** Removing of the shell (usually the complete piece; this greatly speeds up the opening procedure and reduces deposition of debris). **D)** The shell membrane is visible. The yolk sac vasculature and/or the embryo are located underneath this membrane. **E** A drop of Locke-solution is placed on the exposed shell membrane. **F** Incision in the membrane with a forceps; Locke-solution flows into the egg and embryo sinks down (only if air chamber was punctured). You should no longer see blood vessels like in **D**. **G, H** Removing of the shell membrane: carefully introduce the forceps to one corner and tear off the membrane. **I** Shell membrane removed. Do not try to manipulate too much the borders. You should see the embryo; in this case at day 3. **J** At higher magnification, check for normal development of the embryo and for absence of bleeding or debris which may have fallen on the embryo. The embryo (day 4) lies on its left side. **K** Sealing of the egg using Durapore™ tape. **L** Principal equipment needed for opening eggs and manipulating embryos (from left to right): graphite pen, iris scissors, surgical forceps, scalpel, syringe with needle, sterile plastic Pasteur pipette, small metal saw, Durapore™ tape.

▓ Outline

An example of the working schedule for growth factor studies and for tumor cell – CAM interactions is given in Table 1.

Table 1. Timing of CAM experiments

Day	Procedure
Timetable for growth factor studies	
0	Incubation of chick eggs starts
	– Clean with ethanol
	– Clearly mark each egg with graphite pen; note date of incubation on egg
3/4	Sterilization and preparation of tools
	Preparation of working place with stereomicroscope, lamp, alcohol burner, hot water
	Opening of eggs
	– Prepare Locke solution
	– Mark position of embryo
	– Make window into eggs
	– Seal and re-incubate eggs
<13	Prepare Thermanox™ rings
13	Prepare growth factors and inhibitors
	Dialyze growth factors against $_{dd}H_2O$, if necessary
	Deposition of growth factors/inhibitors
	– Deposit Thermanox™ rings on CAM
	– Pipette growth factor into Thermanox™ rings
	– Seal and re-incubate eggs until day 16/17
16/17	Analysis of growth factor/inhibitor experiments
	– Study in vivo blood flow
	– Fix in vivo with glutaraldehyde/formaldehyde
	– Dissect and photograph under stereomicroscope
19	Further processing for electron or confocal microscopy
20/21	Hatching of chicken
	– Keep in mind when planning experiments! Sacrifice chicks before hatching
Timetable for tumor growth studies	
< 10	Prepare sufficient amounts of tumor cells
	Prepare Thermanox™ rings
10	Deposition of tumor cells
	– Collect and count tumor cells
	– Deposit Thermanox™ rings on CAM
	– Pipette cells into Thermanox™ rings
	– Seal and reincubate eggs until day 17/18
11–17/18	Monitoring of tumor growth
>12–17	Treatment of tumors with anti-tumor drugs, if desired
17/18	Analysis of tumor experiments
	– Fixation of tumors in vivo
	– Photo-documentation of results

▥ Procedures

General recommendations

1. Upon arrival, put eggs in a refrigerator (ca. 10–14 °C). They remain viable up to 10–14 days.

2. Place eggs in incubator. First day is day 0. Note date and time on eggshell.

3. Mark eggs only with a graphite pen.

4. Make window at day 3 (72 h) or day 4 (96 h). Later the allantois will stick to the shell membrane.

5. Always wipe eggs with 70 % ethanol before manipulation.

6. Examine eggs on egg lamp prior to manipulation, in order to determine air chamber position, embryo size, position, and viability (see Fig. 1).

7. Prepare nests. Petri dish (50 mm) with cotton swaps.

8. The manipulations can be carried out in an ordinary laboratory room, but frequent passage of people and sitting under air conditioner should be avoided. Do not perform simultaneously animal experiments with mice, rats, or rabbits in the same room.

9. When re-opening the window, do not detach Durapore tape which is collated directly on the shell. Excise the tape above the window with fine scissors.

10. Considerable amounts of salt present in the growth factors applied on the CAM can cause strong non-specific reactions.

11. Finish the experiments around day 17/18 (chicks start breathing at day 19 and hatch on day 20/21).

Opening of eggs (day 3/4): the false air chamber technique

The goal of this technique is to produce an air chamber right above the embryo. The natural air chamber is situated at the blunt end of the egg. The technique first makes the embryo and yolk sac vasculature accessible, and later it exposes the rapidly expanding CAM to the researcher. For an experienced manipulator, the opening procedure can be completed in less then 5 min per egg.

1. Prepare glass recipient with pre-warmed Locke solution, close with aluminum foil, put sterile Pasteur pipette inside.

2. Verify viability of embryos on the egg-lamp; mark the position of the air chamber and of the embryo with a graphite pen.

3. Wipe eggs carefully with 70 % ethanol.

4. Place the egg in a nest; the position of the embryo facing the researcher.

5. Make an incision at the blunt end of the egg with the metal saw to open the natural air chamber. When you see the shell membrane, make a small hole with a forceps or scalpel blade.

6. Saw a window above the embryo. It should be big enough to allow insertion of the Thermanox ring, but not too big. Be careful not to cut through the shell membrane. Wipe away the shell dust with a cotton swap. Do not blow it away: infection! Keep in mind that eggs show a great variability regarding the thickness and the stability of the shell.

7. Under the stereomicroscope: Tear off the shell with a forceps, leaving the shell membrane underneath intact (small holes and ripples in the shell may be still ok). Wipe away the dust. Put some droplets of sterile Locke solution on the exposed shell membrane to soak and clean it.

8. Make an incision from about 12 o'clock to 6 o'clock into the shell membrane with a forceps, letting the solution slip inside the egg. The embryo will sink down immediately; the albumen filling the natural air chamber. This creates a new air chamber above the embryo and is essential for further successful preparation.

9. Remove most of the shell membrane in the window. Verify normal development of the embryo. It should lie on its left side. Discard abnormal embryos. Also verify developmental stage [classification according to Hamburger and Hamilton (1992)] and yolk sac blood vessels.

10. Wipe remaining dust and Locke solution off the egg and seal both the air chamber incision and the window with Durapore tape. Put nests immediately back into the incubator. Do not roll.

Application of growth factors or tumor cells on the CAM

1. Preparation of Thermanox™ rings
   Use a metal paper clip heated over a gas flame. Melt a ca. 7 mm wide round hole into the centre of a 13 mm Thermanox disc. Reduce the outer diameter of the disc and re-sterilize under UV-light or with isopropanol.

2. Dialyzation of growth factors
   (Step can be omitted if the preparation contains not too much salt). Inject sufficient amount of growth factor into a Slide-A-Lyzer™ dialyses box using a 22-gauge needle. Dialyze 45 min against 500 ml of sterile $_{dd}H_2O$ using a magnetic stirrer. Collect desalted growth factor and keep on ice.

3. Preparation of tumor cells
    Make sure you have expanded a sufficiently high number of cells before. Detach cells by trypsination, spin down, and collect cells in as little volume of culture medium as possible. Estimate the number of cells per μl using a Coulter counter or a hemocytometer.

4. Deposition of rings
    Carefully place the ring in the centre of the CAM using a stereomicroscope. Avoid deposition on bigger CAM vessels. Once lying on the surface, the ring should not be removed due to the high adhesiveness of the CAM surface.

Application of growth factors, drugs, and tumor cells

1. Pipette an appropriate volume of growth factor solution into the centre of the ring (usually <20 μl). The solution is absorbed by the CAM within 1 h. To avoid displacement of the solution, be careful when putting the eggs back to the incubator.

2. If you test an anti-angiogenic factor, either premix with an angiogenic factor (VEGF-A) prior to application, or successively apply the agent. The treatment should be repeated the next day (VEGF and inhibitor).

3. For tumor growth, place $1-5 \times 10^6$ tumor cells into the ring (cell number necessary to initiate tumor growth is dependant on the tumor type). It might be necessary to scratch the surface of the chorionic epithelium with a sharp tool in order to facilitate contact of tumor cell with the underlying vascular network. Close eggs with Durapore tape and put back into the incubator. Tumor growth can be monitored regularly.

In vivo fixation of CAMs for immunohistology

1. Pipette approx. 3 ml of fresh 4 % paraformaldehyde onto the CAM. Incubate 15–30 min at RT.

2. Cut out the CAM area of interest under a stereomicroscope. Transfer to a petri dish with phosphate buffered saline (PBS) and proceed with photo documentation.

3. CAM specimens can be kept in PBS in the refrigerator for several days prior to procession for immunohistochemistry.

4. Remove Thermanox ring carefully.

5. Embed tumors in tissue freeze medium and store at –20 °C. Prepare 8–20 μm thick sections with a cryostat, transfer to glycerine-coated slides and store at –80 °C.

In vivo fixation for semi- and ultra-thin sectioning

1. Pipette ca. 3 ml of glutaraldehyde/formaldehyde solution onto the CAM. Incubate 15–30 min at RT.

2. Cut out the CAM area of interest under a stereomicroscope. Transfer to a Petri dish with 0.12 M sodium-cacodylate buffer and proceed with photo documentation.

Immunohistology

1. Cut approx. 5 mm^2 pieces out of the CAM area of interest (or use cover slips with thin or thick sections of tumors).

2. Permeabilize with PBS-T for 30 min at RT.

3. Add antibody of interest or lectin SNA-1/FITC for 45 min at RT (incubate in the dark if fluorescence-labeled markers or antibodies are used).

4. Wash 3 × 15 min in PBS.

5. Add appropriate secondary antibody (if applicable).

6. Wash 3 × 15 min in PBS.

7. Rinse once with $_{dd}H_2O$.

8. Embed with a fluorescence protection reagent (e.g., Prolong Anti-fading Kit, Molecular Probes).

Semi- and ultra-thin sectioning

1. Rinse in Hank's solution.

2. Postfix in Dalton's solution.

3. Rinse in Hank's solution.

4. Dehydrate in a graded series of ethanol (increasing stepwise from 50 % to 100 %).

5. Transfer into propylenoxide.

6. Transfer into propylenoxide/Epon; 1:1.

7. Transfer into Epon; change Epon twice.

8. Polymerize in embedding moulds at 60 °C for 3 days.

9. Cut semi- or ultra-thin sections with an ultramicrotome.

▨ Expected Results

CAM assays are performed with individual outbreed organisms and may therefore generate a higher variability of responses to manipulations, growth factors and drugs as compared to results obtained in murine models using inbred strains. This seems to be an advantage when studying the efficacy of anti-angiogenic or anti-tumor treatments, because the broad genetic background will result in a broader spectrum of responses, which may better reflect the human situation. Therefore, the CAM assay is a very reliable method to discover molecules with strong and consistent angiogenic or anti-angiogenic effects. Molecules with moderate or weak effects in animals with attenuated genetic backgrounds might not give reproducible results in the CAM.

Angiogenesis

3 µg of recombinant human VEGF-A$_{165}$ applied in the centre of a plastic ring on the surface of the CAM induces strong capillary formation (Figs. 2B, 3B). The effect of VEGF-A is also present around the site of application, due to diffusion. Water alone has no effect (Figs. 2A, 3A). When pre-mixed with the growth factor, a prominent anti-angiogenic effect of a peptide derived from platelet factor 4 (PF-4$^{47-70}$DLR) becomes visible inside the ring (Fig. 3C).

Lymphangiogenesis

3 µg of human VEGF-C applied on the CAM induce lymphangiogenesis (Fig. 3G,H). In semi-thin sections, numerous dilated lymphatics can be seen growing towards the chorionic epithelium. The lymphatic endothelial cells express the transcription factor Prox1. Intra-lymphatic injection of Merkox-blue demonstrates the great amount of lymphatics in the application area.

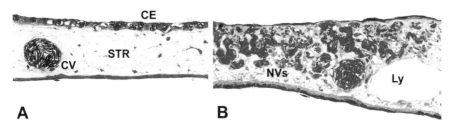

Fig. 2. A Normal day 17 CAM tissue. *CE* chorionic epithelium with dense vascular network; *CV* conduit vessel; *STR* CAM stroma without blood capillaries. **B** VEGF-A-induced angiogenesis in the CAM. *NVs* new blood capillaries in the stroma; *Ly* normal lymphatic vessel

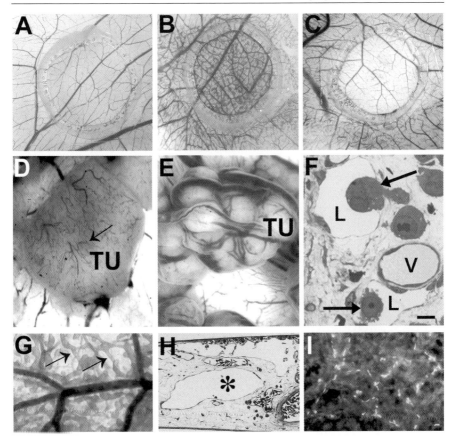

Fig. 3. A Normal CAM at day 17 with a Thermanox ring deposited at day 13. Note that the plastic ring does not cause any vascular irregularities. **B** Deposition of recombinant human VEGF-A$_{165}$ causes a reproducible and strong angiogenic response. **C** Co-application of an angiogenesis inhibitor, PF-4$^{47-70}$DLR, together with VEGF-A completely abolishes VEGF-A effects inside the ring. **D** C6 rat glioma cells, *TU* grown on the CAM. Note induction of tumor angiogenesis (*arrow*). **E** Rat 10AS pancreatic carcinoma cells develop a solid tumor (*TU*) on the CAM. **F** Human A375 melanoma. Tumor cells (*arrows*) are entering lymphatic vessels (*L*), *V* blood capillary. **G, H** Application of VEGF-C on the CAM. The great amount of lymphatics (*arrows*) in the application area is demonstrated by intra-lymphatic injection of Merkox-blue (**G**). Formation of huge lymphatic sinusoids (*asterisk*) (**H**). **I** Anti-Prox1 staining of lymphatic endothelial cells in a VEGF-C expressing A375 melanoma on the CAM

Tumor growth and angiogenesis

Almost all tumor cell lines tested so far have been able to migrate through the chorionic epithelium and to make contact with the vascular network of the CAM. The cell suspensions start forming solid tumors after a few days. Intra-tumoral blood vessels can be seen in large as well as small tumor nodules (Fig. 3D-F), showing that the tumor cell lines have progressed to an angiogenic phenotype. Vascular patterns, capillary morphology, and interactions with pericytes greatly vary between the tumor types. VEGF-C expressing tumor cell lines induce lymphangiogenesis both in the tumor periphery and within the tumor. The lymphatics are positive for VEGFR-3 and Prox1 (Fig. 3I).

▨ Troubleshooting

▸ A major percentage of eggs are damaged upon arrival
  – Tell hatchery to put not more than 15 eggs in a 30-egg carton, egg cartons should be mounted on a wooden palette.
▸ Embryos die during incubation time
  – If >20 % of embryos die during the 17–18 day incubation time, make certain that humidity (>70 %) and temperature (37.8 °C) are constant.
  – Use only fresh Locke solution when opening eggs. Wipe eggs with ethanol prior to opening. Practice opening technique.
  – Remove eggs with dead or moribund embryos from the incubator.
▸ CAM surface is dry
  – Increase humidity; avoid frequent opening of the incubator
  – Do not use plastic tape for closing eggs
  – Avoid deposition of egg shell debris on the CAM during opening
▸ CAM surface is covered with liquid
  – CAM has been ruptured in the periphery during opening. Discard egg.
  – Practice opening technique.
▸ Fine fissures appear in the egg shell during opening
  – Seal fissures with Durapore tape.
▸ No effect of growth factors deposited on the CAM (e.g., VEGF)
  – This happens in a small portion of eggs, most likely due to displacement of the growth factor solution in the ring by embryo movements.
▸ No tumor growth
  – Dispersion of the tumor cells in the ring by embryo movements.
  – Some tumor cell lines may be not invasive enough to penetrate the chorionic epithelium; scratch CAM surface with a scalpel prior to application of cells.
  – Amount of tumor cells deposited was not sufficient; use several million cells per ring.
▸ Thermanox ring disappeared
  – During the growth of the CAM, the ring has been displaced to the periphery. Discard egg; always place Thermanox rings in the centre of the CAM.

References

Armstrong PB, Quigley JP, Sidebottom E (1982) Transepithelial invasion and intramesenchymal infiltration of the chick embryo chorioallantois by tumor cell lines. Cancer Res 42:1826–1837

Ausprunk DH, Knighton DR, Folkman J (1974) Differentiation of vascular endothelium in the chick chorioallantois: a structural and autoradiographic study. Dev Biol 38:237–248

Brooks PC, Clark RA, Cheresh DA (1994) Requirement of vascular integrin alpha v beta 3 for angiogenesis. Science 264:569–571

DeFouw DO, Rizzo VJ, Steinfeld R, Feinberg RN (1989) Mapping of the microcirculation in the chick chorioallantoic membrane during normal angiogenesis. Microvasc Res 38:136–147

Folkman J (1974) Proceedings: Tumor angiogenesis factor. Cancer Res 34:2109–2113

Hagedorn M, Zilberberg L, Wilting J, Canron X, Carrabba G, Giussani C, Pluderi M, Bello L, Bikfalvi A (2002) Domain Swapping in a COOH-terminal Fragment of Platelet Factor 4 Generates Potent Angiogenesis Inhibitors. Cancer Res 62:6884–6890

Hamburger V, Hamilton HL (1992) A series of normal stages in the development of the chick embryo. Dev Dyn 195:231–272

Jakob W, Jentzsch KD, Mauersberger B, Heder G (1978) The chick embryo choriallantoic membrane as a bioassay for angiogenesis factors: reactions induced by carrier materials. Exp Pathol (Jena) 15:241–249

Leeson TF, Leeson CR (1963) The chorioallantois of the chick. Light and electron microscopic observations at various times of incubation. J Anat 97:585–595

Murphy JB, Rous P (1912) The behavior of chicken sarcoma implanted in the developing embryo. J Exp Med 15:119–132

Oh SJ, Jeltsch MM, Birkenhager R, McCarthy JE, Weich HA, Christ B, Alitalo K, Wilting J (1997) VEGF and VEGF-C: specific induction of angiogenesis and lymphangiogenesis in the differentiated avian chorioallantoic membrane. Dev Biol 188:96–109

Papoutsi M, Siemeister G, Weindel K, Tomarev SI, Kurz H, Schachtele C, Martiny-Baron G, Christ B, Marme D, Wilting J (2000) Active interaction of human A375 melanoma cells with the lymphatics in vivo. Histochem Cell Biol 114:373–385

Papoutsi M, Tomarev SI, Eichmann A, Prols F, Christ B, Wilting J (2001) Endogenous origin of the lymphatics in the avian chorioallantoic membrane. Dev Dyn 222:238–251

Willer BH (1924) The endocrine glands and the development of the chick I. The effects of thyroid grafts. Am J Anat 33:67–103

Wilting J, Christ B, Bokeloh M (1991) A modified chorioallantoic membrane (CAM) assay for qualitative and quantitative study of growth factors. Studies on the effects of carriers, PBS, angiogenin, and bFGF. Anat Embryol 183:259–271

Woodruff AM, Goodpasture EW (1931) The susceptibility of the chorio-allantoic membrane of chick embryos to infection with the fowl-pox virus. Am J Pathol 7:209–222

Chapter 24

Corneal Angiogenesis Assay

Lucia Morbidelli, Silvia Cantara, and Marina Ziche

▨ Introduction

Continuous monitoring of angiogenesis in vivo is required for the development and evaluation of drugs acting as suppressors or stimulators of angiogenesis. In this respect, there are concerted efforts to provide an animal model for quantitative analysis of in vivo angiogenesis (Jain et al. 1997). The cornea assay consists in the placement of an angiogenesis inducer (tumor tissue, cell suspension, growth factor) into a micropocket molded in the cornea stroma in order to induce vascular outgrowth from the peripherally located limbal vasculature. In contrast to other in vivo assays, this assay has the advantage of measuring only new blood vessels, since the cornea is physiologically avascular.

The corneal assay performed in New Zealand white rabbits was first described by Gimbrone et al. (1974). It was chosen to investigate the absence of a vascular pattern and for the easy manipulation and monitoring of neovascular growth. This technique, extensively used during the years, has been substantially modified to fulfil different experimental requirements:

- Characterization of angiogenesis inducers
- Evaluation of angiogenesis inhibitors
- Interaction between different factors
- Study of the cellular, biochemical, and molecular mechanisms of angiogenesis.

Springer Lab Manual
H. Augustin (Ed.)
Methods in Endothelial Cell Biology
© Springer-Verlag Berlin Heidelberg 2004

Materials

Animals

- New Zealand albino rabbits of approx. 2 kg (Charles River Laboratories, www.criver.com)

Cells

- Cell suspension (e.g., tumor cells) in medium with 10 % FCS at the dilution of 2–5×10^5 cells/5 µl

Tissue samples

- Fresh tissue fragments (2–3 mg each) isolated not longer than 2 h ago from surgery specimen of patients or animals and kept at 4 °C in medium

Reagents

- Recombinant growth factors or drugs in water, phosphate buffered saline (PBS), ethanol, or methanol in highly concentrated solutions (0.1–1 mg/ml)
- Ethylen-vinyl-acetate copolymer (Elvax-40) (Dupont, www.dupont.com). Elvax-40 preparation and testing:
  1. Weigh 1 g of Elvax-40, wash it in 100-fold excess of absolute alcohol at 37 °C, and dissolve in 10 ml of methylen-chloride to prepare 10 % casting stock solution. Leave Elvax-40 in methylen-chloride at 37 °C for 30–60 min to dissolve it faster.
  2. Test the solution prepared for its biocompatibility (Langer et al. 1976). The casting solution is eligible for use if none of the implants performed with this preparation induces the slightest reaction in the rabbit cornea tissue.
- *Sodium pentothal in saline solution (0.1 g/ml)*
- Benoxinate 0.4 %
- Fixative: 4 % paraformaldehyde in PBS, pH 7.4

Equipment

- Sterile surgical room
- Disposable scalpel for ocular microsurgery (no. 10/11, Aesculap, www.aesculap.com)
- Sterile forceps, silver spatula, microsurgery scissors, teflon plate (10 × 10 cm), micro spatula
- 6 cm glass Petri dishes
- Latex dental dam for endodontic procedures (Dentaltrey, www.dentaltrey.com)

- Insulin syringes
- Slit lamp stereomicroscope equipped with a digital camera

Procedures

Sample preparation

The test material can have the form of slow-release pellets incorporating recombinant growth factors, cell suspensions, or tissue samples.

- Preparation of slow release pellet: Recombinant growth factors are prepared as slow-release pellets by incorporating the test substance in Elvax-40. For testing, a pre-determined volume of Elvax-40 casting solution is mixed with a given amount of the compound to be tested dried on a flat teflon surface and the polymer is allowed to dry under a laminar flow hood. After drying, the film sequestering the compound is cut into $1 \times 1 \times 0.5$ mm pieces. Empty pellets of Elvax-40 are used as negative controls. Vascular endothelial growth factor (VEGF)-containing pellets (200 ng) are considered as positive controls.
- Preparation of cell suspension: Prepare a cell suspension by trypsinization of confluent cell monolayer or concentrated cell suspension to a final dilution of $2–5 \times 10^5$ cells in 5 µl.
- Preparation of tissue sample: When tissue samples are to be tested, samples of 2–3 mg are obtained by cutting fresh tissue fragments under sterile conditions.

Surgical procedure

1. Anaesthetize animals with sodium pentothal (30 mg/kg, i.v.).

2. Segregate the eye with a dental dam. Wash with few drops of local anaesthetic (benoxinate 0.4 %).

3. Produce a micropocket (1.5×3 mm) under aseptic conditions using a pliable silver spatula 1.5 mm wide in the lower half of the cornea (Fig. 1). A small amount of the aqueous humor can be drained from the anterior chamber with an insulin syringe if reduced corneal tension is required.

4. Locate the implant at 2.5–3 mm from the limbus to avoid false positives due to mechanical procedure and to allow the diffusion of test substances in the tissue, with the formation of a gradient for the endothelial cells of the limbal vessels (Fig. 2). Implants sequestering the test material and the controls are coded and implanted in a double-blinded manner.

5. Introduce 5 µl containing $2–5 \times 10^5$ cells in medium supplemented with 10 % serum into the corneal micropocket of each eye by using a

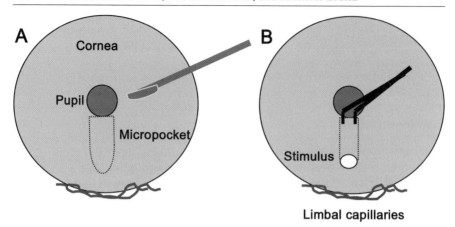

Fig. 1A,B. Schematic representation of the corneal micropocket assay. **A** A micropocket is surgically molded in the corneal stroma of anaesthetized animals by a surgical scalpel and a pliable spatula. **B** The test substance is inserted into the micropocket

micropipette. When the overexpression of growth factors/inhibitors by stable transfection of specific cDNA is studied, one eye is implanted with transfected cells and the other with the wild type or vector-transduced cells. Suitable cell lines for these experiments are Mammary Carcinoma cells (MCF-7), Chinese Hamster Ovary cells (CHO), Lymphoma Burkitt's cells (DG75) (Ziche et al. 1997, Marconcini et al. 1999, Cervenak et al. 2000). It may be necessary to evaluate the angiogenic potential of drug-treated cells. In these experiments, cell monolayers are treated before implantation (18–24 h). One eye is implanted with treated cells and the control eye is treated with control cells.

6. Tissue fragments are inserted in the corneal micropocket with small forceps. The angiogenic activity of tumor samples may be compared with macroscopically healthy tissue (Gallo et al. 1998).

7. If drug solutions incompatible with Elvax polymerization, or genes transduced by viral vectors have to be locally tested, microinjection of concentrated solutions is performed by the use of insulin syringes equipped with 30G needles. A volume of up to 10 μl is injected within the corneal stroma in the space between the limbus and the pellet implant after the removal of aqueous humor.

Quantification of corneal angiogenesis

1. Subsequent daily observation of the implants is made with a slit lamp stereomicroscope without anaesthesia. Angiogenesis, edema, and cellular infiltrations are recorded and images are taken.

Fig. 2. Newly formed vessels grow from the limbal vasculature and progress towards the implanted stimulus (representative positive implant at seven days).

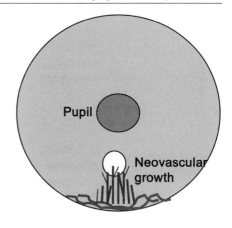

2. An angiogenic response is scored positive when budding of vessels from the limbal plexus occurs after 3–4 days and capillaries progress to reach the implanted pellet in 7–10 days (Fig 2). Implants that fail to produce neovascular growth within 10 days are considered negative, whereas implants showing an inflammatory reaction are discarded (Ziche et al. 1989).

3. The number of positive implants over the total number of implants is scored during each observation.

4. The potency of angiogenic activity is evaluated on the basis of the number and growth rate of newly formed capillaries, and an angiogenic score is calculated by the formula [vessel density × distance from limbus] (Ziche et al. 1997, Ziche et al. 1994). A density value of 1 corresponds to 0 to 25 vessels per cornea, 2 from 25 to 50, 3 from 50 to 75, 4 from 75 to 100 and 5 for more than 100 vessels. The distance from the limbus (in mm) is graded with the aid of an ocular grid.

5. Computerized image analysis: Images are taken from each eye at each observation. Images are digitized and analyzed by an ad hoc software after extracting of the newly formed vessels from the background. The total number of vessels, the area occupied by the vessels, and the branching of the neovascular network are measured and statistically analyzed (Fig. 3).

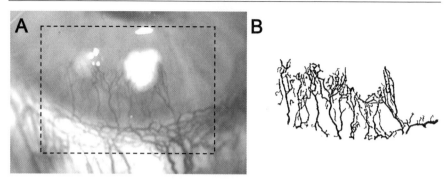

Fig. 3. A Computerized image analysis of an angiogenic response induced by the implantation of two adjacent pellets releasing two different angiogenic stimuli. **B** Following digitized image analysis, the neovessels are extracted from the background. The image, converted in black and white, is analyzed for the number of vessels, the area occupied by vessels, and the degree of branching

Histological examination

1. Sacrifice the animal with a bolus of the anaesthetic drug.

2. Remove the cornea and fix it in paraformaldehyde.

3. If required, freeze the cornea in cooled isopentane for 10 s and store at −80 °C or in liquid nitrogen.

4. Newly formed vessels and the presence of inflammatory cells are detected by hematoxylin/eosin staining or specific immunohistochemical procedure [e. g., detection of macrophages (RAM11) or endothelium (anti-CD31 antibody] (Ziche et al. 1997). Double–staining (i.e., anti-CD-31 for vascular endothelium and specific markers for tumor cells) is useful to label newly formed vessels of the host and proliferating tumor cells implanted into the cornea.

Critical aspects of the rabbit cornea micropocket assay are summarized in Table 1.

Mouse corneal micropocket

The mouse cornea micropocket assay was firstly described by Muthukkaruppan and Auerbach (1979).

1. Anaesthetize animals with methoflurane.

2. Make a corneal micropocket in both eyes reaching within 1 mm of the limbus. Implant pellets containing substances to be tested coated with Hydron (Interferon Science, New Brunswick, NJ).

Table 1. Summary of critical aspects of the rabbit cornea assay

Operator	– Some practice for pellet manipulation, surgery, and monitoring of angiogenesis is required.
Animals	– Body weight: Animals should be in the range of 1.8–2.5 kg for convenient handling and prompt recovery from anaesthesia. – Immobilisation during anaesthetic procedure and observation is important to avoid self-inflicted injury.
Sterility	– Sterility of materials and procedures is crucial to avoid non-specific responses. – Elvax-40 beads should be carefully washed in absolute alcohol as indicated to avoid inflammatory reactions.
Polymers	– Polyvinylalcohol and Hydron can be used instead of Elvax-40. In our experience, a polymer of hydroxyethyl-metacrylate, gives less satisfactory results than Elvax-40.
Surgical procedure	– Make the cut in the cornea in accordance with the pupil and orient the micropocket toward the lower eyelid. – If two factors are tested, make two independent micropockets. – Drain a small amount of the aqueous humor when implanting cells or tissue fragments to reduce corneal tension.

3. Use Hydron as a casting solution [12 % (w/v)], prepared by dissolving the polymer in absolute alcohol at 37 °C (Langer et al. 1976). When peptides are tested, sucralfate (sucrose aluminium sulphate, Bukh Meditec, Copenhagen, Denmark) is added to stabilize the molecule and to slow its release from Hydron (Chen et al. 1995, Voest et al. 1995).

4. The vascular response measured as the maximal vessel length and number of clock hours of neovascularization is scored at fixed time point (usually on postoperative day 5 and 7) using a slit-lamp biomicroscope and photographed. To quantify the section of the cornea in which new vessels are sprouting from the pre-existing limbal vessels, the circumference of the cornea is devided into the equivalent of 12 clock hours. The measurement of the number of clock hours of neovascularization for every eye is performed during each observation.

Rat corneal assay

1. Purified growth factors are combined 1:1 with Hydron as described by Polverini and Leibovich (1984).

2. Pellets are implanted 1–1.5 mm from the limbus of the cornea of anaesthetized rats (sodium pentobarbital, 30 mg/kg, i.p.).

3. Neovascularization is assessed at fixed time points (usually day 3, 5 and 7): animals are perfused with colloidal carbon solution to label vessels, eyes are enucleated and fixed in 10% neutral buffered formalin overnight. The following day, corneas are excised, flattened, and photographed. A positive neovascularization response is recorded only if sustained directional ingrowth of capillary sprouts and hairpin loops toward the implant is observed. Negative responses are recorded if either no growth is observed or if only an occasional sprout or hairpin loop showing no evidence of sustained growth is detected.

Advantages and Disadvantages in Different Species

Species

The rabbit's size (2–3 kg) enables to an easy manipulation of both the whole animal and the eye to be easily removed from its location and to be surgically manipulated.

Rabbit cornea has been found avascular in all strains examined so far. In some strains of rats the presence of preexisting vessels within the cornea and the development of keratitis are serious disadvantages. Furthermore, rabbits are more docile and amenable to handling and experimentation than mice and rats. Inflammatory reactions are easily detectable in rabbits by stereomicroscopic examination as corneal opacity.

Measurements

It is possible to obtain time-point results in mice and rats. The evolution of the angiogenic response in the same animal is not recommended because each time the cornea is observed the animal has to be anaestesized since it is not easy to keep it quiet. Experiments are made with a large number of animals and vessel growth during time can be visualized by perfusion with colloidal carbon solution in individual animals. In contrast, multiple observations are possible in rabbits. The use of slit lamp stereomicroscope and of animals which are not anaestesized allows the observation of newly formed vessels during time with long time monitoring, even for 1–2 months.

Different experimental procedures

Thanks to its size, the rabbit eye offers a large area for the placement of stimuli in different forms. In particular the activity of specific growth factors can be studied with slow-release pellets (Cervenak et al. 2000, Ziche et al. 1982, Taraboletti et al. 2000, Parenti et al. 2001) or tumor or non-tumor cell lines stably transfected for the over-expression of angiogenic factors (Ziche et al. 1997, Cervenak et al. 2000). The modulation of the angiogenic responses by different stimuli can be assessed in the rabbit cornea assay through the implantation and/or removal of multiple pellets placed in parallel micropockets produced in the same cornea. The implantation of tumor samples from different locations can be performed both in corneal micropockets and in the anterior chamber of the eye to monitor angiogenesis produced by hormone-dependent tissues or tumors (i.e., human breast or ovary carcinoma in female rabbits), and it allows the detection of both the iris and the corneal neovascular growth (Ziche et al. 1997, Ziche et al. 1982).

Treatment with drugs

The effect of local drug treatment on corneal neovascularization can be studied with ocular drops or ointment (Presta et al. 1999) or microinjection into the corneal stroma. The effect of systemic drug treatment on corneal angiogenesis can be also evaluated (Ziche et al. 1997, Gallo et al. 1998, Ziche et al. 1994, Ziche et al. 1998). However, systemic drug treatment of rabbits requires a higher amount of drugs than the treatment of smaller animals.

Interestingly, the use of nude mice allows the study of angiogenesis modulation in response to effectors produced and released by tumors or tumor cell lines of human origin growing subcutaneously. Treatment of mice with antiangiogenic or antitumor drugs allows the simultaneous measurement of tumor growth and metastasis and corneal angiogenesis.

References

Cervenak L, Morbidelli L, Donati D, Donnini S, Kambayashi T, Wilson J, Axelson H, Castanos-Velez E, Ljunggren HG, De Waal Malefyt R, Granger HJ, Ziche M, Bejarano MT (2000) Abolished angiogenicity and tumorigenicity of Burkitt lymphoma by Interleukin-10. Blood 96:2568–2573

Chen C, Parangi S, Tolentino MT, Folkman J (1995) A strategy to discover circulating angiogenesis inhibitors generated by human tumors. Cancer Res 55:4230–4233

Gallo O, Masini E, Morbidelli L, Franchi A, Fini-Storchi I, Vergari WA, Ziche M (1998) Role of nitric oxide in angiogenesis and tumor progression in head and neck cancer. J Natl Cancer Inst 90:587–596

Gimbrone M Jr, Cotran R, Leapman SB, Folkman J (1974) Tumor growth and neovascularization: An experimental model using the rabbit cornea. J Natl Cancer Inst 52:413–427

Jain RK, Schlenger K, Hockel M, Yuan F (1997) Quantitative angiogenesis assays: Progress and problems. Nat Med 3:1203–1208

Langer R, Folkman J (1976) Polymers for the sustained release of proteins and other macromolecules. Nature 363:797–800

Marconcini L, Marchio S, Morbidelli L, Cartocci E, Albini A, Ziche M, Bussolino F, Oliviero S (1999) c-fos-induced growth factor/vascular endothelial growth factor D induces angiogenesis in vivo and in vitro. Proc Natl Acad Sci USA 96:9671–9676

Muthukkaruppan V, Auerbach R (1979) Angiogenesis in the mouse cornea. Science 206:1416–1418

Parenti A, Morbidelli L, Ledda F, Granger HJ, Ziche M (2001) The bradykinin/B1 receptor promotes angiogenesis by upregulation of endogenous FGF-2 in endothelium via the nitric oxide synthase pathway. FASEB J 15:1487–1489

Polverini PJ, Leibovich SJ (1984) Induction of neovascularization in vivo and endothelial cell proliferation in vitro by tumor associated macrophages. Lab Invest 51:635–642

Presta M, Rusnati M, Belleri M, Morbidelli L, Ziche M, Ribatti D (1999) Purine analog 6-methylmercaptopurine ribose inhibits early and late phases of the angiogenesis process. Cancer Res 59:2417–2424

Taraboletti G, Morbidelli L, Donnini S, Parenti A, Granger HJ, Giavazzi R, Ziche M (2000) The heparin binding 25 kDa fragment of thrombospondin-1 promotes angiogenesis and modulates gelatinases and TIMP-2 in endothelial cells. FASEB J 14:1674–1676

Voest EE, Kenyon BM, O'Really MS, Truitt G, D'Amato RJ, Folkman J (1995) Inhibition of angiogenesis in vivo by interleukin 12. J Natl Cancer Inst 87:581–586

Ziche M, Donnini S, Morbidelli L, Parenti A, Gasparini G, Ledda F (1998) Linomide blocks angiogenesis by breast carcinoma vascular endothelial growth factor transfectants. Br J Cancer 77:1123–1129

Ziche M, Alessandri G, Gullino PM (1989) Gangliosides promote the angiogenic response. Lab Invest 61:629–634

Ziche M, Jones J, Gullino PM (1982) Role of prostaglandinE1 and copper in angiogenesis. J Natl Cancer Inst 69:475–482

Ziche M, Morbidelli L, Choudhuri R, Zhang HAT, Donnini S, Granger HJ, Bicknell R (1997) Nitric oxide-synthase lies downstream of vascular endothelial growth factor but not basic fibroblast growth factor induced angiogenesis. J Clin Invest 99:2625–2634

Ziche M, Morbidelli L, Masini E, Amerini S, Granger HJ, Maggi CA, Geppetti P, Ledda F (1994) Nitric oxide mediates angiogenesis in vivo and endothelial cell growth and migration in vitro promoted by substance P. J Clin Invest 94:2036–2044

Part IV:
Detection of Endothelial Cells

Chapter 25

Basic Blood Vessel Immunohistochemistry

CYNTHIA OBODOZIE and JENS KROLL

Introduction

The vascular system is the first functional organ system that forms during the development of the mouse embryo. It is essential to provide the growing embryo with nutrition and oxygen to ensure a normal development and proper function of the organs. During later stages of development and in adulthood, blood vessels regulate vascular homeostasis (Carmeliet 2000, Risau 1997).

The development and function of the vascular system is under intense investigation since it became clear that several human diseases are related to a malfunction of the vascular system. Extensive growth of blood vessels is a hallmark during progression and metastasis of tumors. On the other hand, dysregulation of vascular function leads to severe human diseases like myocardial infarction and peripheral artery disease (Conway et al. 2001).

Blood vessels mainly consist of an inner layer of endothelial cells and of surrounding smooth muscle cells or pericytes. In this chapter, we focus on basic methods to detect endothelial cells in culture and in mouse and human tissues. We will provide protocols for the detection of endothelial cells using the endothelial specific markers CD31, CD34, and von Willebrand Factor (vWF), and a method for the quantification of microvessel density (see Table 1).

Materials

Cells and specimens

Immunocytochemistry
- Cultured endothelial cells (e.g., HUVEC) grown in 24-well plates or in chamber slides.

Immunohistochemistry on cryostat sections
- Tissue sections (5 μm) of shock frozen, O.C.T. embedded tissue samples

Springer Lab Manual
H. Augustin (Ed.)
Methods in Endothelial Cell Biology
© Springer-Verlag Berlin Heidelberg 2004

Table 1. Summary of widely used markers and methods to detect endothelial cells in culture dishes and in tissues (frozen and paraffin-embedded sections).

	cultured cells			cryosections			paraffin sections		
	CD31	CD34	vWF	CD31	CD34	vWF	CD31	CD34	vWF
mouse				X		X	X	X	X
human	X	X	X	X		X	X	X	X

- Tissue sections (5 μm) of 4 % PFA fixed, O.C.T. embedded frozen tissue samples.

Immunohistochemistry on paraffin sections
- Tissue sections (5 μm) of 4 % PFA fixed, paraffin-embedded tissue samples

Primary antibodies

Immunocytochemistry on cultured human endothelial cells
- Mouse anti-human CD31, clone JC/70A (DAKO, www.dakocytomation.com)
- Mouse anti-human CD34, clone QBEND/10 (Novocastra, www.novocastra.co.uk)
- Rabbit anti-human von Willebrand Factor, polyclonal (DAKO)

Immunohistochemistry on cryostat sections
- Rat-anti-mouse CD31, clone MEC 13.3 (BD Pharmingen, www.bdbiosciences.com)
- Mouse anti-human CD31, clone JC/70A (DAKO)
- Rabbit anti-human von Willebrand Factor, polyclonal (DAKO)

Immunohistochemistry on paraffin sections
- Rat anti-mouse CD31, clone MEC 13.3 (BD Pharmingen)
- Mouse anti-human CD31, clone JC/70A (DAKO)
- Rat anti-mouse CD34, clone MEC 14.7 (HyCult, www.hbt.nl)
- Mouse anti-human CD34, clone QBEND/10 (Novocastra)
- Rabbit anti-human von Willebrand Factor, polyclonal (DAKO)

Secondary antibodies

Immunocytochemistry on cultured human endothelial cells
- Goat anti-mouse IgG-ALEXA 488 (Molecular Probes, www.probes.com)
- Goat anti-mouse IgG-Cy3 (Dianova, www.dianova.de)
- Goat anti-rabbit IgG-Cy3 (Dianova)

Immunohistochemistry on cryostat sections (light microscopy)
- Goat anti-rat Ig-Biotin (BD Pharmingen)
- Goat anti-mouse Ig-Biotin (Zymed, www.zymed.com)
- Goat anti-rabbit Ig-Biotin (Zymed)

Immunohistochemistry on cryostat sections (fluorescence microscopy)
- Goat anti-rat IgG-ALEXA 488 (Molecular Probes)

Immunohistochemistry on paraffin sections (light microscopy)
- Rabbit anti-rat Ig-Biotin (DAKO)
- Goat anti-mouse Ig-Biotin (Zymed)
- Goat anti-rabbit Ig-Biotin (Zymed)

Immunohistochemistry on paraffin sections (fluorescence microscopy)
- Goat anti-rat IgG-ALEXA 488 (Molecular Probes)
- Goat anti-mouse IgG-ALEXA 488 (Molecular Probes)
- Goat anti-rabbit IgG-Cy 3 (Dianova)

Blocking solutions

- Non-immune goat serum, 10 % (Zymed)
- Non-immune rabbit serum, 10 % (Zymed)
- Biotin blocking system (DAKO)
- 3 % BSA/PBS (Sigma, www.sigmaaldrich.com)

Chromogenic enzyme-conjugates

- Strepavidin-peroxidase conjugate (Zymed)

Chromogens

- Hoechst No. 33258 (Sigma) in PBS (1:5,000)
- DAB brown (Sigma)
- Hemalum solution (e.g., MAYER, Merck, www.vwr.com), 1:10 in distilled water

Solutions and buffers

- Methanol 100 % (Merck)
- Ethanol 99 %, 80 %, 70 %
- Xylene (Roth, www.carl-roth.de)
- Distilled water
- Tap water
- PBS
- PBS containing 0.05 % Tween-20 (Sigma)

- KAISER'S glycerin gelatine (Merck)
- 10 % PFA: dissolve 50 g PFA in 400 ml distilled water by heating in microwave – (40–50 °C, 2 min), dissolve 5 tabs of NaOH (Merck) and adjust the pH to 7.4 with HCl, add final volume to 500 ml with distilled water
- PFA-Fix containing 0.02 mM $CaCl_2$, 1.36 % sucrose, 4 % PFA in PBS
- 3 % H_2O_2 (Roth, www.carl-roth.de) in PBS
- Xylene-soluble mounting medium (e.g., pertex, medite, www.medite.de)
- 18 % sucrose/PBS
- 10 % Triton X-100 (Sigma)
- 4 % PFA/PBS
- Vectashield (Vector Laboratories, www.vectorlabs.com) or Fluorescent mounting medium (DAKO)
- Trypsin (Sigma), 0.1 % in PBS
- Distilled water containing 0.1 % Trypsin (Sigma) and 0.1 % $CaCl_2$
- Proteinase K (Sigma), 0.1 % in PBS

Equipment and supplies

- Light microscope (e.g., Zeiss, www.zeiss.de, type Axiophot)
- Inverted reflected fluorescent microscope (Olympus, type IX-70, www.olympus.com) with a 100 W mercury-lamp equipped with an ultra-violett (360–370 nm ex., 420 nm em.), blue (460–490 nm ex., 510–550 nm em.) and green (510–550 nm ex., 590 nm em.) filter (Olympus)
- Coplin jars, slide racks, coverslips
- Immunostaining pen (DAKO)
- Humid chamber, oven, freezer (–20 °C)
- Software analySIS® (Soft Imaging System, www.soft-imaging.de)

Procedure

Immunocytochemistry on cultured human endothelial cells

Fluorescent microscopic detection of CD31, CD34, and vWF (cultured in 24-well plates or in chamber slides)

1. Wash cells with ice-cold PBS, 5 min on ice.

2. Fix cells in methanol, 10 min at –20 °C.

3. Wash cells with PBS-Tween (0.05 %), 3 × 5 min at RT.

4. Block unspecific antibody binding sites with 10 % non-immune goat serum, 10 min at RT.

5. Remove blocking solution, do not wash.

6. Incubate cells with primary antibody, 30 min at RT
   - Mouse anti-human CD31, 1:50 in PBS
   - Mouse anti-human CD34, 1:25 in PBS
   - Rabbit anti-human vWF, 1:500 in PBS

7. Wash cells with PBS-Tween (0.05 %), 3 × 5 min at RT.

Note: From now on, protect from light.

8. Incubate cells with secondary antibody, 1 h at RT.
   - Goat anti-mouse Cy3, 1:40 in PBS (for human CD31)
   - Goat anti-mouse IgG-ALEXA 488, 1:400 in PBS (for human CD34)
   - Goat anti-rabbit IgG Cy3, 1:400 in PBS (for vWF)

9. Wash cells with PBS-Tween (0.05 %), 3 × 5 min at RT.

10. Counterstain cell nuclei with Hoechst No. 33258, 2 min at RT.

11. Wash cells with PBS-Tween (0.05 %), 3 × 5 min at RT.

12. Cover cells with KAISER's glycerin gelatine.

13. Examine immunostained endothelial cells under a fluorescent microscope.

Immunohistochemistry on frozen sections

Detection of CD31 and vWF in mouse and human tissues (light microscopy)

1. Dry frozen slides, 30 min at RT.

2. Fix slides with the following fixation solutions:
   - For mouse/human CD31: PFA-Fix, 30 min at RT
   - For mouse/human vWF: Methanol, 10 min at −20°C

3. Wash slides with PBS, 3 × 3 min at RT in a coplin jar.

4. Quench endogenous peroxidase activity with 3 % H_2O_2 in PBS, 5 min at RT in a coplin jar.

5. Wash slides with PBS, 3 × 3 min at RT in a coplin jar.

6. Mark sections using an Immunostaining Pen.

7. Cover slides with avidin solution of biotin blocking system to block endogenous avidin for 10 min at RT in a humid chamber.

8. Remove solution, do not wash slides.

9. Cover slides with biotin solution of biotin blocking system to block endogenous biotin for 10 min at RT in a humid chamber.

10. Wash slides with PBS, 3 × 3 min at RT in a coplin jar.

11. Cover slides with 10 % non-immune goat serum to block unspecific antibody binding sites for 10 min at RT in a humid chamber.

12. Remove blocking solution, do not wash.

13. Incubate slides with primary antibody, 1 h at RT in a humid chamber.
    – Rat anti-mouse CD31, 1:50 in PBS
    – Mouse anti-human CD31, 1:20 in PBS
    – Rabbit anti-human vWF, 1:500 in PBS (for mouse and human vWF)

14. Wash slides with PBS, 3 × 3 min at RT in a coplin jar.

15. Incubate slides with secondary antibody, 1 h at RT in a humid chamber.
    – Goat anti-rat Ig-Biotin, 1:50 in PBS (for mouse CD31)
    – Goat anti-mouse Ig-Biotin, ready to use (for human CD31)
    – Goat anti-rabbit Ig-Biotin, ready to use (for mouse and human vWF)

16. Wash slides with PBS, 3 × 3 min at RT in a coplin jar.

17. Incubate slides with streptavidin-peroxidase, 30 min at RT in a humid chamber.

18. Wash slides with PBS, 3 × 3 min at RT in a coplin jar.

19. Add DAB brown solution at RT for about 30 sec until a specific brown color has developed (check under microscope).

20. Stop color reaction by washing slides in distilled water, 1 min at RT in a coplin jar.

21. Wash slides with PBS, 3 × 3 min at RT in a coplin jar.

22. Counterstain in hematoxylin solution, 1 min at RT in a coplin jar.

23. Wash slides in running tap water, 5 min at RT in a coplin jar.

24. Dehydrate slides in ethanol 70 %, 80 %, 99 %, 5 min each step at RT in a coplin jar.

25. Immerse slides in xylene, 2 × 10 min at RT in a coplin jar.

26. Mount coverslip using xylene-soluble mounting medium.

27. Examine under light microscope.

Detection of CD31 in mouse tissues (frozen sections, fluorescence microscopy)

1. Dry sections, approximately 30 min at RT.

2. Mark sections using an Immunostaining Pen.

3. Soak sections in PBS, 5 min at RT in a coplin jar.

4. Fix tissues in 4 % PFA/PBS, 10 min at RT.

5. Wash sections with PBS, 3 × 5 min.

6. Block sections with 3 % BSA/PBS for 30 min at RT in a humid chamber.

7. Remove any blocking solution outside from the marked sections.

8. Add rat anti-mouse CD31 antibody in 3 % BSA/PBS (1 µg/ml) for 1 h at RT in a humid chamber.

9. Wash sections with PBS, 3 × 5 min.

10. Incubate slides with goat anti-rat ALEXA 488 conjugated antibody in 3 % BSA/PBS (1 µg/ml), 1 h at RT in a humid chamber.

11. Wash sections with PBS, 3 × 15 min.

12. Mount slides using Vectashield mounting medium.

13. Examine under fluorescence microscope.

Immunohistochemistry on paraffin-embedded sections

Detection of CD31, CD34, and vWF in mouse and human tissues (light microscopy)

1. Immerse slides in xylene, 3 × 10 min at RT in a coplin jar.

2. Rehydrate slides in ethanol 99 %, 80 %, 70 %, 5 min each concentration at RT in a coplin jar.

3. Immerse slides in distilled water, 1 min at RT in a coplin jar.

4. Retrieve antigenic determinants.
   - 0.1 % Proteinase K, 10 min at 37°C (for mouse CD31)
   - No Proteinsase K treatment is necessary for human CD31 and mouse CD34
   - 0.1 % Trypsin for 30 min at 37°C (for human CD34)
   - 0.1 % Trypsin/0.1 % $CaCl_2$ for 1 h at 37°C (for mouse and human vWF)

5. Wash slides with PBS, 3 × 3 min at RT in a coplin jar.

6. Quench endogenous peroxidase activity with 3 % H_2O_2 in PBS, 5 min at RT in a coplin jar.

7. Wash slides with PBS, 3 × 3 min at RT in a coplin jar.

8. Mark sections using an Immunostaining Pen.

9. Add avidin solution of biotin blocking system to block endogenous avidin, 10 min at RT in a humid chamber.

10. Remove solution, do not wash slides.

11. Cover slides with biotin solution of biotin blocking system to block endogenous biotin for 10 min at RT in a humid chamber.

12. Wash slides with PBS, 3 × 3 min at RT in a coplin jar.

13. Add non-immune serum to suppress non-specific binding of immuno-globulin for 10 min at RT in a humid chamber.

14. Remove blocking solution, do not wash.

15. Incubate slides with primary antibody, at RT in a humid chamber.
    – Rat anti-mouse CD31, 1:20 in PBS for 2 h
    – Mouse anti-human CD31, 1:25 in PBS for 2 h
    – Rat anti-mouse CD34, 1:20 in PBS for 2 h
    – Mouse anti-human CD34, 1:25 in PBs for 2 h
    – Rabbit anti-human vWF, 1:400 in PBS for 1 h

16. Wash slides with PBS, 3 × 3 min at RT in a coplin jar.

17. Incubate slides with secondary antibody, at RT in a humid chamber.
    – Rabbit anti-rat Ig-Biotin. 1:50 in PBS for 1 h
    – Goat anti-mouse Ig-Biotin, ready to use for 1 h
    – Rabbit anti-rat Ig-Biotin, 1:50 in PBS for 1 h
    – Goat anti-mouse Ig-Biotin, ready to use for 1 h
    – Goat anti-rabbit Ig-Biotin, ready to use for 1 min

18. Wash slides with PBS, 3 × 3 min at RT in a coplin jar.

19. Incubate slides with streptavidin-peroxidase, 30 min at RT in a humid chamber.

20. Wash slides with PBS, 3 × 3 min at RT in a coplin jar.

21. Add DAB brown solution at RT for about 30 s until a specific brown color has developed (check under microscope).

22. Stop color reaction by washing slides in distilled water, 1 min at RT in a coplin jar.

23. Wash slides with PBS, 3 × 3 min at RT in a coplin jar.

24. Counterstain in hematoxylin solution, 1 min at RT in a coplin jar.

25. Wash slides in running tap water, 5 min at RT in a coplin jar.

26. Dehydrate slides in ethanol 70 %, 80 %, 99 %, 5 min each step at RT in a coplin jar.

27. Immerse slides in xylene, 2 × 10 min at RT in a coplin jar.

28. Mount coverslip using xylene-soluble mounting medium.

29. Examine under light microscope.

Detection of CD31, CD34, and vWF in mouse and human tissues (paraffin sections, fluorescence microscopy)

1. Immerse slides in xylene, 3 × 10 min at RT in a coplin jar.

2. Rehydrate slides in ethanol 99 %, 80 %, 70 %, 5 min each concentration at RT in a coplin jar.

3. Immerse slides in distilled water, 1 min at RT in a coplin jar.

4. Retrieve antigenic determinants
   - 0.1 % Proteinase K, 10 min at 37°C (for mouse CD31)
   - No Proteinase K treatment is necessary for human CD31 and mouse CD34
   - 0.1 % Trypsin 30 min at 37°C (for human CD34)
   - 0.1 % Trypsin/0.1 % $CaCl_2$, 1 h at 37°C (for mouse and human vWF)

5. Wash slides with PBS, 3 × 3 min at RT in a coplin jar.

6. Mark sections using an Immunostaining Pen.

7. Add 10 % non-immune goat serum to suppress non-specific binding of immunoglobulin for 10 min at RT in a humid chamber.

8. Remove blocking solution, do not wash.

9. Incubate slides with primary antibody, at RT in a humid chamber.
   - Rat anti-mouse CD31, 1:20 in PBS for 2 h
   - Mouse anti-human CD31, 1:25 in PBS for 2 h
   - Rat anti-mouse CD34, 1:20 in PBS for 2 h
   - Mouse anti-human CD34, 1:25 in PBs for 2 h
   - Rabbit anti-human vWF, 1:400 in PBS for 1 h

10. Wash slides with PBS, 3 × 3 min at RT in a coplin jar.

Note: From now on, protect from light.

11. Incubate slides with secondary antibody, at RT in a humid chamber.
   - Goat anti-rat IgG ALEXA 488, 1:200 in PBS for 1 h (for mouse CD31)
   - Goat anti-mouse IgG ALEXA 488, 1:200 in PBS for 1 h (for human CD31)
   - Goat anti-rat IgG ALEXA 488, 1:200 in PBS for 1 h (for mouse CD34)
   - Goat anti-mouse IgG ALEXA 488, 1:200 in PBS for 1 h (for human CD34)
   - Goat anti-rabbit IgG Cy3, 1:500 in PBS for 1 h (for mouse and human vWF)

12. Wash slides with PBS, 3 × 3 min at RT in a coplin jar.

13. Counterstain cell nuclei with Hoechst No. 33258, 2 min at RT.

14. Wash cells with PBS, 3 × 5 min at RT.

15. Cover cells with fluorescent mounting medium.

16. Examine immunostained endothelial cells under fluorescent microscope.

Expected Results

Immunocytochemistry of cultured human endothelial cells

CD31, CD34, or vWF labeled human endothelial cells show a distinct staining at the cell surface. An example is demonstrated in Figure 1A that shows HUVEC stained with mouse anti-human CD31.

Immunohistochemistry of frozen tissue sections and paraffin-embedded tissue sections

Immunohistochemistry is established on frozen tissue sections with pan-endothelial markers CD31 and vWF on mouse and human tissues. Exemplified staining of mouse heart tissue with a CD31 antibody on cryosections shows small capillaries (green dots) and endothelial cells of larger vessels (Fig. 1B).

PFA-fixed, paraffin-embedded tissue sections allow immunolabeling with pan-endothelial markers CD31, CD34, and vWF on mouse and human tissues. Visualization of vessels is performed using DAB chromogen or fluorescent dye, respectively (Fig. 1C, D).

Determination of vessel size and shape

Determination of vessel diameter yields information about the distribution of large (e.g., > 200 μm), small (e.g., 25–50 μm) and capillary-like vessels (e.g., 0–25 μm) in a tissue. This can easily be performed with image analysis software (e.g., analySIS®).

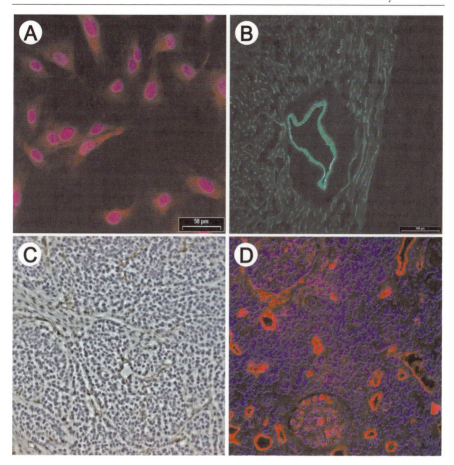

Fig. 1. A Immunolabeling of human umbilical vein endothelial cells (HUVEC) using mouse anti-human CD31 antibody. Nuclei are stained with Hoechst No. 33258 (scale bar 50 μm). **B** Staining of endothelial cells in the adult mouse heart using rat anti-mouse CD31 antibody (scale bar 100 μm). **C** Staining of mouse vessels using anti-human/mouse von Willebrand Factor antibody in human breast R30C tumors grown in nude mice. Nuclei are stained with hematoxylin. **D** Staining of human vessels using anti-human CD34 antibody labeled with Cy3 in renal carcinoma. Counterstaining is performed with Hoechst No. 33258

Determination of microvessel density (MVD)

Microvessel density can be a prognostic marker in some tumors and is often used to evaluate the efficiency of antiangiogenic therapies (Hlatky et al. 2002). The classical method to assess MVD in human tumor specimens is counting pan-endothelial marker-positive cells in vascular hotspots (Weidner et al. 1991). Hotspots are identified at low magnification and microvessels are counted in 1–3 hotspots at high magnification (200×). In contrast, in experimental tumor sections blood vessels are counted in at least 5–10 fields of view and the mean number of vessels is calculated. Apart from this, many other methods are used (Vermeulena et al. 2002). The choice of any methods depends on the posed question.

Troubleshooting

The protocols described here are routinely used in our lab and should be easily adaptable to other labs. Generally, the outcome of any immunostaining of cells and tissues requires proper handling of samples, fixation, dilution of antibodies, blocking reagents, incubation times, and temperatures. Specific problems (e.g., weak/no staining, overstaining, high background) are addressed in Chapter 26 or on the webpage: www.ihcworld.com/trouble shooting.htm.

References

Carmeliet P (2000) Mechanisms of angiogenesis and arteriogenesis. Nat Med 6:389–395

Conway EM, Collen D, Carmeliet P (2001) Molecular mechanisms of blood vessel growth. Cardiovasc Res 49:507–521

Hlatky L, Hahnfeldt P, Folkman J (2002) Clinical application of antiangiogenic therapy: microvessel density, what it does and doesn't tell us. J Natl Cancer Inst. 94:883–893

Risau W (1997) Mechanisms of angiogenesis. Nature 386:671–674

Vermeulen PB, Gasparini G, Fox SB, Colpaert C, Marson LP, Gion M, Belien JA, de Waal RM, Van Marck E, Magnani E, Weidner N, Harris AL, Dirix LY (2002) Second international consensus on the methodology and criteria of evaluation of angiogenesis quantification in solid human tumours. Eur J Cancer 38:1564–1579

Weidner N, Semple JP, Welch WR, Folkman J (1991) Tumor angiogenesis and metastasis – correlation in invasive breast carcinoma. N Engl J Med 324:1–8

Chapter 26

Double-Staining Immunohistochemistry of Blood Vessels

VALENTIN GOEDE

Introduction

Protocols for basic blood-vessel immunohistochemistry for individual pan-endothelial cell-marker molecules are outlined in detail in the previous chapter. Such techniques are widely applied to different areas of ongoing biomedical research areas, particularly within the scope of tumor angiogenesis research (Vermeulen et al. 2002). In brief, the techniques are based on immunohistochemical labeling of the endothelium by antibodies directed against pan-endothelial markers such as von Willebrand factor, CD31, and CD34, followed by light microscopic evaluation. In addition to the qualitative assessment of the microvascular structures of the tissue, which is limited to a description of vessel size and vessel shape, quantitative parameters such as microvessel density counting and capillary size as well as shape determination have been introduced.

Immunohistochemical analyses of microvascular structures that exceed the assessment of vessel number, size, and shape are naturally limited due to the lower level of resolution of the light microscope compared to electron microscopy. However, distinct structural features of the microvascular network of a tissue such as abluminal pericyte coverage of capillaries or basement membrane composition are assessable by double immunolabeling techniques that have been shown to be highly efficient when the research goal is a description of the structural maturity of microvessels (Kakolyris et al. 2001, Wulff et al. 2001, Eberhard et al. 2000, Goede et al. 1998). For example, double immunolabeling with an antibody directed against a pan-endothelial marker and another antibody recognizing a pericyte marker like α-smooth-muscle actin (α-SMA) enables the assessment of the structural maturity of a vascular network. Furthermore, the maturity of the vasculature can be quantified by counting capillaries with and without pericyte coverage and calculating subsequently the ratio of mature and immature microvessels. Table 1 lists markers of structural microvessel components that define vascular maturity and can be assessed by double-immunolabeling procedures.

Springer Lab Manual
H. Augustin (Ed.)
Methods in Endothelial Cell Biology
© Springer-Verlag Berlin Heidelberg 2004

Table 1. Advanced immunohistochemical analysis of microvascular structure

Double immunolabeling targets

1		2		
Structure	Marker	Structure	Marker	Specific assessment of
endothelium	vWF	pericyte	desmin	quiescent vessels
	CD31		α-smooth muscle-actin (α-SMA)	activated angiogenic vessels
	CD34		platelet-derived growth factor (PDGF) receptor β	angiogenic vessels
			aminopeptidase A	angiogenic vessels
			cell surface 3G5 ganglioside antigen	dermal pericytes
		basement membrane	LH39	mature non-angiogenic vessels

In addition to the detection and quantitation of endothelial cells and sur-rounding mural cells, double-immunolabeling techniques can also be used to determine functional parameters of the endothelium. By analogy to the quantitative assessment of vascular maturity, functional activity of tissue endothelium can be quantified, if activated and quiescent endothelial cells are counted and the ratio of both cell populations is calculated. The selection of a suitable antigen that determines activated endothelium depends pri-marily on the functional activity intended to be investigated in endothelial cells (e.g., angiogenesis, regulation of leukocyte trafficking, regulation of coagulation). Table 2 summarizes potential endothelial cell antigens that may mirror an activated endothelial cell phenotype in context with distinct biological functions of the endothelium.

In the following, two laboratory procedures are described in more detail that both apply double-staining immunohistochemistry of blood vessels and can be used to analyze the structural and functional status of the vasculature in tissue sections. While the first procedure provides an example on the assessment of the structural maturity of microvessels (i.e., detection of microvascular pericyte coverage), the second procedure gives an example of assessing the functional activity of endothelium (i.e., detection of endothe-lial cell proliferation). Both laboratory protocols can be easily modified with respect to particular research interests (see Tables 1, 2, and Section "Trou-bleshooting").

Table 2. Advanced immunohistochemical analysis of endothelial function

Double immunolabeling targets

1		2		
Structure	Marker	Structure	Marker	Specific assessment of
endothelium	vWF CD31 CD34	angiogenesis	proliferating cell nuclear antigen (PCNA)	proliferating endothelial cells
			Ki67	proliferating endothelial cells
			αvβ3, αvβ5	migrating endothelial cells
			VEGFR-1, VEGFR-2	angiogenic endothelial cells
			endoglin (CD105)	angiogenic endothelial cells
		leukocyte traffic	inflammatory cell adhesion molecule (ICAM)-1	adhesive endothelial cells
			vascular cell adhesion molecule (VCAM)-1	adhesive endothelial cells
			E-selectin, P-selectin	adhesive endothelial cells
		coagulation	vWF	procoagulative endothelial cells
			plasminogen activator (tPA, uPA)	thrombolytic endothelial cells

Materials

Specimen

- Tissue sections (4 μm) of formalin-fixed, paraffin-embedded tissue samples, or alternatively:
- Tissue sections (4 μm) of cryostat-medium-embedded, shock-frozen tissue samples

Primary antibodies

- Rabbit anti-human von Willebrand factor (vWF)-IgG, polyclonal (final dilution 1:200 in PBS, e.g., DakoCytomation, www.dakocytomation.com)
- Mouse anti-human CD34-IgG, monoclonal (clone QBEnd/10, final dilution 1:25 in 1% normal goat serum, e.g., Novocastra, www.novocastra.co.uk)
- Mouse anti-human α-smooth-muscle actin (α-SMA)-IgG, monoclonal (clone 1A4, final dilution 1:400 in 10% normal goat serum, e.g., Sigma, www.sigmaaldrich.com)

- Mouse anti-human proliferating-cell nuclear antigen (PCNA)-IgG, monoclonal (e.g., Dako, clone PC10, final dilution 1:100 in 10 % normal goat serum)

Secondary antibodies

- Goat anti-rabbit-immunoglobulin IgG, polyclonal, biotinylated
- Goat anti-mouse-immunoglobulin IgG, polyclonal, biotinylated (both available as ready-to-use solutions, e.g., Zymed; www.zymed.com)

Blocking solution

- Normal goat serum 10 % (available as ready-to-use solution (e.g., Zymed)

Chromogenic enzyme-conjugates

- Streptavidin-peroxidase conjugate
- Streptavidin-alkaline-phosphatase conjugate (both available as ready-to-use solutions, e.g., Zymed)

Chromogens

- Nitroblue tetrazolium (NBT)
- Aminoethylcarbazole (AEC) (both available as ready-to-use solutions, e.g., Zymed)

Solutions and buffers

- Distilled water containing 0.1 % Trypsin and 0.1 % $CaCl_2$, pH 7.4
- PBS (phosphate-buffered saline)
- PBS containing 0.05 % Tween-20
- Xylene 100 %
- Isopropanol 100 %
- Ethanol 96 %, 80 %, 70 %
- Methanol 100 % containing 3 % H_2O_2/0.3 % H_2O_2
- Double-staining-enhancer solution (available as ready-to-use solution, e.g., Zymed)
- Glycerine gelatine

Equipment and supplies

- Light microscope (Type Axiophot, e.g., Zeiss, www.zeiss.com)
- Coplin jars, slide rack, humid chamber, coverslips
- Microwave, oven

▨ Procedure

The principles of the-double staining immunohistochemistry techniques are summarized in Fig. 1. Briefly, two different primary antibodies that were generated in different species are used for detection of the antigens, e.g., endothelial cells and pericytes. Then, detection of the primary antibodies is performed by enzyme-conjugated secondary antibodies which are specific for the corresponding primary antibodies. Finally, color reactions using different substrates are used to visualize antigen expression.

Protocol 1: Assessment of microvascular pericyte recruitment

Deparaffinization, blocking of endogenous peroxidase, rehydration, and antigen demasking

1. Immerse slides twice in 100 % xylene for 10 min each.

2. Incubate slides in 100 % methanol containing 3 % H_2O_2: 1 × 20 min at RT.

3. Rehydrate slides in 96 %, 80 %, and 70 % ethanol (5 min each).

4. Rinse slides 3 times in distilled water for 2 min each.

5. Incubate slides in 0.1 % Trypsin/0.1 % $CaCl_2$ for 60 min at 37 °C.

6. Rinse slides 3 times in distilled water for 2 min each.

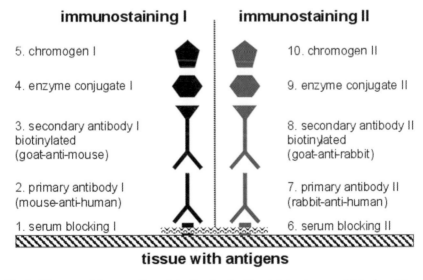

Fig. 1. Diagram of tissue antigen detection by double-staining immunohistochemistry

7. Rinse slides 3 times in PBS/Tween-20 0.05 % for 2 min each.

8. For cryostat sections, air-dry for 60 min, fix in acetone 10 min at −20 °C, rinse slides in distilled water and wash in PBS (3 × 2 min), block endogenous peroxidase in PBS containing 0.3 % H_2O_2, rinse again in PBS/Tween-20 0.05 % 3 × 2 min.

Blocking of unspecific antibody binding

1. Cover slides with 10 % goat serum, incubate in a humid chamber for 30 min at RT.

2. Remove blocking solution, do not rinse.

Immunostaining (labeling of pericytes)

1. Cover slides with mouse anti-human α-SMA antibody, incubate in a humid chamber for 2 h at RT.

2. Rinse slides 3 times in PBS/0.05 % Tween-20 for 2 min each.

3. Cover slides with a biotinylated goat anti-mouse immunoglobulin IgG, incubate in a humid chamber for 10 min at RT.

4. Rinse slides 3 times in PBS/Tween-20 0.05 % for 2 min each.

5. Cover slides with streptavidin-alkaline-phosphatase conjugate, incubate in a humid chamber for 10 min at RT.

6. Rinse slides 3 times in PBS/0.05 % Tween-20 0.05 % for 2 min each.

7. Cover slides with NBT (generates a dark blue color reaction), incubate 8–10 min, stop color reaction by washing slides in distilled water containing 0.05 % Tween-20.

8. Cover slides with double-staining-enhancer solution and incubate in a humid chamber for 30 min at RT.

9. Rinse slides 3 times in PBS/0.05 % Tween-20 0.05 % for 2 min each.

Blocking of unspecific antibody binding II

1. Cover the slides with 10 % goat serum, incubate in a humid chamber for 30 min at RT.

2. Remove blocking solution, do not rinse.

Immunostaining II (labeling of the endothelial cells)

1. Cover slides with rabbit anti-human vWF antibody, incubate in a humid chamber for 60 min at RT.

2. Rinse slides 3 times in PBS/0.05 % Tween-20 for 2 min each.

3. Cover slides with a biotinylated goat anti-rabbit immunoglobulin IgG, incubate in a humid chamber for 10 min at RT.

4. Rinse slides 3 times in PBS/0.05 % Tween-20 for 2 min each.

5. Cover slides with streptavidin-peroxidase conjugate, incubate in a humid chamber for 10 min at RT.

6. Rinse slides 3 times in PBS/0.05 % Tween-20 for 2 min each.

7. Cover slides with AEC (generates a bright red color reaction), incubate 8–10 min, stop color reaction by washing slides in distilled water containing 0.05 % Tween-20.

8. **Note:** Do not counterstain.

9. Embed immunostained tissue sections using glycerine gelatine.

Analysis

1. Evaluate immunostained tissue sections using light microscopy.

2. For quantitation of structural microvascular maturity, count capillary vessels with and without pericyte coverage in at least five representative tissue areas. Calculate the microvessel maturation index (MMI) by dividing the number of pericyte-associated capillaries by the total number of capillaries counted.

Protocol 2: Assessment of endothelial cell proliferation by PCNA/CD34 double-staining

1. Deparaffinize and rehydrate slides following blocking of endogenous peroxidase or prepare cryostat sections as described in Protocol 1.

2. Microwave slides for 1 min (**Note:** Do not microwave when using cryostat sections).

3. Block unspecific antibody binding according to Protocol 1.

Immunostaining (labeling of proliferating cells)

1. Cover slides with mouse anti-human PCNA antibody, incubate in a humid chamber for 60 min at RT.

2. Treat slides with a biotinylated goat anti-mouse immunoglobulin IgG, streptavidin-enzyme conjugate, chromogen, double-staining enhancer, and blocking solution as described in Protocol 1.

Immunostaining (labeling of endothelial cells)

1. Cover slides with rabbit anti-mouse CD34 antibody, incubate in a humid chamber for 2 h at RT.

2. Proceed with a biotinylated goat anti-mouse immunoglobulin IgG, strep-tavidin-enzyme conjugate and chromogen similar to Protocol 1.

3. **Note:** Do not counterstain.

4. Embed immunostained tissue sections using glycerine gelatine.

Analysis

1. Evaluate immunostained tissue under the light microscope.

2. For quantitation of the endothelial proliferative activity, count capillary vessels with and without proliferating endothelial cells in at least five representative tissue areas. Calculate a proliferating capillary index (PCI) by dividing the number of capillaries presenting endothelial cell proliferation by the total number of capillaries.

▨ Expected Results

Both procedures described above have been successfully applied to assess structural maturity and angiogenic activity of the microvasculature in healthy as well as in pathologically altered tissues (Al-Zi'abi et al. 2003, Wulff et al. 2001, Eberhard et al. 2000, Goede et al. 1998, Vartanian and Weidner 1995).

Following double immunolabeling of pericytes and endothelial cells (Protocol 1), analysis of the spatial arrangement of luteal capillaries with and without abluminal pericyte coverage provides qualitative information about the vascular maturation process during luteal development. An example is demonstrated in Fig. 2 showing neovascularization of the growing corpus luteum. Angiogenic endothelial cells invade the growing tissue of the corpus luteum and form new capillaries that have not yet made contact with pericytes. Recruitment of pericytes follows tissue invasion of endothelial cells and results in increasing vascular maturity that can be quantitated by calculating the MMI of the luteal vascular network. Interestingly, the spatial and temporal luteal expression of distinct molecular regulators known to control vascular maturation processes during angiogenesis and angioregression (e.g., Angiopoietin-1/2, VEGF) strictly correlates qualitatively and quantitatively with pericyte recruitment in the course of luteal development (Ramsauer and D'Amore 2002, Goede et al. 1998). This indicates that dual labeling of pericytes and endothelial cells is a valid technique to investigate structural

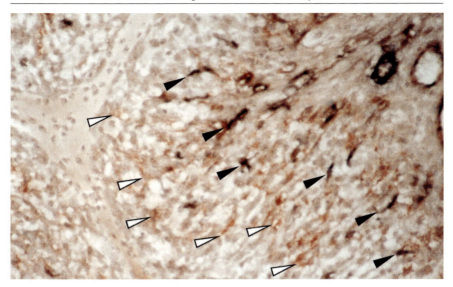

Fig. 2. Double immunolabeling of pericytes (*blue*) and endothelium (*orange-red*) in the growing corpus luteum of the ovary. *White arrows* Immature microvessels without pericyte contact invading the luteal tissue; b*lack arrows* recruitment of a-SMA-positive mural cells to the maturing vasculature. See text for details

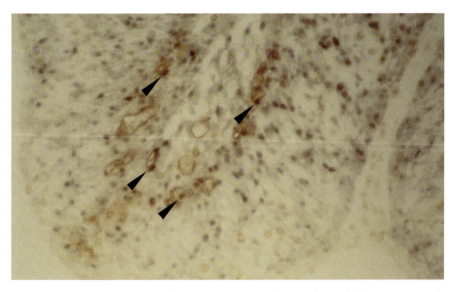

Fig. 3. Double immunolabeling of proliferating cells (*blue*) and endothelium (*orange-red*) in the growing corpus luteum of the ovary. *Black arrows* proliferative endothelial cells invading the luteal tissue. See text for details

maturation of the microvasculature in tissues. Angiogenic activity of the luteal microvasculature is assessable by double immunolabeling of proliferating cells and endothelial cells (Protocol 2). Figure 3 shows as an example the neovascularization of the growing corpus luteum with proliferating endothelial cells that form new capillaries which can be distinguished from surrounding mitotic luteal cells on the basis of the double labeling. Endothelial proliferation quantified by calculating the PCI in different developmental stages of the corpus luteum decreases rapidly during midluteal corpus luteum maturation and further during luteolysis. Endothelial proliferative activity correlates well with the luteal expression of angiogenic growth factors (Goede et al. 1998), demonstrating that dual labeling of mitotic and endothelial cells is a valid technique to examine angiogenic activity in tissues.

Troubleshooting

This section provides additional technical information on the two protocols that may be helpful to establish double-staining immunohistochemistry in order to assess the microvasculature in tissues.

Tissue slides

The double immunolabeling techniques described above can be applied to formalin-fixed, paraffin-embedded tissue sections or alternatively carried out on cryostat sections. However, paraffin-embedded tissue slides are preferred due to frequent loss of tissue integrity occurring during double labeling of cryostat sections. Preserved tissue morphology is a prerequisite to assign staining signals to specific tissue structures like the endothelium or pericytes and to analyze the spatial relationship of staining signals generated by double immunolabeling procedures.

The thickness of sections should not exceed 4 µm. When using thicker sections, staining signals may project on top of each other although actually not colocalizing. This may result in a false positive assessment with subsequent overestimation of the quantified MMI and PCI.

Antigen demasking is crucial for optimal antibody binding to specific tissue structures. Chemical (e.g., tissue digestion by proteases) and physical (e.g., microwave) treatment of tissue slides may vary with respect to the antibodies used and should be evaluated systematically prior to applying double-staining immunohistochemistry on larger quantities of samples.

Primary antibodies

The primary antibodies should be tested separately for high-quality staining by a single immunolabeling technique prior to double staining immunohistochemistry. This step is also necessary to determine optimal antibody concentrations.

As also previously mentioned, it is strongly recommended to take individual research goals into consideration when selecting primary antibodies to visualize specific structural or functional properties of the tissue microvasculature. For example, several antibodies recognizing distinct pericyte subpopulations have been identified (Protocol 1). Most pericytes express desmin, whereas pericytes within an activated angiogenic microvasculature may express additional markers such as α-SMA, PDGFRβ, and aminopeptidase A (Abramsson et al. 2002, Morikawa et al. 2002, Schlingemann 1996). Thus, when analyzing microvascular maturity by assessing pericyte recruitment, the functional status of the tissue must be critically considered (e.g., tissue of healthy organs, neovascularized tissue, inflamed tissue) in order to select the optimal pericyte-specific antibody for the immunolabeling reaction.

Detection system

Some frequent problems that may arise when detecting tissue antigens by double-staining immunohistochemistry are summarized in Table 3. Classic double-staining immunohistochemistry sequentially applies two primary antibodies that come from different animal species (e.g., mouse and rabbit). As the secondary antibodies are derived from a third animal species (e.g., goat) and recognize either mouse immunoglobulin or rabbit immunoglobulin, cross-interactions within the detection system do not occur (see Fig. 1). However, as indicated in Protocol 2, it is also possible to utilize two primary antibodies that both derive from a single animal species (e.g., both from mouse). These primary antibodies can be detected by the same secondary antibody (e.g., goat-anti-mouse immunoglobulin) without significant cross-interaction within the detection system if slides are extensively rinsed between the incubation steps.

To facilitate microscopic analysis, the chromogenic substrates should produce an intense color contrast (e.g., NBT: dark blue, AEC: bright red). Since most chromogens are alcohol-soluble, slides should not be dehydrated and embedded in alcohol-soluble mounting medium subsequent to the staining procedure. As the tissue morphology is sufficiently visualized by the faint background staining, a counterstaining is not recommended.

Instead of utilizing chromogenic enzyme conjugates and chromogens, double-staining immunohistochemistry can also be carried out with fluorescent secondary antibodies that may generate well-contrasting fluorescent

Table 3. Troubleshooting in double staining immunohistochemistry

Problem	Presumed cause	Solution
negative staining	– improper tissue fixation – destruction of labile antigens – skipping of incubation steps	– fix and embed new samples – try alternative antigen demasking – repeat staining procedure
weak staining	– poor titer of primary antibodies	– try different primary antibody titers
	– short incubation times	– prolong incubation times
	– old or improper chromogen preparation	– check chromogen preparation, replace if necessary
	– inadequate liquid retention on slides after rinsing	– carefully remove liquid after washing steps
background staining	– inadequate blocking of endogenous peroxidase	– completely block endogenous peroxidase
	– non-specific binding of primary antibody	– prolong incubation time with blocking serum
	– high titer of primary antibodies	– try different primary antibody titers
	– long color reaction times	– shorten color reaction times
	– drying-out of sections	– keep sections wet

signals (e.g., fluorescent green and red). However, compared to basic light microscopy, it is not possible to localize fluorescent signals and simultaneously visualize tissue morphology with fluorescence microscopy. Further, due to fading of the dye's fluorescent signal, tissue slides cannot be stored for longer periods of time.

References

Abramsson A, Berlin O, Papayan H, Paulin D, Shani M, Betsholtz C (2002) Analysis of mural cell recruitment to tumor vessels. Circulation 105:112–117

Al-Zi'abi MO, Watson ED, Fraser HM (2003) Angiogenesis and vascular endothelial growth factor expression in the equine corpus luteum. Reproduction 125:259–270

Eberhard A, Kahlert S, Goede V, Hemmerlein B, Plate KH, Augustin HG (2000) Heterogeneity of angiogenesis and blood vessel maturation in human tumors: implications for antiangiogenic tumor therapies. Cancer Res 60:1388–1393

Goede V, Schmidt T, Kimmina S, Kozian D, Augustin HG (1998) Analysis of blood vessel maturation processes during cyclic ovarian angiogenesis. Lab Invest 78:1385–1394

Kakolyris S, Giatromanolaki A, Koukourakis M, Kaklamanis L, Kouroussis CH, Bozionelou V, Georgoulias V, Gatter KC, Harris AL (2001) Assessment of vascular maturation in lung and breast carcinomas using a novel basement membrane component, LH39. Anticancer Res 21:4311–4316

Morikawa S, Baluk P, Kaidoh T, Haskell A, Jain RK, McDonald DM (2002) Abnormalities in pericytes on blood vessels and endothelial sprouts in tumors. Am J Pathol 160:985–1000

Ramsauer M, D'Amore PA (2002) Getting Tie(2)d up in angiogenesis. J Clin Invest 110: 1615–1617

Schlingemann RO, Oosterwijk E, Wesseling P, Rietveld FJ, Ruiter DJ (1996) Aminopeptidase a is a constituent of activated pericytes in angiogenesis. J Pathol 179:436–442

Vartanian RK, Weidner N (1995) Endothelial cell proliferation in prostatic carcinoma and prostatic hyperplasia: correlation with Gleason's score, microvessel density, and epithelial cell proliferation. Lab Invest 73:844–850

Vermeulen PB, Gasparini G, Fox SB, Colpaert C, Marson LP, Gion M, Belien JA, de Waal RM, Van Marck E, Magnani E, Weidner N, Harris AL, Dirix LY (2002) Second international consensus on the methodology and criteria of evaluation of angiogenesis quantification in solid human tumours. Eur J Cancer 38:1564–1579

Weidner N (2000) Angiogenesis as a predictor of clinical outcome in cancer patients. Hum Pathol 31:403–405

Wulff C, Dickson SE, Duncan WC, Fraser HM (2001) Angiogenesis in the human corpus luteum: simulated early pregnancy by HCG treatment is associated with both angiogenesis and vessel stabilization. Hum Reprod 16:2515–2524

Chapter 27

In Situ Hybridization Analysis of Vascular Endothelium

Regina Heidenreich and Georg Breier

Introduction

In situ hybridization with RNA probes is a valuable technique to determine the expression pattern of endothelial genes in tissue and allows the spatial localization of the corresponding mRNA at the cellular level. Labeled single-stranded complementary RNA probes (cRNA) are used to detect the mRNA of interest on histological sections, i.e., in situ. In comparison to the immunological detection of gene products, in situ hybridization is more time-consuming and has a lower resolution at the cellular level. However, it is the method of choice if suitable antibodies to detect the protein itself are not available. Moreover, the detection of mRNA allows to unambigously determine the site of synthesis of endothelial receptors, of secreted molecules such as angiogenic factors, or of molecules whose extracellular domains can be proteolytically processed, and which, upon release from their producer cells, may travel to distant locations in the tissue (e.g., Breier et al. 1995, Breier et al. 1996, Steffen et al. 1996, Vajkoczy et al. 2002).

RNA probes can be labeled with radioactive isotopes or with non-radioactive digoxygenin or biotin, respectively. Although the non-radioactive detection of mRNA on tissue sections in principle enable a rapid detection, it tends to be less sensitive on histological sections than radioactive techniques. This protocol describes a procedure for in situ hybridization of frozen tissue sections with radioactive cRNA probes. Cryosections are quickly and relatively easily prepared and show acceptable morphology although the histological preservation of tissue is not optimal. Better histological results at the expense of a higher background and longer exposure times may be obtained using sections of paraformaldehyde fixed tissue embedded in paraffin. This technique has been described in detail elsewhere (Hogan et al. 1994).

The result of the in situ hybridization protocol is largely dependent on the use of a suitable hybridization probe. Templates may range in size between less than 100 and more than several thousand base pairs. Probes should be

Springer Lab Manual
H. Augustin (Ed.)
Methods in Endothelial Cell Biology
© Springer-Verlag Berlin Heidelberg 2004

derived from the coding region of a protein. Repetitive sequences (which may be contained in non-coding mRNA regions or in introns) must be excluded. It is also advisable to avoid sequences encoding conserved protein domains since these might hybridize unspecifically. Sequences with extremely high or low GC contents should not be used. Two or more different probes derived from a single gene may be required if alternative splicing occurs.

Materials

Equipment

- Plastic molds of different sizes for tissue embedding
- Dry ice/isopentane bath
- Cryostat
- Silane-coated glass slides for mounting tissue sections. These can be purchased ready-to-use (e.g., Roth; www.carl-roth.de) or prepared from standard glass slides as follows: clean slides by wiping with 70 % ethanol. In a fume hood, immerse slides in glass staining jars in a 2 % solution of 3-aminopropyl-trimethoxysilane in acetone for 5 min. Wash slides 3 × 5 min in acetone. Air-dry and store in a glass container.
- Thermostat heating plate for drying tissue sections
- Plastic staining boxes with racks for slides. These should be cleaned with 70 % ethanol before use.
- Microcentrifuge
- Liquid scintillation counter
- Incubator at 48–60 °C
- Hybridization chamber with racks for orienting slides in a horizontal position.
- Parafilm pieces
- Shaking water bath
- Large light-tight box with support for drying slides in an upright position
- X-Ray films or Phospho-Imager
- Microscope with bright field and dark field illumination and (digital) camera

Reagents

Tissue embedding

- Autoclaved phosphate buffered saline (PBS) solution.
- Tissue Tek O.C.T (Miles Scientific, e.g., Ted Pella, www.tedpella.com)

Sectioning and postfixation

- 4 % Paraformaldehyde (PFA) solution: Dissolve PFA in 1× PBS at 60–80 °C with constant stirring in a fume hood (takes approx. 1 h). PFA fumes are toxic. Cool to RT. Should be made fresh before use.
- PBS solution
- Absolute ethanol. Graded ethanols: 30 %, 60 %, 80 %, 95 %, 100 % EtOH. Use autoclaved RNase-free ddH$_2$O for making all dilutions.
- Silica gel. Wrap in paper wipers.

Preparation of RNA probes

- ddH$_2$O. Autoclave. Treatment with diethylpyrocarbonate (DEPC) is optional.
- Restriction enzymes. Store at –20 °C.
- Phenol/chloroform/isoamylalcohol (PCI; 50:50:1) saturated with TE buffer (10 mM Tris-HCl pH 7.5, 1 mM EDTA). Store at –20 °C. Phenol is very toxic upon inhalation and skin contact, and should be handled with care in a fume hood.
- 3.3 M Na-acetate, pH 5.5. Autoclave.
- Absolute ethanol. 80 % ethanol.
- T3-, T7- or SP6 RNA polymerase (e.g. Stratagene) and 5× transcription buffer. Store at –20 °C. RNA polymerases are used at 10 units/μl.
- 1 M DTT and 0.2 M DTT. Dissolve in water. Filter sterilize and store in aliquots at –20 °C. Avoid repeated freeze-thaw cycles.
- RNase inhibitor (e.g., RNAguard, Pharmacia, www.pfizer.com). Store at –20 °C.
- 10 mM mix of ATP, CTP, and GTP solutions (e.g., Pharmacia) in ddH$_2$O. Store at –20 °C.
- Uridine 5′-[α-^{35}S]thiotriphosphate ([^{35}S]UTPαS, e.g., Amersham Biosciences, www5.amershambiosciences.com), specific activity >1,000Ci/mmol.
- 1 M MgCl$_2$. Autoclave.
- DNase I (10 U/μl). Must be RNase free (e.g., Roche, www.roche.com). Store at –20 °C.
- Yeast tRNA (50 mg/ml in sterile water). Store at –20 °C.
- 6 M Ammonium acetate. Filter sterilize.
- Deionized formamide (e.g., GibcoBRL, www.gibcobrl.com). Store at –20 °C.
- 2 M NaOH
- 2 M Acetic acid
- Liquid scintillant

Section pretreatment

- 20× SSC (3 M NaCl, 0.3 M NaCitrate; pH 7.0) stock solution. Autoclave. Dilute to 2× SSC.
- Pronase (e.g., Roche, www.roche.com), Nuclease-free. Store in aliquots at 40 mg/ml. At 20 °C.
- 0.2 % Glycine in PBS. Autoclave.
- 4 % PFA solution in PBS (see above).
- 2 M Triethanolamine-HCl pH 8.0 stock solution. Autoclave.
- Acetic anhydride in 0.1 M triethanolamine: Dilute triethanolamine-HCl stock solution to 0.1 M in 1× PBS. Add 1/400 volume of acetic anhydride in a fume hood immediately before use.
- Graded ethanols: 30 %, 60 %, 80 %, 95 %, 100 % EtOH (see above).

Hybridization

Hybridization buffer is made from stock solutions:
- Deionized formamide (e.g., GibcoBRL)   10 ml
- 20× SSC   2 ml
- 1 M Tris-HCl pH 7.5   200 µl
- 0.2 M Na-phosphate pH 6.8   1 ml
- 0.5 M EDTA pH 8.0   200 µl
- 50 mg/ml yeast tRNA   60 µl
- 100 mM UTP solution   20 µl
- Dextran sulfate   2 g
- ddH$_2$O to a total volume of   20 ml

Dissolve dextrane sulfate (e.g., Pharmacia or Sigma) with heating in dd H$_2$O. Hybridization buffer is stored in 1 ml aliquots at –70 °C. Before use, add 1/100 volume each of 100 mM ADPβS, 1 mM ATPγS, 1 M DTT, 1 M 2-mercaptoethanol for stabilization of the RNA probe. These compounds are stored in aliquots at –70 °C.

Post-hybridization washes and RNAse digestion

- Wash buffer: 50 % formamide, 2× SSC, 10 mM 2-mercaptoethanol.
- RNase buffer: 0.5 M NaCl, 10 mM Tris pH 7.5, 1 mM EDTA pH 8.0.
- 1 M DTT: Dissolve in water. Filter sterilize. Store in aliquots at –70 °C.
- RNase A (10 mg/ml) in H$_2$O. Store in aliquots at –20 °C.

Autoradiography

- Kodak NTB-2 emulsion
- Silica gel

Developing and staining of slides

- Kodak D19 developer. Store protected from light.
- 1 % Acetic acid
- 30 % Sodium thiosulfate. Freshly prepared solutions are very cold and must be brought to RT before use.
- 0.1 % Toluidine blue in 0.2 M Na-acetate pH 4.2.
- 0.2 M Na-acetate pH 4.2
- Glass Coverslips
- Mounting fluid (e.g., Entellan).

Procedure

Tissue embedding

1. Dissect out tissue and wash in PBS (see Notes 1,2).

2. Carefully blot off excess liquid and orient tissue in a plastic mold filled with Tissue Tek. Avoid air bubbles. Cover tissue with Tissue Tek.

3. Transfer mold onto isopentane/dry ice and hold steady until frozen. Store at –70 °C.
   Blocks frozen for several months have given satisfactory results.

Sectioning and postfixation

1. Bring tissue blocks from –70 °C to cryostat temperature (approx –20 °C). Wait until tissue has adapted to cryostat temperature (approx. 1 h). Mount onto cryostat chuck using Tissue Tek and cooling in the cryostat. Orient block and trim with a razor blade.

2. Cut sections at approximately –20 °C at 8–10 µm (see Notes 3,4).

3. Collect sections onto silane-coated glass slides at RT (2 sections per slide: one each for antisense and for sense control probe).

4. Dry slides on a hot plate at 50 °C for approximately 5 min.

5. Fix sections in 4 % PFA solution (20 min at RT).

6. Wash slides in PBS (5 min at RT)

7. Dehydrate sections in graded ethanols: 30 %, 60 %, 80 %, 95 %, 100 % EtOH (2 min each)

8. Air dry slides (approx. 1 h) and store dissicated in a box containing silica gel at –70 °C (see Note 5).

Preparation of RNA probes

The plasmid vectors used for cloning template DNA contain promoters for bacteriophage RNA polymerases such as T3-, T7-, or SP6 RNA polymerase on either side of the insert, e.g., pBluescript (Stratagene). T3 and T7 polymerases are preferable over SP6 RNA polymerase which is less active. Plasmids are linearized by restriction enzyme digest (Maniatis et al. 1989) on either side of the insert in order to generate templates for in vitro transcription of antisense RNA probe or of control sense RNA probe, respectively. The latter does not hybridize with the corresponding mRNA. Linearized templates are then incubated with the respective RNA polymerase to generate antisense and sense RNA probes.

Preparation of the linearized DNA template

1. Digest 10 μg of plasmid DNA 3´ to the DNA insert using 20 units of restriction enzyme in a final volume of 50 μl. This template is used for generating antisense RNA probe. In a second reaction, digest plasmid DNA 5´ to the DNA insert. This DNA serves as a template for generating control sense RNA probe.

3. Extract with PCI, 50:50:1

3. Add 0.1 vol. of 3.3 M Na-acetate, pH 5.5, mix and precipitate DNA with 2.5 vol. of ethanol (30–60 min at –20 °C). Spin 20 min in a microcentrifuge at maximum speed and pipette off supernatant. Wash precipitate as follows: add 100 μl 80 % EtOH without disturbing the pellet, spin 5 min and pipette off supernatant.

4. Dissolve DNA in 20 μl of ddH$_2$O (RNase-free), i.e., at 0.5 μg/μl. Check a 1 μl aliquot of the DNA by gel electrophoresis in a 1 % agarose gel. Store DNA at –20 °C.

In vitro transcription

Radioactivity security guidelines for working with [^{35}S] must be followed. Mix at RT in an eppendorf tube:

5× Transcription buffer	4 μl
200 mM DTT	1 μl
RNAse inhibitor (33 U/ml)	1 μl
10 mM GTP, ATP, and CTP	1 μl
Template DNA (0.5 μg/μl)	2 μl
[^{35}S] UTPαS (10 mCi/ml)	10 μl
RNA polymerase (10 U/μl)	1 μl

Incubate at 37 °C for 60–90 min.

Removal of the DNA template

Add 80 µl ddH$_2$O, 1 µl of 1 M MgCl$_2$, 1 µl DNase I (10 U/µl). Incubate at 37 °C for 10 min.

Probe purification

All centrifugations are made at maximum speed.

1. Mix with 100 µl PCI by vortexing. Spin for 5 min in a benchtop centrifuge. Transfer the upper aqueous phase to a fresh eppendorf tube.

2. To the aqueous phase, add 5 µl yeast tRNA (50 mg/ml) and 50 µl of 6 M ammonium acetate. Mix. Add 500 µl ethanol. Mix.
   Incubate at –20 °C for at least 60 min. Spin 20 min in a microcentrifuge. Pipette off supernatant (see Note 6).

3. Dissolve RNA pellet in 100 µl ddH$_2$O.
   Add 50 µl of 6 M ammonium acetate. Mix. Add 500 µl ethanol. Mix.
   Incubate at –20 °C for at least 60 min. Spin 20 min in a microcentrifuge and pipette off supernatant. Take care not to disturb the pellet.

4. Add 100 µl 80 % ethanol. Spin 5 min and pipette off supernatant.
   For efficient hybridzation, probes should be less than 800 nucleotides in length. Longer in vitro transcripts are subjected to alkaline hydrolysis. If hydrolysis of probes is required, omit the 80 % ethanol washing step and proceed as described below ("Hydrolysis of RNA probe").

5. Dissolve RNA pellet in 100 µl ddH$_2$O. Add 2 µl of 1 M DTT and 100 µl deionized formamide.

6. Measure incorporation of radioactive nucleotide by counting a 1 µl aliquot in liquid scintillant. Expected radioactive concentration is 500,000 to 1,000,000 cpm/µl.

7. Store probes at –20 °C. Probes can be used for up to four weeks.

8. Alternative purification method: Probes can also be purified by using the PCR purification kit from Quiagen according to the manufacturers instructions. However, this purification method sometimes leads to a greatly decreased radioactive recovery. If hydrolysis of RNA probes is necessary, continue as described below.

Hydrolysis of RNA probe (for transcripts longer than 800 nucleotides)

1. Chill probe (dissolved in 100 µl ddH$_2$O) on ice (see Note 7).

2. Add 10 µl ice-cold 2 M NaOH. Incubate on ice for the required time period (5–20 min). Incubation time varies depending on the original length of the transcript. As a rule of thumb, incubation time in min is probe length (number of nucleotides) divided by 200, e.g., 10 min for a 2,000-nucleotide transcript.

3. Neutralize by mixing with 10 µl 2 M acetic acid.

4. Precipitate RNA with 750 µl ethanol at –20 °C for at least 60 min. Spin 20 min in a microcentrifuge. Pipette off supernatant. Take care not to disturb the pellet.

5. Add 100 µl of 80 % ethanol. Spin for 5 min. Pipette off supernatant.

6. Dissolve RNA pellet in 100 µl ddH$_2$O. Add 2 µl of 1 M DTT and 100 µl formamide.

7. Measure incorporation of radioactive nucleotide and store probes at –20 °C as described above.

Section pretreatment

Bring slides (still dissicated in the box) to RT (this takes at least 1 h). Do not open the boxes before the slides are at RT. Slides are transferred to racks and the following incubations are performed in plastic staining boxes at RT unless otherwise stated:

ddH$_2$O	(1 min)
2× SSC	(30 min at 70 °C in a water bath)
ddH$_2$O	(1 min)
40 µg/ml Pronase in 1× Pronase buffer	(10 min) (see Note 8)
0.2 % glycine in 1× PBS	(30 s to block Pronase digestion)
1× PBS	(30 s)
4 % PFA in PBS	(20 min)
1× PBS	(3 min)
Acetic anhydride (1:400 in 0.1 M triethanolamine)	(10 min)
1× PBS	(2 min)
30 %, 60 %, 80 %, 95 %, 100 % EtOH	(2 min each)

Air dry for 30 min or longer. Use for hybridization.

Hybridization

1. Prewarm slides in a hybridization chamber in an incubator at 60 °C.

2. Dilute probe to 50,000 cpm/µl in hybridization buffer. Mix well. Calculate 20 µl for each section.

3. Denature probe in hybridization buffer at 95 °C for 2 min. Do not chill on ice.

4. Apply probe. In general, antisense and sense probes are applied to serial sections on the same slide. This allows to compare signal of sense and antisense probes on serial sections. Depending on the size of the section, 10–30 µl of hybridization probe is used.

5. Lower a piece of parafilm precut to the appropriate size and press gently with forceps. Avoid trapping air bubbles (see Note 9).

6. Hybridize in a humid chamber overnight (12–20 h) at 48 °C (see Note 10). A humid atmosphere in the chamber is generated with a paper wiper soaked with 50 % formamide.

Post-hybridization washes and RNAse digestion

Washing and RNase A digestion are important procedures for achieving specific signals. RNase A will remove all single-stranded RNA probe that does not form a hybrid with mRNA. All washing steps are performed in plastic staining jars at 37 °C in a water bath with gentle agitation. A separate set of boxes must be used for these treatments because these will be contaminated with RNase. Prewarm all washing buffers.

1. Wash in wash buffer for 2–3 h until the coverslips float free (see Note 11).

2. Wash in RNase buffer for 15 min.

3. Digest with RNAse (20 µg/ml in RNase buffer) for 15 min.

4. Wash in RNase buffer for 15 min.

5. Wash in wash buffer supplemented with 5 mM DTT overnight.

6. Dehydrate sections in graded ethanols at RT: 30 %, 60 %, 80 %, 95 %, 100 % EtOH (2 min each). Air dry slides for at least 1 h.

Autoradiography

1. Expose slides on X-ray film for 2–3 days at RT.

2. Develop X-ray film. The image should give an impression of whether the experiment was successful and how long the slides should be exposed with the less sensitive photographic emulsion.

3. In a dark room, dilute Kodak NTB-2 emulsion 1:1 with dH_2O and melt in a water bath at 42 °C. The emulsion is applied and evenly distributed with a plastic pipette tip. Avoid air bubbles.

4. Air dry slides in an upright position in a light-tight box for at least 2 h.

5. Expose slides dissicated in a black plastic box with silica gel at 4 °C. Seal box in a plastic bag or aluminium foil. The standard exposure time is 2 weeks.

Developing and staining of slides

1. Bring slides to RT.

2. In a dark room, develop in Kodak D19 developer for 3 min at RT. Stop developing in dH_2O. Fix in 30 % sodium thiosulfate for 3 min. Rinse with dH_2O and several times in tap water.

3. Stain with 0.2 % Toluidine blue (in 0.2 M Na-acetate, pH 4.2) in a staining jar for 5 min at RT. Destain for several min in tap water. Rinse with 1× PBS, tap water and finally with dH_2O. Air dry (at least 1 h).

4. Mount coverslips using mounting medium (e.g., Entellan).

5. View the localization of silver grains under the microscope. Dark field illumination is very helpful for detecting weak signals.

■ Notes

▶ Extreme care must be taken to avoid contamination of all glassware and solutions with RNase. Wear gloves during all steps of the procedure. It is also recommended to have a separate clean bench space that is used only for RNA in situ hybridization and other RNA techniques.

▶ All aqueous solutions should be sterilized by autoclaving or by filtration trough a 0.45 μm filter. Solutions that cannot be sterilized (e.g., hybridization buffer) are made up of sterile solutions. Use sterilized tubes and freshly autoclaved pipette tips throughout the whole procedure.

▶ Optimal sections are a prerequisite for a successful experiment. For best preservation of tissue, it is essential that the tissue is frozen slowly on isopentane/dry ice. Do not use liquid nitrogen for freezing tissue blocks. Also, the tissue should be frozen soon after dissection to avoid proteolysis by endogenous proteases. Stretching or excessive bending of the tissue during dissection can lead to tissue damage and therefore to poor morphology.

▶ For best results, the knife has to be in an excellent condition. The optimal cutting temperature must be determined empirically. For most tissues, temperatures between –14 °C and –20 °C are optimal. The tissue temperature must be raised in one degree steps if the tissue ruptures. Soft tissues are cut at lower temperatures than more solid ones. The optimal temperature of the knife is often several degrees lower than the temperature of the tissue. Lower the temperature if the section adheres to the knife.

▶ Slides can repeatedly be thawed and frozen without loss of quality provided that great care is taken to avoid humidity. Boxes for storing slides must contain silica gel wrapped in wiping paper. Before freezing slides, check that the silica gel is still dissicated, as indicated by the blue color.

▶ It is advisable to monitor the radioactivity contained in supernatants and in pellets with a sensitive radioactivity contamination monitor. This helps to minimize unintended loss of ^{35}S-labeled probe. In general, approximately 60 % of ^{35}S-labeled nucleotide should be contained in the pellet after the first ethanol precipitation. No significant loss of radioactivity should occur in all subsequent ethanol precipitation and washing steps.

▶ It is extremely important that alkaline hydrolysis of in vitro transcripts is performed on ice and that all reagents are pre-cooled on ice because the reaction proceeds faster at higher temperatures, resulting in too small fragments.

▶ Optimal Pronase concentration varies for different tissues. Incomplete Pronase digestion may result in high hybridization background levels. In this case, increase Pronase concentration to 80–120 µg/ml.

▶ Removal of air bubbles trapped under parafilm is easier following incubation of the slides for a few minutes in the hybridization incubator.

▶ While applying the hybridization probe, cooling of slides to RT should be avoided. Return slides to the incubator from time to time.
The optimal hybridization temperature is depending on the GC content of the probe. Probes with lower or higher than average GC content require lower or higher hybridization temperatures, respectively. Optimizing the hybridization temperature can greatly improve results.

▶ Ideally, antisense probe generates a strong specific hybridization signal whereas sense control probe give only weak background levels. However, some sense probes generate relatively high unspecific background which can only be judged under the microscope. To decrease unspecific background the first washing step after hybridization can be extended up to 4–5 h and washing temperature can be raised up to 42 °C. Furthermore, RNase A digestion can be extended up to 30 min.

References

Breier G, Clauss M, Risau W (1995) Coordinate expression of vascular endothelial growth factor receptor-1 (flt-1) and its ligand suggests a paracrine regulation of murine vascular development. Dev Dyn 204:228–39

Breier G, Breviario F, Caveda L, Berthier R, Schnürch H, Gotsch U, Risau W, Dejana E (1996). Molecular cloning and expression of murine vascular endothelial cadherin in early stage development of cardiovascular system. Blood 87:630–641

Hogan B, Beddington R, Costantini F, Lacy E (1994) Manipulating the mouse embryo: A laboratory manual. Cold Spring Harbor Laboratory Press, New York

Sambrook J, Fritsch EF, Maniatis T (1989) Molecular Cloning: A Laboratory Manual. Cold Spring Harbor Laboratory Press, New York

Steffen B, Breier G, Butcher EC, Schulz M, Engelhardt B (1996) ICAM-1, VCAM-1 and MAd-CAM-1 are expressed on choroid plexus epithelium but not endothelium and mediate binding of lymphocytes in vitro. Am J Pathol 148:1819–1838

Vajkoczy P, Farhadi M, Gaumann A, Heidenreich R, Erber R, Wunder A, Menger MD, Breier G (2002) Microtumor growth initiates angiogenic sprouting with simultaneous expression of VEGF, VEGF-receptor-2, and angiopoietin-2. J Clin Invest 109:777–85

Chapter 28

High-Resolution in Situ Confocal Analysis of Endothelial Cells

HOLGER GERHARDT and CHRISTER BETSHOLTZ

Introduction

The analysis of endothelial cells in situ differs in many ways from the study of cultured endothelium. Endothelial cells in situ are embedded in their natural environment – the tissue, which extends in three dimensions and exhibits some properties that are far from ideal for imaging techniques. These properties vary also within the organism. For example, the capillary plexus in the skin has a more planar dimension as compared to that of the lung, liver or brain. In general, the capillary plexus is patterned according to the dimensions and demands of the surrounding tissue. Consequently, the imaging techniques must be adapted to the three-dimensional properties of the tissue. Similarly, the optical properties differ between different tissues. Structures like hair, pigment, fat, or muscle fibers differ tremendously, for example, in transparency and autoflourescence. Accordingly, imaging techniques must be adapted to include bleaching of pigment or quenching of autoflourescence. In general, the maximum resolution will be achieved by imaging structures of minimal 3D dimensions and maximal transparency. Therefore, the choice of the tissue to be used as a model system for high-resolution analysis of endothelial cells is important and already sets the limits for the final resolution. Although physical sectioning increases the optical properties of a given tissue, it is often unwanted since it disrupts spatial relationships and, thus, limits overview and 3D reconstruction possibilities.

Laser scanning confocal microscopy (LSCM) of fluorescently labeled structures provides excellent possibilities to achieve both high-resolution and 3D imaging. Simplistically, LSCM is based on the principle to suppress all image contribution from structures that are out of focus. Rather than bathing the whole specimen in excitation light, illumination is achieved through a laser source that is focused on the specimen and scans the region of interest. The emission light passing through the lens is focused to a spot where an adjustable diaphragm (pinhole) blocks all light from structures lying above or below the level of focus. Thereby blur is minimized and the resolution is

Springer Lab Manual
H. Augustin (Ed.)
Methods in Endothelial Cell Biology
© Springer-Verlag Berlin Heidelberg 2004

ideally only limited by the numerical aperture of the lens (more details on the microscopic technique and applications can be found at www.leica.com, www.zeiss.com, or in more specialized literature [e.g., Paddock 1999]). With this technique, the tissue can be sectioned optically and the stacks of sections can be used for 3D imaging. The numerical aperture of the lens is crucial for the thickness of the optical section. However, the size of the pinhole will determine the final outcome by increasing or reducing contribution from structures that are out of focus, thereby adjusting the thickness of the optical section. For 3D reconstructions, the resolution in the Z dimension is dependent on the number and the overlap of sections. However, even under ideal conditions, the Z dimension will always provide less resolution as compared to the xy dimension. This is due to a basic characteristic of optical lenses known as the point-spread function, which describes how a perfectly round point will be distorted by projection through a lens. The result is that the projected point will be close to round in the xy dimension but distorted into an elongated ellipse with diffuse extensions in the upper and lower periphery. Although new developments aim to tackle this problem by digital processing, there is still considerably less resolution in the Z dimension.

The considerations above only briefly outline some of the technical possibilities and limitations that apply even if the stained samples are of excellent quality. However, the staining intensity and signal-to-background ratio have the strongest impact on the imaging possibilities, and failure to achieve strong labeling contrast for your structure of interest will dramatically reduce the imaging quality. For the analysis of vascular endothelium, it is therefore crucial to select a marker molecule that gives a strong staining and high specificity. Below, we describe a simple and robust protocol that performs well on different mouse tissues. We will provide examples from early post-natal retina and embryonic hindbrain (see also Ruhrberg et al. 2002).

▪ Outline

Day 1

Collecting specimen, fixation:
- Collecting embryos for hindbrain material
- Collecting eyes for retina material
- Fixation over night

Day 2

Preparing specimen, permeabilization:
- Dissecting hindbrain
- Dissecting retinas

- Removal of fixative
- Tissue permeabilization overnight

Day 3

Biotinylated isolectin incubation:
- Washes in PBS
- PBlec, Isolectin incubation

Day 4

Primary antibody incubation (double labeling, optional):
- Washes in PBS
- Incubation for double labeling

Day 5

Fluorescence conjugated Streptavidin/secondary antibody incubation, mounting, observation:
- Washes in PBS
- Incubation with secondary antibodies
- Washes
- Mounting
- Analysis

Materials

Day 1: Collecting specimen

Animals for retina specimen
- Mouse pups, P1 to P7 (for analysis of sprouting angiogenesis); P5 to P21 (for analysis of vascular remodeling)

Animals for hindbrain specimen
- Mouse embryos, E10.5 to E12.5

Equipment
- Dissection tools including forceps, strong and fine scissors, a fine round-tipped spatula, and two pairs of fine tweezers (watchmaker tweezers, e.g., Dumont no. 5)
- Stereomicroscope (e.g., Nikon SMZ-U, www.nikon-instruments.com) and fiberoptic illumination (e.g., F-Q1000)
- Orbital or horizontal shaker (e.g., Heidolph polymax 1040 or GFL 3006, www.heidolph.com) preferably in a coldroom

Supplies
- Phosphate buffered saline (PBS), pH 7.2 (prepare from 10 × stock)
- 4 % freshly depolymerized paraformaldehyde in PBS (PFA-fixative)
- 10 cm petri dishes
- 2 ml round bottom Safe-lock eppendorf tubes
- Ice

Day 2: Preparing specimen

Equipment
- Fine feather iris scissors (e.g., Lawton, www.lawton.de),
- Two pairs of fine tweezers (watchmaker tweezers, e.g., Dumont no. 5, www. fullam.com)
- Angled tweezers (e.g., Dumont 5/45, www. fullam.com)
- Preferably a tungsten needle
- Stereomicroscope and illumination as above.

Supplies
- PBS
- 1 % bovine serum albumin (BSA), 0.5 % Triton-X100 in PBS (block/permeabilization buffer)
- 35 mm petri dishes

Day 3: Isolectin incubation

Reagents
- Biotinylated Isolectin B4 (lectin from *Griffonia simplicifolia*, Sigma L2140, www.sigmaaldrich.com; make 200 ng/µl stock)

Supplies
- PBS
- PBS pH6.8 containing 1 mM $CaCl_2$ (prepare 1 M stock solution), 1 mM $MgCl_2$ (make 1 M stock solution), 0.1 mM $MnCl_2$ (prepare 1 M stock solution) and 1 % Triton-X100 (PBlec)

Day 4: Primary antibody incubation for multi-labeling (optional)

Reagents
- Primary antibody of choice [e.g., rabbit anti-glial fibrillary acidic protein (GFAP) for detection of Astrocytes; DAKO Z 0334, www.dakocytomation.com]

Supplies
- PBS
- 0.5 % BSA, 0.25 % Triton-X100 in PBS, pH 7.2 (incubation buffer)

Day 5: Streptavidin-conjugate incubation (and secondary antibody for multi-labeling, optional), mounting, and analysis

Equipment
- Confocal Laser Scanning Microscope, upright or inverted setup, equipped with lenses up to 63× magnification and laser lines for excitation at 488 nm, 568 nm and 633 or 647 nm.

Reagents
- Streptavidin-fluorochrome conjugate (e.g., Streptavidin Alexa-Fluor 568, Molecular Probes, www.probes.com)
- Secondary antibody (for multi-labeling, optional; e.g., goat-anti-rabbit Alexa-Fluor 488, Molecular Probes)
- ToPro-3 (Molecular Probes) for nuclear counterstain (optional)

Supplies
- PBS
- Incubation buffer (see Day 4)
- Fluorescence compatible mounting media (e.g., Moviol plus anti-bleach DABCO or phenylendiamine)
- Microscopic slides (e.g., Menzel, www.menzel.de)
- Coverslips (e.g., Menzel)

Procedure

Day 1: Collecting specimen

Retina
1. Sacrifice pups by decapitation (strong scissors).

2. Remove skin and eyelid (forceps and fine scissors).

3. Carefully enucleate the eyes using the fine rounded spatula.

4. Fix eyes overnight in cold PFA-fixative (1 ml or more per pair of eyes in Eppendorf tube).

Hindbrain
1. Sacrifice pregnant female by cervical dislocation.

2. Dissect uterus and place in ice-cold PBS in a 10 cm Petri dish (keep on ice).

3. Separate and dissect the embryos free of deciduas.

4. Fix embryos overnight in cold PFA-fixative (1 ml or more per embryo in Eppendorf tube).

Day 2: Preparing specimen

1. Remove PFA fixative and wash in PBS.

2. Transfer to ice-cold PBS in 35 mm Petri dish for dissection under stereomicroscope.

Retina

1. Open the eye by cutting carefully along the ora serrata using the fine scissors while supporting the eye with the angled tweezers.

2. Remove anterior parts including the lense.

3. Remove remaining hyaloid vessels by carefully detaching them from the retina at the very periphery and subsequently pulling them out directly at their insertion at the optical disc.

4. Strip the retina from the sclera, choroids, and retinal pigment epithelium using two fine tweezers: hold the sclera at the very periphery and carefully slide the second pair of tweezers between the retinal pigment epithelium and the neural retina. Detach the neural retina by carefully sliding the tweezers sideways while keeping the pointed end of the tweezers facing away from the retina (to avoid damage). Subsequently, grasp the sclera/choroids/RPE firmly with both tweezers and rip open. The sclera will flip and turn inside out, thereby detaching completely from the retina. Finally, free the retina by pulling the sclera with one pair of tweezers through the second pair of tweezers which is held flat and slightly open over the entry point of the optic nerve in order to hold down the retina (a similar dissection technique is graphically illustrated in Chan-Ling, 1997).

5. Collect retina with a rounded spatula and place in round-bottom eppendorf tube containing PBS (Note: in case long-term storage is wanted before labeling, the retinas can be collected in –20 °C methanol and stored at –20 °C. Rehydrate in PBS before labeling).

Hindbrain

1. Cut off the head of embryos at the medulla oblongata (border between hindbrain and spinal cord).

2. Detach the hindbrain from surrounding mesenchyme using fine tweezers.

3. Cut at the midbrain/hindbrain border and carefully remove remaining mesenchyme and the transparent dorsal roof of the hindbrain using tweezers and tungsten needle (if available).

4. Collect hindbrain with a rounded spatula and place into round-bottom eppendorf tube containing PBS (alternatively store in −20 °C methanol, rehydrate in PBS before labeling).

All following steps are identical for retina and hindbrain specimen

1. Wash three times for 5 min in PBS.

2. Incubate in block/permeabilization buffer overnight at 4 °C.

Day 3: Isolectin incubation

1. Wash three times for 5 min in PBS.

2. Change buffer to Pblec and equilibrate by washing in Pblec three times 20 min.

3. Incubate in isolectin diluted 1:10 in Pblec, overnight at 4 °C.

Day 4: Primary antibody incubation for multi-labeling (optional)

1. Wash five times for 10 min in PBS.

2. Incubate in primary antibody (e.g., rabbit anti-GFAP, 1:75) diluted in incubation buffer overnight at 4 °C.

Day 5: Streptavidin-conjugate incubation (and secondary antibody for multi-labeling, optional), mounting and analysis

1. Wash 5 times for 10 min in PBS.

2. Incubate in streptavidin-Alexa Fluor 488 (1:100; for visualizing biotin-isolectin; e.g., www.probes.com), goat anti-rabbit Alexa Fluor 568 (1:100; for visualizing rabbit anti-GFAP; e.g., www.probes.com) and ToPro-3 (1:1,000; for nuclear counterstain) in incubation buffer no longer than 2 h at RT (in the dark; use aluminum foil to wrap or other appropriate light protection. Keep protected from light also during the following steps).

3. Wash three times for 10 min in PBS.

4. Post-fix in 4 % PFA in PBS for 5 min (optional, recommended if intended to analyze after longer storage periods).

5. Wash twice in PBS.

6. Mount in mounting media including anti-bleach: Put a drop of PBS onto the slide and transfer the retinas to this drop using the spatula. Make 4 to 5 radial incisions into each retina under the stereomicroscope (use low illumination to avoid extensive bleaching). Remove most of the PBS using a pipette: The retinas will flatten onto the slide – if not, carefully

flatten out using tweezers. Place a drop of mounting media on the coverslip (not on the retinas, because otherwise they will fold up due to the high surface tension of the mounting media!!), and quickly turn the coverslip upside down with the mounting media now hanging as a drop. Lower the coverslip slowly onto the retinas while observing under the stereomicroscope. After mounting, add more mounting media slowly at the edge of the coverslip, if needed. In case too much has been added, tip the edge of the slide on Kleenex tissue. Hindbrains do not need incisions and can be mounted by adding mounting media directly onto the specimen and placing coverslips on top. If wanted, several layers of adhesive aluminum tape can be used as spacers to avoid deformation of the tissue. Note: Make sure to mount the coverslips completely planar. Failure to do so will impair the optical qualities during analysis.

7. Let mounting media settle for a couple of hours at RT in the dark.

8. Analyze in the LSCM.

Expected Results

The isolectin B4 from *Griffonia simplicifolia* binds with high specificity to alpha-galactosylated glycoproteins. The staining is confined to the membrane of labeled cells and not detected in basement membranes or other extracellular matrices. For example, double staining with extracellular matrix proteins, such as fibronectin, shows that regressing blood vessels loose the epitopes recognized by the isolectin whereas the remaining basement membrane is still detectable by anti-fibronectin antibodies. Isolectin B4 generally gives a very strong staining of the entire endothelial cell surface enabling to visualize the three-dimensional extensions of the endothelium (Fig. 1). The staining is usually slightly stronger at endothelial junctions (Fig. 1A). Additionally, surrounding pericytes are weakly positive. In the CNS, Isolectin B4 furthermore stains microglial cells, however, these can easily be distinguished by their morphology, position, and punctuate staining pattern (Fig. 1B). The intensity of endothelial staining enables also to observe all extensions of the endothelial cells. For example, the leading cells in the periphery of the retina show long filopodial extensions that are labeled by Isolectin B4 (Fig. 1A). The same is true for the hindbrain in regions where sprouting and fusion occurs. The strong labeling enables to perform 3D analysis of thick specimen at high resolution (Fig. 1C-E). The example of the hindbrain shows a 3D transparency projection based on a stack of 200 optical sections taken at 0.5 μm interval. This mode of data presentation is well suited to study the organization of the vasculature in tissues that do not possess a very planar vascular plexus. The resolution is optimal in the XY-view, looking directly on top (Fig. 1C) or on the bottom of the stack (Fig. 1D), and somewhat reduced when observing from the side (Fig. 1E).

The study of different stages of vascular development reveals that the epitopes stained by Isolectin are much more abundant on angiogenic endothelium as compared to quiescent endothelium. Regressing vascular processes also loose their staining, indicating that Isolectin B4 may be considered as an excellent marker for active endothelium. The epitopes are very stable and appear to be insensitive to strong fixation, detergents, and to some degree even to enzymatic degredation by proteinase K. Therefore, Isolectin

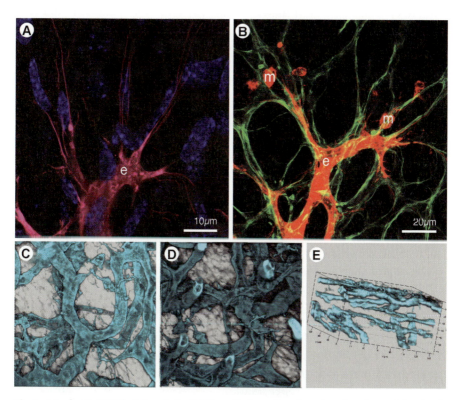

Fig. 1. A Isolectin B4 labeled endothelial tip-cells (*e*, *red*) at the migrating front of the developing retinal vasculature of a five day old mouse pup. The image represents a single optical section taken with a 63× lens and digital zoom factor 2 (pinhole setting: 1 airy unit) at 1024 × 1024 pixel resolution. Detector noise was reduced by averaging 8 scans of the same frame. B represents a maximal projection of 8 optical sections taken at 0.5 µm interval using a 40× lens and digital zoom factor 2. Astrocytic intermediate filaments (GFAP) are labeled in *green*. Note microglia are also labeled (*m*). Nuclei are labeled using Topro-3. **C-E** Three-dimensional transparency projections of Isolectin B4 labeled vasculature in the embryonic mouse hindbrain at E12.5. The images **C**, **D**, and **E** represent top, bottom, and side view, respectively, of 200 partially overlapping optical sections taken at 0.5 µm interval (pinhole settings 1 airy unit) using a 63× lens at 512 × 512 pixel resolution (no averaging)

B4 labeling works also very well after in situ hybridization procedures. Long-term storage of material in either PBS or methanol does not affect the staining significantly. However, the preservation of the endothelial morphology is dependent on fixation. For example, the filopodia are best visualized after 4 % PFA fixation and are partially disrupted by methanol.

In many cases multiple labeling is desired, both by antibody labeling techniques and by combinations with GFP-tagged protein expression. We only provide one example for double labeling of the endothelium and the astrocytes in the retina. However, many other combinations can be performed and enable to study the spatial relationship between endothelial cells and their neighbors in the tissue as well as cellular distribution of molecules and, in special cases, even molecular interactions (for example by Fluorescence Resonance Energy Transfer analysis). With the advent of numerous new and brightly fluorescent dyes and new laser lines for excitation, as well as data processing that now enables unmixing of overlapping fluorescence emission spectra (for examples, see information on websites by Zeiss [www.zeiss.com] and Leica [www.leica.com]), multiple-labeling has lost many limitations.

Troubleshooting

The isolectin labeling is very robust and only rarely gives problems. However, when applying this staining to other tissues than the CNS, the specificity for endothelium may be lost. This is, for example, true for the skin and for some tumors including fibrosarcoma. In our hands, Isolectin also works well on heart and skeletal muscle.

The retinal staining can be troublesome if the hyaloid vasculature and larger parts of the vitreal matrix are not removed. This occasionally reduces the penetration of antibodies, leads to high levels of background and, thus, diminishes contrast and resolution dramatically. A solution to the problem can be to dissect the retinas already after short fixation (max. 1 h). This facilitates detachment of the hyaloids. However, after dissecting, the retina should be post-fixed to ensure proper fixation.

Some intracellular epitopes, like cytoskeletal proteins, may require special protocols including extra permeabilization. Permeabilization can be facilitated by methanol. A modification that works well in our hands is to dissect the retinas after short PFA fixation, to remove all PBS, and to flatten the retina by radial incisions (as described for mounting, see above), and to add –20 °C methanol directly onto the retinas. They can now be stored in methanol until staining. Before staining, the retinas should be post-fixed briefly in 4 % PFA.

General considerations

The above protocol can be adapted according to the requirements of the antibodies in question. For some antibodies low PFA fixation is required. In other cases, as for staining of some junctional molecules, fixation with PFA should be totally avoided and replaced by ethanol – acetone fixation.

High background can usually be reduced by lowering antibody concentrations, increasing wash steps and time, adding more detergent to the wash steps, or raising the salt concentration. Quenching of autofluorescence may be achieved by brief incubation with 0.1 % sodium borohydride (NaBH$_4$) in PBS after fixation. Bleaching of endogenous pigment can be achieved by hydrogen peroxide incubation.

More general protocols for antibody staining, whole mount techniques, and possible adaptations and solutions may be found in Harlow et al. (1998) and Hogan et al. (1994).

References

Chan-Ling T (1997) Glial, vascular, and neuronal cytogenesis in whole-mounted cat retina, Microscopy research technique 36:1–16

Harlow E, Lane D, Harlow, E (1998) Using antibodies: a laboratory manual: Portable protocol No. 1, Cold Spring Harbor, NY, Cold Spring Harbor Laboratory Press

Hogan BL, Beddington R, Costantini F, Lacy E (1994) Manipulating the mouse embryo: A laboratory manual, 2nd edn, Cold Spring Harbor, NY, Cold Spring Harbor Laboratory Press

Paddock WS, ed (1999) Confocal microscopy methods and protocols, Totowa, NJ, Humana Press

Ruhrberg C, Gerhardt H, Golding M, Watson R, Ioannidou S, Fujisawa H, Betsholtz C, Shima DT (2002) Spatially restricted patterning cues provided by heparin-binding VEGF-A control blood vessel branching morphogenesis, Genes Dev 16:2684–2698

Chapter 29

Whole Mount Analysis of the Embryonic Vasculature

JODY J. HAIGH and ANDRAS NAGY

▦ Introduction

Similar to highways, roads, paths and trails reaching the different communities of a country, vessels get to all parts of an organism and they are spatially organized for trafficking capacity to achieve efficient circulation of blood. In vertebrates, this network is so important that no solid tissue survives above 1–2 mm in diameter without being vascularized and connected to the functioning cardiovascular system. This applies to embryonic development as well. Therefore, it is not surprising that in embryos of many species the cardiovascular system is the first functional system to be developed. In the mouse just four days after implantation, the embryo has to have a functional heart with connected vessel structure to the embryo, yolk sac, and placenta in order to survive. The vasculature is developing from early embryonic stages until the end of juvenile phase, as long as an individual is growing (Gerber et al. 1999). After this growth phase, vessel formation ceases except in organs subject to periodical remodeling, such as ovary and uterus, and in physiological/pathological conditions, such as wound healing and tumor growth.

The "blueprint" of the embryonic vasculature is followed with amazing precision during development. The determinants of this organization, however, are not very well understood, but it is certain that there are a large number of players. Disturbances in these processes could have serious consequences, frequently leading to embryonic lethality. This is also true for embryonic lethality associated with many targeted gene mutations.

For characterization of cardiovascular phenotypes, good imaging technology of the entire vasculature is pivotal. In this chapter, two separate global vessel visualization methods are given. The first method involves the injection of India ink into the circulation whereas the other method uses whole mount immunohistochemistry to visualize endothelial cells. The difficulty of this task is obvious, since the vascular system quickly develops into a high-density network with enormous complexity. In addition, the size of the embryo quickly exceeds the maximum area, which can be covered and pen-

Springer Lab Manual
H. Augustin (Ed.)
Methods in Endothelial Cell Biology
© Springer-Verlag Berlin Heidelberg 2004

etrated by most of the high resolution imaging techniques. Interestingly, a "classic" embryology method provides one of the best and simplest solutions for visualizing fine details of vessel structure normally hidden inside the embryo. India ink injection into the circulatory system of freshly recovered chicken embryos was used since the early years of developmental studies to increase the visual contrast of vessels. This classic chicken embryo technique can be adapted to mouse embryos as well. Clearing the tissue after injection to glass equivalent transparency reveals the three-dimensional locations of the ink carbon particles trapped inside the vessels. This adaptation provides a unique opportunity to visualize the three-dimensional vessel structure of even late gestation stage embryos, which opens the possibility of detecting global malformations in cardiovascular system development.

The task of adapting this methodology to the mouse embryo lies in the need of injecting the ink into the still operating circulation of the embryo. For this, the embryos have to be removed from the uterus, thereby limiting their survival for not more than 15–20 min depending on the embryonic stage.

India ink injection

▨ Outline

Day 1

Dissection of embryos from the uterus. Immediate injection of India ink into the still active circulation. Fixing and dehydrating the embryos with ethanol

Day 2–3

Clearing the embryos with a mixture of benzyl benzoate and benzyl alcohol

Day 4 and on

Observation of the vasculature and documentation

▨ Materials

For day 1

- Timed pregnant females
- Fine dissecting tools
- Drawn out Pasteur pipette

- Mouse pipetting tube, mouthpiece, and adaptor to connect to the glass pipette (e.g., a blue tip for P1000 pipette man)
- India ink (for example Higgins waterproof drawing ink, Black India 4415)
- 0.45 µm syringe filter
- Eppendorf tubes
- 100 mm Petri dishes
- 26G needles
- PBS with Ca^{++} and Mg^{++} pre-warmed at 37 °C
- Dissecting scope
- 30 %, 50 %, 70 %, and 100 % ethanol

For day 2

- Glass scintillation vials (20 ml)
- 2:1 mixture of benzyl benzoate and benzyl alcohol
- Fume hood

For day 3

- Dissecting scopes
- Compound microscope with normal light optics, long working distance objectives
- Digital camera
- Fine surgical tools
- Glass slides or depression glass slides

Procedure

Preparation of India ink for injection (day 1)

1. Dilute 1 ml of India ink with PBS in a 1:3 ratio.

2. Load the mixture into a 3 ml syringe and filter is through a 0.45 µm syringe filter to fill up a 1.5 ml Eppendorf tube.

3. Spin down the filtered carbon particles and carefully remove the supernatant.

4. Resuspend pellet in 500 µl of PBS.

5. Using the mouth pipetting apparatus tip-load the drawn out Pasteur pipette with the carbon particle suspension just prior to the surgical recovery of the embryos.

Surgical procedures (day 1)

1. Kill pregnant female by cervical dislocation and dissect uterus as quickly as possible (Nagy et al. 2003).

2. Place the uterus into a 100 mm Petri dish with 10 ml of warm PBS.

3. Quickly cut along the uterus wall avoiding rupturing the yolk sac membrane of the embryos or damaging any other part.

4. Place 2–3 embryos into the a new 100 mm Petri dish with 10 ml warm PBS.

5. Leave the remaining embryos inside the uterus and place the plate on a warm (37 °C) surface.

Ink injection (day 1)

1. Observe the embryos subject to ink injection under a dissecting scope first with low magnification and turn on that side which reveals the yolk sac viteline vein descending to the embryo.

2. Then using high power on the dissecting scope, observe the direction of the blood flow and decide about the site of injection. This must be in the lower segment of the vein, at one of the last merging branches.

3. At this site, right above the vessel, poke through the yolk sac with the needle to rupture the vessel (Fig. 1A).

4. Pick up the Pasteur pipette loaded with the ink and insert the pipette into the ruptured vessel.

5. Slide the capillary deep into the vessel, down to the branching point, and then slowly blow a puff of India ink into the circulation (Fig. 1B).

6. Wait until the blood coming to the merging point from the intact branch flushes most of the ink into the embryo.

7. Expel a small volume of ink several times and wait 20–30 s in between. This way, the embryo will keep circulating the blood-ink mixture for longer time.

8. Observe the change in the color of the embryo from pink to grayish as the ink disperses into different areas.

9. Withdraw the injection pipette and place the embryos into fresh warm PBS until the circulation stops.

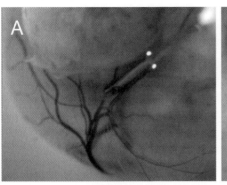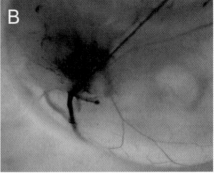

Fig. 1A,B. 14.5 dpc embryo prepared for India ink injection into its circulatory system. A The embryo with intact yolk sac is oriented and the 26G needle is punched through cutting the yolk sac vein at the desired spot. B The ink filled capillary is inserted into the lumen of the vessel and the ink is slowly blown into

Dehydration (day 1 and overnight)

1. Wash the embryos through increasing concentrations of ethanol (30 %, 50 %, 70 %, 100 %) by exposing them to each concentration for 2–3 h.

2. Leave the embryos in 100 % ethanol overnight.

Clearing (day 2)

1. Using only glass tools and glassware, prepare an adequate amount of clearing solution in a fume hood. The volume should be 30–50 times higher than the volume of the embryos.

2. Place the completely dehydrated embryos into the clearing solution kept in glass (e.g., glass scintillation vials) for 48 h or more. If dehydration was complete, this treatment should clear the embryos to glass like transparency, with a little yellowish coloration remaining in the liver.

Observation and documentation (day 4)

The global overview of the vessel structure of a late stage embryo (Fig. 2A) requires a low resolution microscope which could be a dissecting scope or compound microscope equipped with low magnification and large working distance objective. The embryos should be kept in the above clearing solution in a glass container (dish). Using high magnification of the dissecting scope, it is possible to optically penetrate deeply into the embryo and reveal amazing details of vessel structures (Fig. 2B–D). Microscopes with Z dimension optical sectioning capabilities may allow for deconvolution and flattening the 3D structure into 2D without carrying over out-of-focus areas.

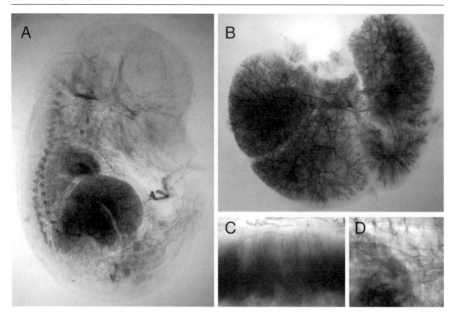

Fig. 2A–D. Cleared 14.5 dpc embryo injected with India ink into its circulatory system. A Global view of the cardiovascular system. Vessel structure of the liver (**B**), embryonic vessels in the placenta (**C**), and connecting vessels to the heart (**D**)

Deconvoluted Z dimension optical sections may also allow 3D reconstruction of the vasculature in an exceptional depth due to the extremely good transparency of the embryo.

▪ Comments

Due to the filtration of the ink, the size of the carbon particles is at least ten times smaller than the diameter of a red cell. These small particles can pass through the capillaries and enter into the veins as well. If unfiltered India ink is injected, the larger-sized carbon particles quickly clot the capillaries and retain the ink in the arterial vessels. In theory, this could be useful if the arteries are the focus of the study. However, since the circulation usually stops much faster in this case, often before the ink distributes into all arteries, there is the high variability in ink distribution.

The longer there is an active circulation after injection, the better the ink distributes. Therefore, the best results are obtained if an embryo is injected immediately after dissection from the mother. From one pregnant female, one person usually can inject 2–6 embryos, depending on experience.

The tip diameter of the injection needle is critical and should be determined experimentally, depending on the embryonic stage under investigation. The proper diameter is best determined by the size of the vessel being injected. If it is much smaller than the inner lumen, a backflow of the ink could occur. If the injection pipette is too large, there could be a problem injecting it into the vessel.

Injecting ink into the viteline vein can be very challenging with embryos younger than 12.5 dpc. In these embryos, direct heart injection could give good results as well.

If the embryo has a particular organ or area of interest, one might dissect this region from the already cleared embryo before documentation. After clearing, the embryo tissue is very hard. The dissection is more like breaking the non-essential parts away.

29.2 Whole mount CD31 (PECAM-1) staining

Introduction

Unlike the India ink injection protocol described in the previous section of this chapter, the whole mount CD31 staining protocol described below utilizes an antibody and immunohistochemical methodology to detect and label the vascular endothelium. CD31 or PECAM-1 (Platelet Endothelial Cell Adhesion Molecule-1) is a 130-kDa integral membrane protein that mediates cell-to-cell adhesion (DeLisser et al. 1994). CD31 is expressed constitutively on the surface of adult and embryonic endothelial cells and is weakly expressed on peripheral leukocytes, platelets, hematopoietic and embryonic stem cells (DeLisser et al. 1997, DeLisser et al. 1994, Ling et al. 1997, Vecchi et al. 1994). CD31 is involved in the transendothelial migration of neutrophils and endothelial cell-cell interactions involved in angiogenesis (DeLisser et al. 1997, Duncan et al. 1999).

Through the use of a biotinylated anti-CD31 antibody and an immunoperoxidase signal amplification system (Hsu et al. 1981), it is possible to selectively label the vascular endothelium and obtain a global overview of the complex vasculature network in whole mount embryos from 8.5–12.5 days p.c. (E8.5-E12.5). For a complete description of the immunoperoxidase "ABC" methodology and reagents see the VECTOR Laboratories website at: www.vectorlabs.com/products.asp?catID=42&locID=609342.

Solutions and Reagents

- Fixation solution # 1: 4 % paraformaldehyde (PFA; to be prepared on the day of use).

For 50 ml:

- 2 g paraformaldehyde in 45 ml of water
- 40 µl 5 N NaOH
- Heat at 65 °C with occasional mixing to dissolve PFA .
- Once dissolved, add 20 µl of concentrated HCl.
- Add 5 ml 10× PBS and cool on ice before use.
- Fix solution # 2: 4 % PFA/0.2 % glutaradehyde (to be prepared on the day of use).
- Make up 4 % PFA as described above and supplement with 400 µl of 25 % aqueous solution of glutaradehyde.
- PBT = PBS with 0.1 % Triton-X
- PBSMT = PBS with 0.1 % Triton-X and 2 % skim milk powder
- DMSO
- 30 % H_2O_2
- Methanol
- Proteinase K (Roche Applied Sciences; Cat. No. 1 092 766, www.roche-applied-science.com)
- Glycine
- Biotin-conjugated rat anti-mouse CD31 (PECAM-1) monoclonal antibody (BD Biosciences-Cat. No. 553371, www.bdbiosciences.com)
- Vectastain Elite ABC kit (standard, Vector Laboratories, Cat. No. PK-6100, www.vectorlabs.com)
- DAB substrate kit (Vector Laboratories, Cat. No. Sk–4100)

All microscopes and dissecting instruments needed for this protocol have been described in the previous section.

Procedure

Day 1

1. Dissect E8.5-E12.5 embryos in PBS free from the maternal decidua and reflect or remove yolk sacs (depending on whether or not yolk sac vasculature will be analyzed).

2. Fix embryos in 50 ml of fresh fixative # 1 (4 % PFA in PBS) overnight at 4 °C gently rocking on an orbital platform or nutator.

Day 2

1. Wash extensively for 3 × 15 min in 50 ml PBT. After the last PBT wash, embryos may be transferred to a 10 cm dish containing PBT and the anterior hindbrain (neuropores) gently punctured with forceps to allow flow of solutions into and out of the embryos.

2. Dehydrate embryos with sequential solutions of 25%, 50%, 75% MeOH/PBT, then 100% MeOH twice, for 15 min each. The embryos can be stored at –20 °C in 100% methanol at this point. Embryos can be processed the next day or stored for up to one month without significant degeneration of the embryos or loss of protein signal. The embryos can be dehydrated as a group in 10 ml volumes or they can be separated and dehydrated separately in 2 ml/well of a 24-well tissue culture dish (E8.5-E11.5) or 3 ml/well of a 12-well culture dish (E12.5). Separation of the embryos may be needed if genotyping is required. For genotyping of E8.5-E10.5 embryos, parts of the yolk sac are dissected and PCR geno-typed. For E11.5-E12.5 embryos, small pieces of tail or yolk sac are taken for genotyping.

Day 3

1. Bleach embryos and block endogenous peroxidases by incubating embryos in a 10 ml 4:1:1 mixture of MeOH:DMSO:30% H_2O_2 for 5–6 h at RT. Embryos are rocked by hand 1–2 times per hour to avoid excessive damage to the embryos. Embryos of the same genotype can be pooled prior to this bleaching stage.

2. Rehydrate embryos in 75%, 50%, 25% MeOH/PBT at RT for 15 min each and give a final 100% PBT wash for 10 min.

3. Treat embryos with 10 µg/ml proteinase K solution in PBT for the following times:

   E8.5:   4 min
   E9.5:   6 min
   E10.5:   12 min
   E11.5:   13 min
   E12.5:   15 min

4. The embryos are quite fragile at this time. So they should not be excessively shaken and care should be taken when removing the liquid.

5. Wash 2 × 5 min in freshly prepared 2 mg/ml glycine in PBT followed by a 3 × 5 min wash in PBT.

6. Refix in fresh fixative # 2 (0.2 % glutaraldehyde/4 % paraformaldehyde in PBT) for 20 min with minimal shaking. Wash away PFA with 3 × 5 min washes in PBT.

7. Incubate 2 × 1 h in PBSMT supplemented with 2 % BSA.

8. Nutate in PBSMT with 2 % BSA overnight at 4 °C with a 1/500 or 1/1,000 dilution of biotin-conjugated rat anti-mouse CD31 (PECAM-1) monoclonal antibody. To save on the amount of antibody used, incubation can be performed in 1.5 ml screw cap tubes.

Day 4

1. Wash embryos 5–7 times in 10 ml of PBSMT on nutator at 4 °C for 1 h each wash.

2. During the final wash, set up the Vectastain ABC reagents by gently mixing a 1/50 dilution of reagent A with PBSMT followed by a 1/50 dilution of reagent B and gentle mixing. Incubate ABC complexes on ice for 30 min. This step can also be performed in a 1–1.5 ml volume to save on reagents.

3. After the final wash, add preformed ABC complexes to the embryos and nutate overnight.

Day 5

1. Wash embryos 5–7 times in 10 ml of PBSMT on nutator at 4 °C for 1 h each wash. Wash a final 1 × 20 min at RT in PBT.

2. Incubate embryos in the dark in 0.3 mg/ml DAB in PBT (1 DAB tablet in 2.3 ml PBT; DAB is a carcinogen – handle with care!) and 0.03 % H_2O_2 at RT until staining is satisfactory. The color can be enhanced by adding 0.5 % NiCl2 to the DAB solution. Rinse in PBT after staining is finished.

3. Document staining results and examine vascular patterning.

Observation and documentation

Optional:
1. Dehydrate through methanol series (25 %, 50 %, 75 %, 100 %) for 5 min each.

2. Embryos may be cleared in benzyl alcohol:benzyl benzoate (1:2) as described previously.

Comments

The whole mount CD31 protocol is limited up to E12.5 in development due to the fact that background staining due to trapping of reagents becomes problematic in larger, later stage embryos. Adult tissues that are around the same size as E12.5 embryos can also be dissected and subjected to the whole mount CD31 protocol. The main benefits of this whole mount protocol are that it is not dependent on a functional circulatory system to label the vascular endothelium and can be combined with whole mount mRNA in situ protocols. If mRNA in situ analysis is to be performed, then all solutions before this analysis should be made up with DEPC-treated water to avoid mRNA degradation.

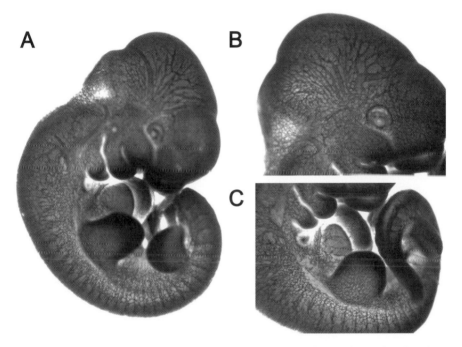

Fig. 3. A Depicts an 11.5 dpc embryo before clearing that has been subjected to the whole mount CD31 staining protocol. This figure demonstrates the complexity of the vascular system in the head (**B**) and throughout the embryo (**C**).

References

DeLisser HM, Christofidou-Solomidou M, Strieter RM, Burdick MD, Robinson CS, Wexler RS, Kerr JS, Garlanda C, Merwin JR, Madri JA, Albelda SM (1997) Involvement of endothelial PECAM-1/CD31 in angiogenesis. Am J Pathol 151:671–677

DeLisser HM, Newman PJ, Albelda SM (1994) Molecular and functional aspects of PECAM-1/CD31. Immunol Today 15:490–495

Duncan GS, Andrew DP, Takimoto H, Kaufman SA, Yoshida H, Spellberg J, Luis de la Pompa J, Elia A, Wakeham A, Karan-Tamir B, Muller WA, Senaldi G, Zukowski MM, Mak TW (1999) Genetic evidence for functional redundancy of Platelet/Endothelial cell adhesion molecule-1 (PECAM-1): CD31-deficient mice reveal PECAM-1-dependent and PECAM-1-independent functions. J Immunol 162:3022–3030

Gerber HP, Hillan KJ, Ryan AM, Kowalski J, Keller GA, Rangell L, Wright BD, Radtke F, Aguet M, Ferrara N (1999) VEGF is required for growth and survival in neonatal mice. Development 126:1149–1159

Hsu SM, Raine L, Fanger H (1981) Use of avidin-biotin-peroxidase complex (ABC) in immunoperoxidase techniques: a comparison between ABC and unlabeled antibody (PAP) procedures. J Histochem Cytochem 29:577–580

Ling V, Luxenberg D, Wang J, Nickbarg E, Leenen P J, Neben S, Kobayashi M (1997) Structural identification of the hematopoietic progenitor antigen ER-MP12 as the vascular endothelial adhesion molecule PECAM-1 (CD31). Eur J Immunol 27:509–514

Nagy A, Gertsenstein M, Vintersten K, Behringer R (2003). Manipulating the Mouse Embryo: A Laboratory Manual: Third Edition: Cold Spring Harbour Laboratory Press

Vecchi A, Garlanda C, Lampugnani MG, Resnati M, Matteucci C, Stoppacciaro A, Schnurch H, Risau W, Ruco L, Mantovani A et al. (1994) Monoclonal antibodies specific for endothelial cells of mouse blood vessels. Their application in the identification of adult and embryonic endothelium. Eur J Cell Biol 63:247–254

Methods for Visualizing Intact Blood Vessels Using Immunofluorescent Localization of PECAM (CD31) and α-Smooth Muscle Actin

Gavin Thurston and Danielle Jean-Guillaume

▓ Introduction

The vasculature is a complex and beautiful system that is organized on several levels. On one level, blood vessels are organized into a hierarchical network of arteries, capillaries, and veins. Within this network, the structure, function, and molecular properties of each segment are highly specialized. On another level, blood vessels are organized around the structure and function of the organ that they invest. The blood vessels of the kidney, for example, are patterned on a macroscopic scale very differently than the blood vessels of the brain. Similarly, on a microscopic scale, some of the blood vessels of the kidney are highly specialized and easily distinguishable from brain vessels.

Because of their dramatic heterogeneity and specialization, it is critical to identify the location of the vessels within the vascular network and within the specialized regions of an organ. Unfortunately, many methods to examine the histology and morphology of blood vessels involve cutting thin sections of tissue, or digesting the non-vascular tissue components, with the result that much information on the location of the vessels may be lost.

Over the past several years, researchers have recognized the power of maintaining intact blood vessels in their tissue context. Various methods have been developed to localize cellular and molecular structures of endothelial cells or smooth muscle cells, and then study them in intact blood vessels. Key developments in this process have been the use of reagents that specifically stain blood vessels, and microscopic approaches to visualize and record blood vessels, particularly the confocal microscope.

We have previously used several different approaches to stain and visualize intact blood vessels, including: 1) perfusing fluorescent dyes into lightly-permeabilized cannulated microvessels (Baldwin and Thurston 1995, Thurston and Baldwin 1994); 2) injecting or perfusing iv a fluorescent or biotinylated lectin which binds to the luminal surface of the endothelium

Springer Lab Manual
H. Augustin (Ed.)
Methods in Endothelial Cell Biology
© Springer-Verlag Berlin Heidelberg 2004

(McLean et al. 1997, Thurston et al. 1996, Thurston et al. 1998b, Thurston et al. 1999); 3) rapidly perfusing silver nitrate to stain endothelial cell borders (McDonald 1994, Murphy et al. 1999, Thurston et al. 1998b); 4) injecting iv an antibody that binds to an endothelial-specific cell-surface protein (Corada et al. 1999); 5) injecting iv fluorescently labeled cationic liposomes that bind to the luminal surface of the endothelium (McLean et al. 1997, Thurston et al. 1998a); and 6) staining blood vessels in thick tissues by prolonged immersion in antibodies (Thurston et al. 2000, Thurston et al. 1999).

These different methods have advantages and disadvantages depending on the specific information being sought, but they also require different levels of technical expertise and non-commercial reagents. This chapter describes a simple and robust method to stain endothelial cells and smooth muscle cells in intact blood vessels, with application to several different tissues from mice. The method involves the staining of thick tissue sections or whole tissues by prolonged incubation with vascular specific antibodies and fluorescent secondary antibodies. Suggestions will also be provided on how to visualize and record some of the available information using a confocal microscope.

Materials

Reagents

- Rat anti-mouse CD31 (PECAM-1) monoclonal antibody (Pharmingen, Cat. No. 550274, www.pharmingen.com; various other PECAM-1 antibodies are available)
- Cy3-labeled monoclonal anti-α-smooth muscle actin antibody (Sigma, clone 1A4, Cat. No. C-6198, www.sigmaaldrich.com)
- FITC-labeled goat anti-rat antibody (Jackson ImmunoResearch Laboratories, Cat. No. 112-095–102, www.jacksonimmuno.com)
- Phosphate buffered saline (PBS-Sigma, Cat. No. P-3813, www.sigmaaldrich.com)
- 0.3 % Triton X-100 (Sigma, Cat. No. X-100, www.sigmaaldrich.com) in PBS
- Goat serum (Gibco BRL, Cat. No. 16210-064, e.g., www.invitrogen.com)
- 24-well plates (Nunclon Surface, Nalge Nunc, Cat. No. 143982, www.nalgenunc.com)
- Vectashield (Vector Laboratories, #H-1000, www.vectorlabs.com)
- Glass slides (VWR Scientific, superfrost®, Cat. No. 48311-702, www.vwrsp. com)
- Glass coverslips, 50 mm (Corning, Cat. No. 469750, www.corning.com)
- Paraformaldehyde (Sigma, Cat. No. P6148) made up to 1 % in PBS, pH 7.4

- SeaPlaque low melting temperature agarose (BioWhittaker Molecular Applications, Cat. No. 50100, e.g., www.cambrex.com) made up to 7 % in PBS.

Supplies

- Polystyrene beaker cups (VWR Scientific Products, Cat. No. 13915-985)
- Instant Krazy glue, industrial razor blades (VWR Scientific Products, Cat. No. 55411-050)
- Feather® blades (Ted Pella, Inc, Cat. No. 121-9, www.tedpella.com)

Equipment

- Confocal microscope, Vibratome (The Vibratome Co., www. vibratome. com)

Procedure

Prepare tissue

1. Euthanize mouse by approved procedures. Although not necessary, better results can be obtained by perfusing the vasculature of the mouse with PBS, or, better still, perfusing with fixative (Thurston et al. 1998a, Thurston et al. 1999).

2. Harvest tissue. For obtaining ear skin, remove whole ear at base. Other tissues can also be collected. For example diaphragm, kidney, and intestine.

3. Fixation. Immerse tissues in fixative (1 % paraformaldehyde in PBS, pH 7.4) for 60 min at RT.

Staining of whole mount tissues

Many mouse tissues, such as ear skin, diaphragm, trachea, bladder, etc., can be processed and examined as whole mounts. Working with whole tissue not only preserves the entire vascular architecture, but also saves much time for tissue processing. This section will describe the dissection of mouse ear skin which is a particularly useful vascular bed.

1. Dissection of tissue (use dissecting microscope if necessary). For ear skin, rinse tissue briefly in PBS. Trim base and sides of ear with razor blade to create clean tissue edges. Wet ear with PBS and keep wet during procedure. Using medium-fine forceps, gently grab base of ear skin and peel back one layer of skin, thus exposing dermal layers, auricle muscles, and cartilage. While peeling back skin, try to keep entire cartilage sheaf

attached to one layer of skin (either dorsal or ventral). After skin has been peeled apart, gently remove cartilage by peeling and scraping with medium-fine forceps. Place tissue in a 4-well plate.

Steps 2 to 7 should be done using a rocker at 4 °C

2. Wash ear skin several times with PBS, then permeabilize for 1 h in PBS + 0.3 % Triton X-100.

3. Block with 3 % goat serum in PBS + 0.3 % Triton X-100 for 2–3 h.

4. Primary antibody. Incubate tissue with anti-PECAM antibody (diluted 1:500 to 1:1000) in PBS + 0.3 % Triton X-100 for 24 h.

5. Wash 3 times with PBS + 0.3 % Triton X-100 for 40 min each.

6. Secondary antibodies. Incubate tissue with FITC anti-rat antibody (1:800) and Cy3 monoclonal anti-α-smooth muscle actin antibody (1:800) in PBS + 0.3 % Triton X-100 for 24 h.

7. Wash 3 times with PBS + 0.3 % Triton X-100 for 40 min each.

8. Remove tissue from wells and gently blot on Kimwipe. Mount tissue on glass slides in Vectashield, being carefully to spread tissue and avoid dirt and air bubbles (recommended to use dissecting microscope). For ear skin, mount with dermal aspect toward coverslip.

9. Examine with fluorescence or confocal microscope. See below for suggestions with confocal microscopy. Keep track of location of vessels and whether from dorsal or ventral aspect of ear.

Vibratome sections

Some tissues, such as kidney, lung, and liver, are best prepared by cutting thick (100 µm) sections. These sections are then stained, mounted, and examined.

1. Fix tissues for an additional 2 h to overnight at 4 °C in PBS + 1 % paraformaldehyde.

2. Prepare 7 % agarose solution in PBS. Prior to embedding tissue, warm agarose solution to 65 °C to remove all air bubbles.

3. When agarose is ready, pour into small polystyrene cups. Let agarose cool to approx. 40 °C (test by touch to inner wrist, similar to testing baby's milk). Higher temperatures will heat tissue and result in auto-fluorescence. Lower temperatures will make it difficult to embed and orient tissue.

4. Remove tissue from fixative using forceps, blot tissue dry, and then immerse in agarose. Orient tissue such that the surface to be cut first is facing up.

5. Quickly cool and harden agarose, using ice or placing at 4 °C. Allow agarose to thoroughly harden. Do not freeze.

6. Remove agarose block with tissue from polystyrene cup, being careful not to crack agarose. Trim agarose block with razor blade, leaving at least 2 mm of agarose around tissue.

7. Mount agarose/tissue block in chuck of Vibratome, sealing in place using superglue, and being careful to keep orientation of tissue. Let glue dry, then cover specimen with cold (4 °C) PBS so that sections will float after cutting.

8. Insert one half of sharp Feather® razor blade into blade holder of Vibratome. Cut sections at 75–150 μm thickness.

9. Once a number of sections have been cut and are floating in the cold PBS, carefully transfer sections to a 12- or 24-well plate. Keep immersed in PBS. Sections can be left in their agarose support.

10. Stain thick sections, essentially following Steps 2 to 7 above. Shorter incubation times (12 h instead of 24 h) may be applied.

11. Mount tissue in Vectashield. When mounting the tissue sections, agarose can be left on the tissue if it provides support. Use considerable care to keep sections flat and properly oriented.

Suggestions for using confocal microscope

The confocal microscope is very useful for visualizing and documenting intact blood vessels, particularly when the vessels are specifically stained with fluorescent markers. Fluorescence really comes into its own when multiple markers are used: two or three distinct fluorophores can be easily detected and distinguished with fluorescence confocal microscopy, making co-localization studies very powerful. Each make and model of confocal microscope has its own operating procedure, and this section is not intended to be a user's manual for confocal microscopy. Rather, it is intended to provide some tips that we have found useful for imaging intact blood vessels.

1. Get the overview first. Collect low magnification (10× objective lens or lower) images first, and then move to higher magnification.

2. For higher magnification imaging of thick tissue or whole mounts, some excellent 40× high numerical aperture (i.e., oil immersion) lenses are available which also provide very long working distances. Such lenses are available from Zeiss, Leica, and Nikon. Images can be collected up to 120 μm into tissue with these oil-immersion lenses.

3. Collect a series of images, at different focus levels (z-series), which spans the entire vascular structures of interest in the field. Match the step size of the z-motor to the numerical aperture (resolution) of the lens.

4. The easiest and most routine way to visualize the z-series is to compress the images down into a single image, normally called a projection. Although the images can be displayed in a wide variety of ways (rotations, surface contours, transparent structures), these visualizations cannot be customized, and so the process takes a lot of time without necessarily providing much more information.

5. For some applications, particularly when trying to see subcellular structure in endothelial cells, it is often useful to project only one surface of the blood vessel. The image series can be collected such that it spans both the upper and lower surfaces of the vessel, but then a subset of the images can be selected for projection, such that only one surface is shown at a time. This can eliminate confusing overlap of opposing endothelial cells.

6. Strive to get specificity of staining. Although it may be tempting to use the confocal microscope to localize the vascular distribution of a ubiquitous protein, most likely the signal from surrounding tissue will obscure or obliterate the vascular staining. Vessel-specific stains should be used, and, if that is not possible, then consider ways to perfuse the stain into the blood vessels and thereby limit the extent of staining (Baldwin and Thurston 1995, Thurston and Baldwin 1994). For example, nucleic acid stains can be perfused into the vasculature of an animal after perfusion of fix and very brief perfusion of permeabilization buffer. With this approach, staining is largely confined to the nuclei of the blood vessels (Thurston and Baldwin 1994).

7. Get to know the organ system and vasculature that is under study, and keep track of the type, size, and location in the organ and in the network of the vessels that are imaged. Some aspects of vascular biology have been shown to be remarkably specific to a particular organ or to a particular segment of the vasculature. Only with experience and familiarity with the particular organ system can these effects be properly documented.

Expected Results

Figure 1 shows examples of tissues stained according to the above procedures. The blood vessels (PECAM, CD31) are shown in green, and smooth muscle cells (α–smooth muscle actin) are shown in red. These images are projections from confocal z-series that are 20–30 µm thick.

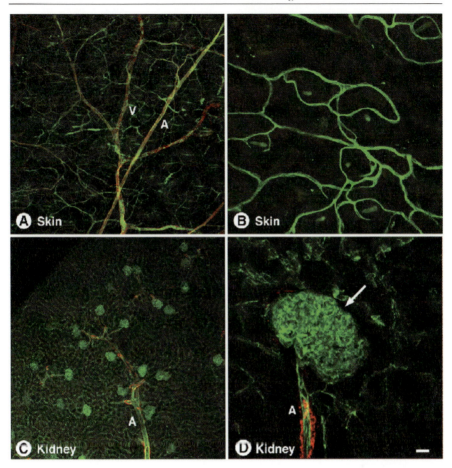

Fig. 1A-D. Confocal micrographs showing blood vessels stained for PECAM (CD31) (*green*) and α-smooth muscle actin (*red*). **A** Low magnification view of ear skin, showing artery (*A*) and vein (*V*). **B** Higher magnification view of skin microvessels around hair follicles. Hair shafts appear as diffuse *green spots* or *ovals* in the middle of capillary loops. **C** Low magnification view of cortex region of kidney, showing artery (*A*) and numerous glomeruli. **D** Higher magnification view of kidney showing glomerulus (*arrow*) and afferent arteriole (*A*) with smooth muscle cell coat

In a low magnification (10× objective lens) image of a whole mount preparation of the skin vessels (Fig. 1A), arterioles are identifiable by their straight course and their organized layer of smooth muscle cells, whereas venules are characterized by a meandering course and a less organized coat of smooth muscle cells. Lymphatic vessels can also be seen occasionally. Lymphatics have fainter PECAM staining than blood vessels, few if any smooth muscle cells, and blind-end channels. In higher magnification (40× oil-immersion objective lens) images of skin vessels (Fig. 1B), the microvascular network around the hair follicles is clearly seen. The pericytes in this tissue show little or no immunoreactivity for α–smooth muscle actin.

In low magnification images of Vibratome sections of the kidney (Fig. 1C), arterioles are again identifiable by their prominent layer of smooth muscle cells. The glomeruli are easily seen as a dense cluster of small vessels. At higher magnification (Fig. 1D), an afferent arteriole with a smooth muscle coat is visible, and the glomerular capillaries form a tightly woven network. Although the glomerular capillaries are somewhat obscured in this projected image, individual optical sections through the glomerulus show much more details of the vasculature (not shown).

Troubleshooting

Below are some potential problems and possible solutions.

▶ Strong non-specific or background staining
  May be a problem with the secondary antibody – check that it has been cross-absorbed against host species.
▶ Too much tissue autofluorescence
  The fixative may be too strong or too old, or tissue may have been in fixative too long. Alternatively, tissue may have been over-heated during agarose embedding.
▶ Tissue will not cut with the Vibratome
  Try using a stronger fixative or longer duration of fixation. Keep the tissue cooled when cutting by putting ice around the chuck and tissue holder. Cutting thicker sections usually helps too. If problems persist, try cutting smaller piece of tissue.
▶ Tissue pops out of agarose block during cutting
  Tough connective tissue may be catching on blade. Alternatively, tissue may not have been dry before embedding in agarose. Trim tissue, gently blot dry, and re-embed in agarose.

References

Baldwin AL, Thurston G (1995) Changes in endothelial actin cytoskeleton in venules with time after histamine treatment. Am J Physiol 269:H1528–1537

Corada M, Mariotti M, Thurston G, Smith K, Kunkel R, Brockhaus M, Lampugnani MG, Martin-Padura I, Stoppacciaro A, Ruco L, McDonald DM, Ward PA, Dejana E (1999). Vascular endothelial-cadherin is an important determinant of microvascular integrity in vivo. Proc Natl Acad Sci USA 96:9815–9820

McDonald DM (1994) Endothelial gaps and permeability of venules in rat tracheas exposed to inflammatory stimuli. Am J Physiol 266:L61-L83

McLean JW, Fox EA, Baluk P, Bolton PB, Haskell A, Pearlman R, Thurston G, Umemoto EY, McDonald DM (1997) Organ-specific endothelial cell uptake of cationic liposome-DNA complexes in mice. Am J Physiol 273:H387-H404

Murphy TJ, Thurston G, Ezaki T, McDonald DM (1999) Endothelial cell heterogeneity in venules of mouse airways induced by polarized inflammatory stimulus. Am J Pathol 155:93–103

Thurston G, Baldwin AL (1994) Endothelial actin cytoskeleton in rat mesentery microvasculature. Am J Physiol 266:H1896-H1909

Thurston G, Baluk P, Hirata A, McDonald DM (1996) Permeability-related changes revealed at endothelial cell borders in inflamed venules by lectin binding. Am J Physiol 271:H2547-H2562

Thurston G, Baluk P, McDonald DM (2000) Determinants of endothelial cell phenotype in venules. Microcirculation 7:67–80

Thurston G, McLean JW, Rizen M, Baluk P, Haskell A, Murphy TJ, Hanahan D, McDonald DM (1998a) Cationic liposomes target angiogenic endothelial cells in tumors and chronic inflammation in mice. J Clin Invest 101:1401–1413

Thurston G, Murphy TJ, Baluk P, Lindsey JR, McDonald DM (1998b) Angiogenesis in mice with chronic airway inflammation: strain-dependent differences. Am J Pathol 153:1099–1112

Thurston G, Suri C, Smith K, McClain J, Sato TN, Yancopoulos GD, McDonald DM (1999) Leakage-resistant blood vessels in mice transgenically overexpressing angiopoietin-1. Science 286:2511–2514

Chapter 31

Intravital Multi-fluorescence Microscopy

MOHAMMAD FARHADI, KARSTEN SCHWINN, and PETER VAJKOCZY

Introduction

Intravital multi-fluorescence microscopy represents an ideal tool for the quantitative assessment of the dynamic process of angiogenesis and microcirculation of normal and neoplastic tissue (Vajkoczy et al. 1998, Vajkoczy et al. 2000, Vajkoczy et al. 2002). In combination with transparent chamber models, it allows for a direct, continuous, and noninvasive in vivo visualization of the tumor microvasculature at the level of individual microvessels. Moreover it allows for a quantitative evaluation of various microhemodynamic parameters. Contrast enhancement with fluorescently labeled dextrans or albumin enables the visualisation of angiogenic sprouts, individual microvessels, and the tumor microvasculature. The use of nuclear dyes such as Rhodamin 6G enables visualisation and sequential analysis of leukocyte-endothelial cell interactions within individual microvessels. Finally, labeling tumor cells prior implantation by either genetic means (eGFP) or fluorescent dyes (e.g., Fast Blue, DiI) allows the simultaneous identification of individual tumor cells, their proliferation, migration behaviour, and their interaction with host and tumor vessels.

We use two transparent chamber models: the dorsal skinfold chamber and chronic cranial window. The dorsal skinfold chamber represents an ectopic model but is easier to prepare and allows observations for up to 28 days. In contrast, the cranial window represents an orthotopic model for brain tumors, but only allows observations for up to 14 days. When performing these microsurgical preparations, it is essential that the preparation remains atraumatic so that surgery by itself will not induce angiogenesis. Therefore, preparations should be performed 2 days prior to tumor cell implantation in order to exclude traumatized, inflamed, or hemorrhaged preparations.

Springer Lab Manual
H. Augustin (Ed.)
Methods in Endothelial Cell Biology
© Springer-Verlag Berlin Heidelberg 2004

Materials

Chamber preparation

- Mice (nu/nu, SCID, or immunocompetent, male/female, 28–32 g)
- Ketamine hydrochloride and xylazine (Hoechst Roussel Vet, www.archive.hoechst.com and Bayer Vital, www.bayervital.com)
- Microsurgical instruments (fine forceps and iris microscissor; Aesculap, www.aesculap.com)
- High-speed drill with a burr tip size of 0.5 mm (Bürklin, www.buerklin.com)
- Stereomicroscope (Leica, www.leica.com)
- Surgical consumables (4-0 sutures, saline, paddies, etc.)
- Rodent stereotactic head holder (for chronic cranial window; Bilaney, www.bilaney.co.uk)
- Titanium skinfold chamber (Institute for Surgical Research, www.icf.med.uni-muenchen.de)
- Glass coverslips (8 mm and 12 mm for the cranial window and skinfold chamber; Plano, www.plano-em.de)
- Histocompatible glue (Kisling Cie AG, http://einzelhandel.freepage.de/lkammer/)

Fluorescence-labeling of tumor cells

- Appropiate culture medium (depending on tumor cells)
- FBS (Fetal Bovine Serum)
- Dulbecco's PBS
- Viralex trypsin/EDTA (0.05 %/0.02 %)
- Sterile glass and plastic pipettes (1 ml, 2 ml, 5 ml, 10 ml)
- Sterile Falcon tubes (15 ml or 50 ml)
- Pump to discard old medium
- Inversion microscope
- Biosafety cabinet (vertical flow preferably)
- 75 cm² culture flasks
- Aliquots of 25 µl of Fast Blue (10 mg/ml, in cryotubes; Sigma, www.sigmaaldrich.com)

Tumor cell implantation

- Micropipet with tips
- Glass coverslips (8 mm and 12 mm for the cranial window and skinfold chamber)
- Stereomicroscope
- Histocompatible glue

Intravital microscopy

- Epi-fluorescence microscope with a 100-W mercury lamp attached to a block with an ultraviolet (340 to 380 nm), blue (450 to 490 nm) and green (520 to 570 nm) filter (Zeiss, www.zeiss.com)
- 2% FITC-Dextran$_{150}$ (or other molecular sizes; Sigma) and 0.2% Rhodamine 6G (Serva, www.serva.de)
- 3.2× long distance, 10× long-distance and 20× water immersion objectives (Zeiss)
- Low-light level charge-coupled device (CCD) video camera (Kappa, www.kappa.de)
- S-VHS video system (Panasonic, www.panasonic.de)
- Computer assisted image analysis system

▨ Procedure

Chamber preparation

Chronic cranial window preparation:
1. Anesthetize animal by s.c. injection of 1.2 mg of ketamine hydrochlorid and 0.16 mg of xylazine per 10 mg of body weight.

2. Fix head in stereotactic head holder.

3. Clean with water and desinfect head with 70% alcohol.

4. Incise skin medially over the frontal and parietal skull and remove periosteum.

5. Create a circular bone flap (6 mm in diameter) between coronal and lambdoid suture using the electrical high-speed drill. Cool the skull by continuous irrigation!

6. Free bone flap from underlying dura and superior sagittal sinus.

7. Remove dura under high magnification, avoiding damage to the sinus or bridging veins.

8. Seal window with glass coverslip using histocompatible glue.

Dorsal skinfold chamber preparation:
1. Anesthetize animal as described above

2. Remove dorsal skin hair, clean with water, and desinfect skin with 70% alcohol

3. Lift skin to form an extended double layer and identify subcutaneous arteries by holding it up against a light source (Fig. 1A). The skinfold has

to be big enough to span between the screws of the titanium frame without compromising the breathing of the animal. This is usually guaranteed for mice >20 g.

4. Fix one titanium frame to the opposite side of the skin using the 4-0 suture and perforating the skinfold with the screws so that the observation window is localized a few millimeters behind the major vessels. Make sure that these vessels are not injured in the course of the chamber preparation.

5. Remove the side of the extended double-layer of skin facing you (preferably the left side for right-handed surgeons) as well as the inner layers of the opposite side in a stepwise fashion using microsurgical techniques (Fig. 1B). The tissue should be removed in a circular way with a diameter of approximately 15 mm, being just a little bit larger than the the size of the observation window. The concept is to visualize the subcutaneous tissue from the inside through the window.

6. Close chamber with the second frame fixed by the three screws and the 4-0 suture. The frames now sandwich the remaining layers within the observation window, consisting of the striated skin muscle, subcutaneous tissue, and epidermis. Make sure not to tighten the screws too much which would compromise blood flow through the chamber´s major feeding and draining blood vessels.

7. Cover the preparation with a glass coverslip and fix it with an elastic ring (Fig. 1C).

Fig. 1A-C. Microsurgical preparation of dorsal skinfold chamber preparation. **A** Identify main feeding arterioles of skinfold (*arrow heads*) by holding it up against a light source. *Circle* indicates area where obervations window should be placed and where the tissue should be removed. **B** One titanium frame has been fixed to the opposite side of the chamber with the screws perforating the skin and the suture. Then remove the layers of the animal´s extended skinfold circularly using microsurgical techniques. **C** The two titanium frames sandwich the extended skinfold and the observation window is covered by a glass coverslip

Fluorescence-labeling of tumor cells

1. Dissolve Fast Blue (25 μl Fast Blue and 30 ml medium in a 50 ml falcon tube).

2. Add approx. 10 ml medium with Fast Blue to each culture flask.

3. Incubate for approx. 2 h (not longer) at 37 °C, 5 % CO_2, discard medium.

4. Pipet 3–4 ml PBS for a quick rinse.

5. Run over the whole surface, discard.

6. Add 2 ml trypsin-EDTA, incubate approx. 3–5 min, stop by adding 2–3 volumes of medium.

7. Pipet into centrifuge tube.

8. Wash trypsinized cells repeatedly.

9. Check cells microscopically. Cells don´t have to be completely free-flowing. Pool them so that tubes contain 5×10^6 cells.

Tumor cell implantation

1. Centrifuge tubes (2,000 rpm, 5 min, 37°) and discard supernatant.

2. Anesthetize animal.

3. Remove the coverslip with a 27G needle.

4. Place a suspension of 5×10^5 cells onto either hemisphere or skin muscle.

5. Seal the window with a new glass coverslip adhered to the bone using glue.

Intravital microscopy

1. Anesthetize animal as described above.

2. Inject 0.1 ml (i.v. via tail vein) of a cocktail mixed by 0.5 ml FITC-Dextran and 0.5 ml Rhodamin 6G.

3. Place the chamber preparation on the microscope stage.

4. Start microscopy with × 3.2 and × 10 objectives for overview, for more details use higher magnifications (×20).

5. Tumor cells and the tumor mass can be identified with ultraviolet epi-illumination.

6. Blood vessels can be visualized with blue light epi-illumination; Red blood cells appear in negative contrast.

7. Leukocyte-endothelial cell interaction can be assessed with green light epi-illumination.

8. For an unbiased analysis of the tumor microvasculature, follow a fixed pattern of microscopy and include central, marginal, as well as peri-tumoral areas.

9. Record microscopic images by means of the CCD camera and the video system.

10. The experiments will be analyzed off-line using the image analysis system.

Analysis of parameters

Quantitative analysis of intravital microscopic observation can be performed by a computer-assisted image analysis system. The following parameters can be analyzed:

- Tumor size (mm^2): area covered by tumor mass within the chamber.
- Total vessel density (cm^{-1}): defined as length of all newly formed microvessels per area, representing the classical vessel count and indicating the angiogenic activity.
- Functional vessel density (cm^{-1}): defined as length of red blood cell-perfused microvessels per area, reflecting the nutritive vessel density.
- Perfusion index (%): calculated from the functional and total vessel density, reflecting the angiogenic efficacy.
- Microvessel diameter (μm)
- Red blood cell velocity ($\mu m/s$)
- Blood flow rate (nL/s): calculated from the red blood cell velocity and microvessel diameter ($Q_v = x(D/2) \times RBCV/K$, where RBCV represents the red blood cell velocity, D the microvessel diameter and K (=1.3) the Baker Wayland factor considering the parabolic velocity profile of blood in microvessels).
- Permeability index: calculated as the ratio between intra- and extravascular contrast and reflecting the extent of fluorescent marker extravasation.
- Tumor cell migratory activity ($\mu m/h$): assessed by daily measuring the distance from the initial cell implantation site to the most distant population of individual cells.

Expected Results

Intravital microscopy can be performed for approximately 14 to 28 days using the chronic cranial window and dorsal skinfold chamber, respectively.

Contrast enhancement with FITC-conjugated Dextran or albumin allows visualization of the host and neoplastic blood vessels. In normal tissue, the arterial and venous system can be differentiated by means of vessel morphology and blood flow direction (Fig. 2). Neoplastic vessels, in contrast, lack an obvious microvascular organization. Neoplastic blood vessels are rather characterized by a heterogeneous and chaotic microangioarchitecture with an irregular branching pattern of vessels, heterogeneous blood flow, high permeability (as assessed by extravasation of the fluorescent dye), and some enlarged sinusoidal blood vessels with either sluggish or no blood flow (Fig. 3). The first signs of tumor-induced angiogenesis can be observed between day 2 and day 4 after cell implantation, characterized by dilated and tortuous host vessels and vascular sprouting from pre-existing capillaries and post-capillary venules. Between day 6 and day 14, the tumor is usually completely vascularized and the tumor begins to grow steadily. At this stage, the microvasculature is characterized by a heterogenous and chaotic microangioarchitecture, a population of microvessels with diameters ranging between 8 and 16 μm as well as a gradual perfusion failure in central areas. Simultaneous in vivo staining of leukocytes with Rhodamine 6G enables an analysis of leukocyte-endothelial cell interaction within individual vessels (Fig. 2). In contrast to normal vessels, neoplastic vessels show only a few number of leukocytes and almost no sticking or rolling cells. Furthermore, fluorescent labeling of the tumor cells prior implantation allows for the study of tumor cell migration and tumor cell/blood vessel interaction. For this purpose it is advisable to use tumor spheroids instead of tumor cell suspensions (Fig. 3). Within hours after tumor implantation, individual tumor cells begin to detach from their initial implantation site and migrate centrifugally in all

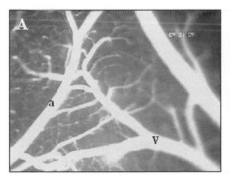

Fig. 2A,B. Intravital multi-fluorescence microscopic overview of brain microcirculation through cranial window preparation. A Visualization of microvasculature after contrast enhancement with FITC-Dextran₁₅₀. B Visualization of leukocyte-endothelial cell interaction for the same vessel segments as in A after i.v. injection of Rhodamin 6G; *a* artery, *v* vein, *arrow heads* leukocytes (Magnification × 216)

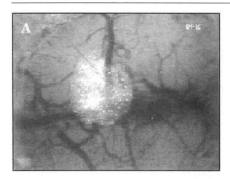
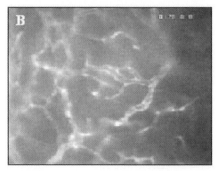

Fig. 3. A Intravital multi-fluorescence microscopy of Fast-Blue labeled tumor-cell suspension after implantation into the cranial window. **B** Tumor microvasculature on day 10 after implantation. Note the heterogeneous microangioarchitecture and the difference in total vessel density, vessel diameter and permeability compared to normal vessel. Contrast enhancement with FITC-Dextran$_{150}$ (Magnification × 216)

directions. Higher magnifications of migrating cells reveal two types of cell morphology: some cells display a bipolar morphology while others appeared round.

▧ Troubleshooting

Problem	Presumed cause	Solution
Cells		
– Cells do not grow after implantation	– Contamination of the cells	– Work as sterile as possible!
– Cells grow too fast (high mitosis)	– Early implantation after thawing	– Never implant frozen cells (directly after thawing). Grow fresh cells (at least passaged once). After 1 or 2 days, the cells are ready for implantation.
– Unbound fluorescent dye	– No or too little washing	– Wash the cells at least three times to remove any excess of dye
Chambers		
– Infection/inflammation	– Contamination during surgery	– Work as sterile as possible!
	– Leaky chamber	– Make sure that the coverslip is completely adherent to the skull by the glue. Give antibiotics prophylactically after cranial window preparation and after cell implantation

Problem	Presumed cause	Solution
Microscopy		
– No image or very dark pictures without visualisation of any vessels – Blurred image	– Paravasal injection of fluorescent dyes – Inflammation, pus formation – Edema	– Reinject via tail vein – In case of inflammation, pus or hemorrhage there is no chance to get clear pictures. Exclude animal from experiment – Edema is encountered in the first days after cell implantation. Try again on later days. – Check technical equipment. Change settings on the camera

References

Vajkoczy P, Farhadi M, Gaumann A, Heidenreich R, Erber R, Wunder A, Tonn JC, Menger MD, Breier G (2002) Microtumor growth initiates angiogenic sprouting with simultaneous expression of VEGF, VEGF receptor-2, and angiopoietin-2. J Clin Invest 109:777–785

Vajkoczy P, Schilling L, Ullrich A, Schmiedek P, Menger MD (1998) Characterization of angiogenesis and micro-circulation of high-grade glioma: an intravital multifluorescence microscopic approach in the athymic nude mouse. J Cereb Blood Flow Metab 18:510–520

Vajkoczy P, Ullrich A, Menger MD (2000) Intravital fluorescence videomicroscopy to study tumor angiogenesis and microcirculation. Neoplasia 2:53–61

Chapter 32

Corrosion Cast Analysis of Blood Vessels

Valentin Djonov and Peter H. Burri

▨ Introduction

The field of angiogenesis and lymphangiogenesis research has rapidly expanded in the last decades, which has led to the development and improvements of numerous investigative techniques. Inspite of the efforts, however, the three-dimensional (3D) visualization of the vasculature remains a methodological challenge. Nuclear magnetic resonance, angiography, computer tomography, and ultrasonics can so far not provide the resolution required for visualizing capillary beds. The successful application of these techniques remains restricted to the investigation of larger vessels. Three-dimensional reconstruction from stacks of serial sections or by laser scanning microscopy are delicate and time consuming approaches, which are limited to small pieces of tissue.

Analysis of vascular corrosion casts is a powerful technique for the investigation of the 3D vessel architecture and of the microvascular phenotype in different organs. Owing to the relatively simple procedure and to the high resolution achieved, the vasculature of whole animals can be investigated in details. The origin of today's methods can be traced back to Leonardo da Vinci (1452–1519) who made wax replica of the heart cavities and of the cerebral ventricles. Since then, the method was repeatedly refined using different casting materials, varying the procedures for removing the surrounding tissues and improving the viewing technique. The quality of this approach improved dramatically with the combination of inviscid methylmethacrylate casting media and scanning electron microscopic cast examination (Murakami 1971). During the last three decades, a large number of papers using the corrosion cast technique have been published. They provided a 3D insight into the phenotype of blood and lymphatic vessels phenotype, and the architecture of vascular beds.

Springer Lab Manual
H. Augustin (Ed.)
Methods in Endothelial Cell Biology
© Springer-Verlag Berlin Heidelberg 2004

◾ Materials

Reagents

Flushing solutions
- 0.9 % sodium chloride containing 1 % Liquemine® and 1 % Procaine®
- PBS buffer

Fixatives

- 0.5–2.5 % glutaraldehyde:
  2.5 % glutaraldehyde in 0.03 M potassium phosphate – buffer, pH 7.4, 350 mOsm
- ½ Karnovsky fixative:
  2 % Paraformaldehyde + 2.5 % Glutaraldehyde in 0.1 M Sodium-cacody-lat solution, pH 7.4, 1,100–1,200 mOsm

Casting media

- Methylmethacrylate (Mercox®, Vilene Company, www.vilene.co.jp)

Equipment

- Syringes and cannulas: different sizes (from 0.3 to 50 ml) and calibers (blunted commercial needles). In case of small capillaries, glass cannulas with tip diameters of 4–10 µm are recommended.
- Micromanipulator (may be needed if cannulating is problematic)
- Water bath or incubator
- Stereomicroscope
- Manometer
- Ventilated hut or gas mask
- Sputtering machine (in our case: MED 020 from BAL-TEC, www.bal-tec.com) with gold targets.
- Scanning electron microscope (SEM). In our case: Philips XL 30 FEG (www.philips.com).

◾ Procedure

Flushing the blood

Complete flushing of the blood from the vascular lumina has been indicated to be crucial in obtaining good corrosion casts (Lametschwandtner et al. 1990). The vascular bed of the region of interest should therefore be rinsed with a solution of 0.9 % sodium chloride containing 1 % Liquemine® and 1 % Procaine® at body temperature to prevent blood clotting and induce vascular relaxation (Caduf et al. 1986).

Alternatively, 0.9 % NaCl alone or PBS could be used without significant loss of quality. In case of very small areas of interest and small tissue specimens with thin supplying vessels (in the range of 20–30 μm), the double step procedure, i.e., rinsing followed by casting may be quite difficult to perform due to the fragile vessels and the complicated micromanipulation. In this case, rinsing could be omitted and the vasculature directly perfused with the casting media.

Perfusion prefixation

The significance of vascular prefixation prior to the casting perfusion is controversial. Some authors suggest that this pre-treatment induces vasospasm and alterations in the intravascular pressure. From our practical experience, we learned that perfusion prefixation with 0.5–2.5 % glutaraldehyde or 0.5 % Karnovsky fixative is appropriate for fine fragile vessels (neoangiogenesis in tumors, VEGF treatment, etc.).

In case of small vessels, fixation and casting could be performed in one step: the first 0.2–1.0 ml of the syringe is filled with fixative and the rest with Mercox. Perfusion with the fixative is done very slowly – about 0.5 ml/min One to two minutes are sufficient for the fixation of the tiny vessels and the following resin perfusion takes place in stable and tight vessels without ruptures and leakage.

Casting media and polymerization

A broad spectrum of casting media such as resins, plastics, araldite, tardoplast, or silicon rubber, etc., have been used for analyses of the vasculature (for more details see the review articles by Lametschwandtner and Lametschwandtner 1992, Ohtani and Murakami 1992). Very important factors determining the casting quality are the viscosity and shrinking behavior of the casting medium. Lametschwandtner and colleagues reported that material viscosity and shrinking properties are inversely related – media with lower viscosity rather suffer from shrinkage and vice versa (Lametschwandtner and Lametschwandtner 1992).

Our experience with different casting media indicates that a solution of methylmethacrylate (Mercox®, Vilene Company) containing 0.1 ml accelerator per 5 ml resin is one of the most appropriate casting media. This mixture allows 5–6 min for the perfusion before getting viscous. Reduction of accelerator concentration to 0.05 ml per 5 ml resin can prolong the perfusion time to 8–9 min, if needed, without significant reduction of quality. Alternatively, a mixture of Mercox with 20 % methylmethacrylate monomer was used frequently as a low viscosity medium (Murakami 1971). Methylmethacrylate is toxic; for this reason, the perfusion and the following cast preparation should be performed under a ventilated hut, or the personnel should wear gas masks.

Perfusion pressure and cannula size

Perfusion pressure and the size of the perfusion cannula are important factors affecting the intravascular flow of the perfusate. Many of the reported casting studies have been performed with a perfusion pressure allegedly equal to the physiological pressure in the cannulated artery – about 70 to 120 mmHg. However, very often the perfusion pressure monitored by a side-manometer does not reflect the pressure in the vascular bed. The latter depends on many factors such as cannula size, perfusion distance, side connections, caliber variation in the syringe-cannula assembly, etc. Often pressure monitoring only puts the researcher's conscience at ease; indeed, measurements cannot reflect at all the real pressure in the vascular bed. Pressure measurements are recommended, however, in case of repetitive casting experiments of the same organ or tissue followed by qualitative or quantitative comparative studies, in order to maintain an inner standard.

As a general rule of thumb, the following recommendations can be made (especially important for tumors and small objects): I) cannula size should approximate the diameter of the perfused vessel; II) slow perfusion with low pressure until Mercox flows out through the veins; III) small syringe with a large cannula causes a high perfusing pressure, while a large syringe in combination with a thin cannula results in a lower perfusing pressure when applying the same force; IV) swelling of tissue or rupture of vessels in the perfused organs should be avoided by all means.

Tissue corrosion

Polymerization of the medium starts immediately after addition of the accelerator. The viscosity increases unnoticeably during about 5 min and then very rapidly. One hour after perfusion, the resin is hardened enough for the organs to be excised and transferred to a water bath at 40–60 °C overnight. During the setting process the perfused organs should be kept moist to prevent desiccation and shrinkage. The immersion in the water bath prevents cast deformation and shrinking during definitive setting of the resin. A water temperature of 40–60 °C is suggested for optimal polymerization and to prevent artefacts (Lametschwandtner and Lametschwandtner 1992).

After casting, the maceration process is very important for a good quality of the cast. Treatment with different acids and alkalis at variable concentrations and combinations were reported (Ohtani and Murakami 1992). Recently, commercially available proteinases such as Prozyme 6® (fungal protease, Amano, www.amano-enzyme.co.jp) have been applied for 2 weeks at 40 °C resulting in very good soft tissue digestion (Matsuo and Takahashi 2002). In case of methylmethacrylate as casting medium, the maceration with concentrated acids is not recommended because it severely damages the surface of the vascular casts.

With two decades of casting experience in our laboratory, the maceration with 15 % potassium hydroxide (KOH) for 3 to 4 weeks at RT yields excellent results. Recently, in an attempt to shorten the procedure, we applied the following treatment protocol after casting: immersion of excised organ in a water bath at 60 °C for 2–3 days, followed by immersion in 15 % KOH at the same temperature for 2–7 days. A disadvantage of all rapid procedures at higher temperature or with higher concentrations of KOH (20 %–25 %) is the risk of surface damage; therefore, daily cast inspections are recommended. With specimens originating from skin, muscle, liver, etc., the casts may be wrapped after digestion with a white frothy material, probably saponified lipids. Those remains can be removed in gently running warm water in a few hours.

Cast dehydration and drying

The casts should be washed gently in running water before dehydration, and the last rinse must be made with distilled water to avoid salt crystallization on the cast surface during drying. After the last wash, casts are dehydrated in ascending ethanol solutions: 70 % for 30 min, 80 % for 60 min, 96 % for 60 min, 100 % for 60 min, and again 100 % for 60 min. The casts are then dried in a vacuum desiccator, at RT, or in oven at 60 °C overnight.

In case of delicate fragile casts, the specimens can be freeze–dried. This method may protect from cast breakages due to surface tension forces. The casts should be frozen slowly in distilled water, and subsequently freeze-dried in a lyophilizor (Lametschwandtner et al. 1990).

Cast dissection and mounting

Dissection of the vascular casts can be performed at several levels during preparation. Usually, we excise the target organs about 1 to 2 h after injection, or after complete polymerization (24 h).

Large specimens can first be cut into smaller pieces of up to 10–15 mm side length. Do not forget to indicate some topographical hallmarks of the probes (caudal, cranial, outer, inner part, etc.). In order to investigate the internal vascular phenotype, the casts are best dissected after maceration. To obtain this effect, the probes should be frozen in distilled water as described above and cut into slices with a fine saw blade, or a mini tool (10,000–15,000 rpm) blade. The final sectioning of the frozen blocks with a cryomicrotome provides a perfect specimen surface and is appropriate for overviews and for high-resolution images (Fig. 1). This method can be combined with the freeze–drying procedure mentioned above.

In case of small specimens with small areas of interest, the final dissection should be performed just before mounting, in order to achieve an optimal cast preservation.

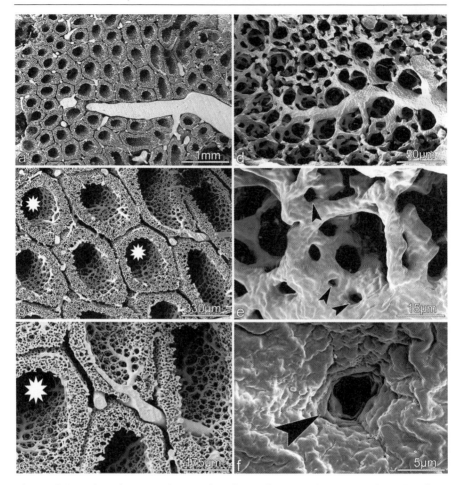

Fig. 1a–f. Scanning electron micrographs of vascular corrosion casts originating from embryonic chick lung at stage 45 (embryonic day 20). Zoom-in demonstrates the vascular phenotype at different magnifications (see *scale bars*). *Asterisks* label parabronchi and *arrowheads* the transluminal tissue pillars represented in the casts as holes (reminiscent of intussusceptive vascular growth)

Similarly, one may also remove casting artefacts and unnecessary vascular structures just before mounting, using for this a dissecting microscope.

The definitive cast specimens are then mounted on metal stubs. These are of different diameters and are commercially available (Plano GmbH, www.plano-em.de). We use LEIT-C as mounting medium (Conductive Carbon Cement, e.g., www.canemco.com). Alternatively, a copper multi-wire fixation, known as "conductive bridging", was proposed with good results (Lametschwandtner and Lametschwandtner 1992).

Sputtering with gold

To allow observing in the SEM, the casts must be covered with a conductive layer for electrons. The classical evaporation of OsO_4 onto the specimen is time consuming (2 to 3 days) and the results are not very satisfying. Sputtering with a gold layer of 10 nm thickness (MED 020 sputtering machine from BAL-TEC) usually yields excellent image rendering. Sputtering should be performed in 2–3 steps allowing rotation of the specimens, so that a nearly overall gold coating can be achieved – a measure, crucial for avoiding electrostatic charging.

Observation in SEM

The evaluation of the casts in the SEM requires a good morphological and technical background. The former is self-evident, the latter shall be briefly commented on. Selection of the appropriate accelerating voltage is important for good quality SEM pictures. Voltages in the range from 7.5 to 15 kV provide sharp "deep" micrographs with good 3D visualization. In contrast, low accelerating voltages (around 2.5 kV) produce less contrasted, "flat" images. Reducing the voltage has the advantage to diminish the electrostatic charging typical for old or insufficiently sputtered specimens. Too high voltages may lead to charging even in adequately sputtered preparations (Fig. 2).

The identification of the different vascular components in normal tissue is not a big challenge at first sight. While the caliber of capillaries ranges normally between 4 and 20 μm, it may reach in some cases, (such as in embryonic development or after pro-angiogenic treatment), up to 30–40 μm. Terminal arterioles and postcapillary and collecting venules have a diameter of about 20–50 μm. Small arteries and veins range from 50 to 100 μm. Distinction between arteries and veins can be difficult. The shape of the endothelial imprints visible on the cast surface is a good differential criterion: elongated nuclear imprints running in parallel to the vessel axis are characteristic of arteries. In contrast, oval or circular imprints are typical for the venous system. This criterion cannot be applied in the vicinity of the capillary plexus where all nuclei look very similar. Vascular typing in casts should always be done in the light of the typical, organ-specific vascular pattern, which is to be verified in the literature. When the vasculature is experimentally challenged (special pro- or anti-angiogenic treatment, tumors, etc.), and the effects are unknown so far, a combined methodological approach is required. The vascular characteristics have to be confirmed additionally by light and/or transmission electron microscopy, immunohistochemistry, confocal microscopy or in vivo investigation, etc. (Djonov et al. 2000a, Djonov et al. 2000b, Djonov et al. 2002). The additional methods will provide supplementary informations about the luminal configuration, the morphology and functional status of the endothelium, the presence of pericytes, basal membranes, etc. – all rel-

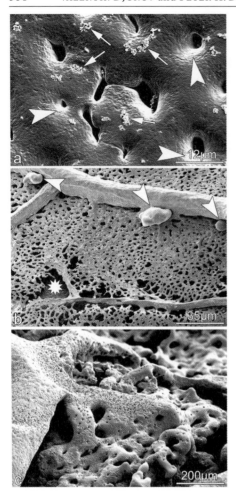

Fig. 2a–c. Different types of artefacts in corrosion cast specimens obtained from chick CAM. a Insufficient digestion is represented by tissue remains on the smooth cast surface (*arrows*). Same panel illustrates high electrostatic charging (*arrowheads*) caused by insufficient gold sputtering in combination with a high accelerating voltage (15 kV). b Globular-like extravasations connected to the vessels (*arrowheads*) and incomplete vascular casting (*asterisk*) are the most frequent vascular filling artefacts. c represents specific surface erosions caused by overdone maceration

evant data for the general appraisal of the investigated vasculature. Combined investigations may be needed also to distinguish between incomplete filling, vascular sprouts and/or obturated vessels in regressing tissue (Figs. 2 and 3).

Short casting protocol

1. Cannulate the heart, thoracic aorta, carotid artery, or regional supplying artery depending on area of interest.

2. Perfuse the animal (organ) with a solution of 0.9 % sodium chloride containing 1 % Liquemine® and 1 % Procaine® (5–8 min) at 37 °C.

3. Prefixation with 0.5–2.5 % glutaraldehyde (only in case of fine, fragile vessels, where Mercox alone failed).

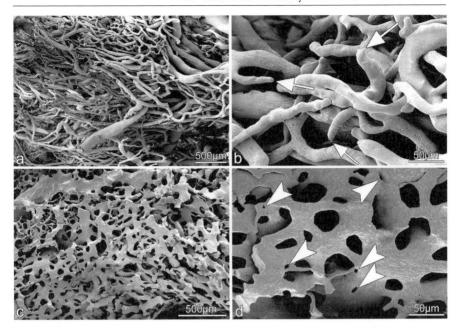

Fig. 3a–d. Scanning electron micrographs of methylmethacrylate corrosion casts of the vasculature in mammary gland tumors of c-neuT transgenic mice. **a, b** sprouting phenotype; capillary sprouts are indicated by *arrows*. **c, d** intussusceptive phenotype; transcapillary pillars are indicated by *arrowheads* [reprinted from Djonov et al. (2001) Microsc Res Tech., with permission]

4. Perfuse with freshly prepared methylmethacrylate (Mercox®, Vilene Company) containing 0.1 ml accelerator per 5 ml resin (perfusion time is limited to 5–6 min).

5. One hour after casting, excise and transfer the specimens to water bath at 60 °C, leave overnight.

6. Maceration with 15 % potassium hydroxide (KOH) for 3 to 4 weeks at RT. Alternative short procedure: 2–3 days water bath at 60 °C, followed by 15 % KOH at the same temperature for 2–7 days; daily cast inspections recommended.

7. Wash the casts gently with running water, 1–3 h.

8. Wash the casts briefly with distilled water.

9. Dissection of the corrosion casts.

10. Cast dehydration in ethanol solutions: 70 % – for 30 min; 80 % – for 60 min; 96 % – for 60 min; 100 % – for 60 min, and again 100 % for 60 min.

11. Dry the casts in a vacuum desiccator or in an oven at 60 °C overnight.

12. Mounting of cast on specimen holder.

13. Sputtering with gold.

14. Observation in SEM.

Specific Approaches

Vascular casting of human specimens

Casting of human material is usually performed in resected human organs or on cadaver material. In both cases, the vascular casting may occur many hours after blood flow stopped. This makes the flushing of blood a crucial step. The flushing procedure should be long and extensive. In case of surgically removed material, immediate clearance of blood is recommended.

Cast analysis of lymph vessels

The field of lymphangiogenic research has exploded in the last few years. 3D visualization of lymphatic vessels, however, is very difficult and casting them is not a trouble-free approach. The limiting factors are the thin fragile vascular walls and the valve system, the caliber alterations along the lymph vessels, the presence of lymph nodes, and the complex delicate and leaky saccular structures at the periphery.

The casting procedure for lymph vessels either allows or requires some specific modifications: i) lumen clearance is not mandatory, ii) prefixation can be very helpful to reinforce the fragile vessel wall and prevent extravasations, iii) low viscosity casting media are to be preferred, iv) interstitial application of ink prior to casting could nicely visualize the draining lymph vessels.

The casting medium can be applied in two different ways – by intraluminal or interstitial injection.

The first mode is recommended for larger collecting vessels. However, the retrograde filling may be severely impaired by the well-developed valve system. Depending on the experimental setup, the technique works well: e.g., in testing new pro-lymphangiogenic strategies retrograde filling was successful: under these conditions, the newly formed lymph vessels were enlarged with or without insufficient valves, which greatly favored their filling.

Many pro-angiogenic and permeability factors applied on the top of the growing chick chorioallantoic membrane (CAM) induce primary or secondary lymphangiogenesis. To investigate their effects, we prepared duplex casts, i.e., simultaneous blood/lymph vessel casting. This can be achieved by separately injecting red Mercox into the blood and blue Mercox into the

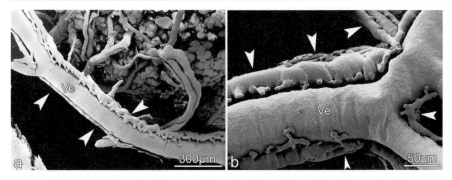

Fig. 4a,b. Combined vascular and lymphatic corrosion cast of chick CAM at day 11 of incubation, 48 h after VEGF$_{121}$ application. **a** Overview. **b** The distended lymphatics (*arrowheads*) escort larger vessels, in this particular case a draining vein (*Ve*)

lymph vessels accompanying the veins or arteries. After maceration, washing and drying, the casts are ready for observation under the stereomicroscope; (important: do not dehydrate in alcohol – the coloration disappears). This procedure is very appropriate for a rough cast evaluation.

For a more detailed high resolution investigation, the probes should be processed for SEM observation. The vascular patterns of lymph and blood vessels are so completely different, that their respective identification in SEM is unproblematic (Fig. 4).

The indirect or interstitial casting media injection has been used for the selective casting of small lymph vessels in their distal parts. Different approaches have been described by Castenholz and colleagues (Castenholz et al. 1998). The rationale of the method is based on the combination of a low viscosity casting medium and its interstitial injection. The resin spreads into the initial lymph vessels and successively fills the collecting vessels. The injection has to be performed with very low pressure (rupture and swelling are to be avoided), and only small volumes (less than 1 ml) of casting media are allowed.

Troubleshooting

▶ Corrosion artefacts
▶ Insufficient maceration is represented by tissue remains on the smooth vessel surface (Fig. 2a). Sometimes additional maceration for a few hours or days followed by a thorough washing procedure can help to remove those artefacts. Evidently renewed drying and sputtering procedures are required.

▶ Overdone maceration results in specific surface erosions and those casts usually are no more appropriate for qualitative high-resolution images (Fig. 2 c).

▶ Incomplete vascular filling
Incomplete casting is obvious if larger defects in the integrity of vascular networks are found. Defects are less conspicuous if they are only represented by blind endings of smaller vessels or capillaries (Fig. 2b). The distinction between incomplete filling, vascular sprouts, and regressing vessels can be difficult (for more details see "Observation in SEM").

▶ Extravasations
Perfusion with high pressure can cause extravasations. Pressures as high as 200 mm Hg have been used and represent the upper limit (Ohtani and Murakami 1992). In a specific situation, only a pilot study allows to determine the optimal pressure and the best compromise between complete filling and leakage. The extravasations in SEM pictures appear as spheres or globular structures connected to the vessels (Fig. 2b). In case of leaky and fragile vessels, short prefixation may help to avoid extravasations. It is to be noted that the resin in the fenestrated capillaries protrudes at the fenestrae and forms spotty elevations on the cast surface. They should not be interpreted as artefacts.

Acknowledgements. We would like to thank D. Stauffer, D. Friis, B. Haenni, K. Sala, B. de Breuyn, K. Babl, and B. Krieger for technical assistance and R. Hlushchuk and O. Baum for the critical reading of the manuscript. This research was supported by the Swiss National Science Foundation (grants no. 3100-055895.98/2; 3100A0-104000/1) and the Bernese Cancer League.

References

Caduff JH, Fischer LC, Burri PH (1986) Scanning electron microscope study of the developing microvasculature in the postnatal rat lung. Anat Rec 216:154–64

Castenholz A (1998) Functional microanatomy of initial lymphatics with special consideration of the extracellular matrix. Lymphology 3:101–18

Djonov V, Galli AB, Burri PH (2000a) Intussusceptive arborization contributes to vascular tree formation in the chick chorio-allantoic membrane. Anat Embryol 202:347–357

Djonov V, Schmid M, Tschanz SA, Burri PH (2000b) Intussusceptive angiogenesis: its role in embryonic vascular network formation. Circ Res 86:286–292

Djonov V, Kurz H, Burri PH (2002) Optimality in the developing vascular system: branching remodeling by means of intussusception as an efficient adaptation mechanism. Dev Dyn 224:391–402

Lametschwandtner A, Lametschwandtner U, Weiger T (1990) Scanning electron microscopy of vascular corrosion casts – technique and applications: updated review. Scanning Microsc 4:889–940

Lametschwandtner A, Lametschwandtner U (1992) Historical review and technical survey of vascular casting and scanning electron microscopy. In: Motta PM, Murakami T, Fujita H (eds) Scanning electron microscopy of vascular casts. Methods and applications. Kluwer Academic Publisher, Boston, pp 1–11

Matsuo M, Takahashi K (2002) Scanning electron microscopic observation of microvasculature in periodontium. Microsc Res Tech 56:3–14

Murakami T (1971) Application of the scanning electron microscope to the study of the fine distribution of the blood vessels. Arch Histol Jpn 32:445–454

Ohtani O and Murakami T (1992) Routine methods for vascular casting and SEM. In: Motta PM, Murakami T, Fujita H (eds) Scanning electron microscopy of vascular casts. Methods and applications. Kluwer Academic Publisher, Boston, pp 13–25

Appendix

Suppliers

Products	Supplier	Internet
Animals	Charles River Laboratories	www.criver.com
	Harlan Sprague Dawley	www.harlan.com
	Hoechst-Roussel-Vet	www.archive.hoechst.com
	Monsanto	www.monsanto.com
	Taconic	www.taconic.com
	The Jackson Laboratory	www.jaxmice.jax.org
Antibodies and histology	DakoCytomation	www.dakocytomation.com
	Dianova	www.dianova.com
	Ladd Research Industries	www.laddresearch.com
	Microbix	www.microbix.com
	Molecular Probes	www.probes.com
	Novocastra	www.novocastra.co.uk
	OEM concepts	www.oemconcepts.com
	R&D Systems	www.rndsystems.com
	Research Diagnostics Inc. (RDI)	www.researchd.com
	Upstate Biotechnologies	www.upstatebiotech.com
	Vector Laboratories	www.vectorlabs.com
	Vibratome	www.vibratome.com
	Zymed	www.zymed.com
Biochemical reagents	Amersham Biosciences	www.amershambiosciences.com
	Becton, Dickinson & Co./	www.bd.com
	Pharmingen/Clontech	www.bdbiosciences.com
	Chemicon International	www.chemicon.com
	Dunnlab	www.dunnlab.de
	Gibco/Invitrogen	www.gibcobrl.com
		www.invitrogen.com
	Helixx Technologies	www.helixxtec.com
	J.T.Baker Chemicals	www.jtbaker.com
	Leo Pharmaceutical Products	www.leo-pharma.com
	Mallinkrodt Chemicals	www.mallchem.com
	MDS Nordion	www.nordion.com
	Merck Biosciences	www.merckbiosciences.de
		www.merck.com
	Pharmacia/Pfizer	www.pfizer.com
	Pierce Biotechnology	www.piercenet.com

Products	Supplier	Internet
	Promega	www.promega.com
	QBIOgene	www.qbiogene.com
	Roth	www.carl-roth.de
	Selectscience	www.selectscience.net
	Serva	www.serva.de
	Sigma-Aldrich	www.sigmaaldrich.com
	VWR	www.vwr.com
Cells, media and plastics	American Type Culture Collection	www.atcc.org
	Biochrom	www.biochrom.de
	Biosource/Biofluids	www.biosource.com
		www.biofluids.com
	C.C.Pro	www.c-c-pro.com
	Cambrex	www.cambrex.com or www.cambrexbioproductseurope.com
	Cascade Biologics	www.cascadebio.com
	Cell Concepts	www.cellconcepts.de
	Cell Systems Biotechnologie	www.cellsystems.de
	Corning	www.corning.com
	Costar	www.labpages.com
	Forma Scientific	www.thermo.com
	Greiner	www.greiner-lab.com
	Nalge Nunc International	www.nalgenunc.com
	PAA	www.paa.at
	Paragon Biosciences Inc	www.paragonbioservices.com
	Promocell	www.promocell.com
Cytokines and enzymes	Amano	www.amano-enzyme.co.jp
	Becton, Dickinson & Co./ Pharmingen/Clontech	www.bd.com or www.bdbiosciences.com
	HyCult	www.hbt.nl
	ProQuinase	www.proquinase.com
	R&D Systems	www.rndsystems.com
	Reliatech	www.reliatech.de
	Sigma-Aldrich	www.sigmaaldrich.com
	Upstate Biotechnologies	www.upstatebiotech.com
Drug target discovery	Neurogenex Inc	www.neurogenex.com
	Sanquin	www.sanquin.nl
Flow cytometry	Baxter Healthcare Corp.	www.baxter.com
	Bayer Vital	www.bayervital.de
	Boehringer-Ingelheim	www.boehringer-ingelheim.com
	Chemicon International	www.chemicon.com
	Dade Behring	www.dadebehringcom
	Dunnlab	www.dunnlab.de
	Dynal Biotech	www.dynal.no
	Eppendorf	www.eppendorf.com
	Fisher Scientific	www.fisherscientific.com

Products	Supplier	Internet
	Gibco/Invitrogen	www.gibcobrl.com
		www.invitrogen.com
	Helixx Technologies	www.helixxtec.com
	Medite	www.medite.de
	Merck Biosciences	www.merckbiosciences.de or
		www.merck.com
	Miltenyi Biotec	www.miltenyibiotec.com
	Pharmacia/Pfizer	www.pfizer.com
	Pierce Biotechnology	www.piercenet.com
	Polysciences	www.polysciences.com
	Promega	www.promega.com
	QBIOgene	www.qbiogene.com
	Selectscience	www.selectscience.net
	StemCell Technologies	www.stemcell.com
	VWR	www.vwr.com
Laboratory equipment	Aesculap	www.aesculap.com
	Alzet	www.alzet.com
	Braintree Scientific	www.braintreesci.com
	Chemicon International	www.chemicon.com
	Dunnlab	www.dunnlab.de
	Dupont	www.dupont.com
	Ehret Laboratory Technology	//manuf.labworld-online.com/ehret
	Eppendorf	www.eppendorf.com
	Ethicon	www.ethicon.com
	Fisher Scientific	www.fisherscientific.com
	FST (Fine Science Tools)	www.finescience.com
	Gibco/Invitrogen	www.gibcobrl.com or www.invitrogen.com
	Gilson	www.gilson.com
	Heidolph	www.heidolph.com
	Helixx Technologies	www.helixxtec.com
	Heraeus	www.heraeus-instruments.de
	IKA Laboratory Technology	www.ika.de
	Kisling	www.kisling.com
	Merck Biosciences	www.merckbiosciences.de or www.merck.com
	Millipore	www.millipore.com
	Oster	www.oster.com
	Pharmacia/Pfizer	www.pfizer.com
	Pierce Biotechnology	www.piercenet.com
	Praxair	www.praxair.com
	Promega	www.promega.com
	QBIOgene	www.qbiogene.com
	Selectscience	www.selectscience.net
	Steris Laboratories	www.steris.com
	The Baker Company	www.bakercompany.com
	Vilene Company	www.vilene.co.jp
	VWR	www.vwr.com

Products	Supplier	Internet
Microscopes: **image analysis** **and equipment**	Bal-Tec	www.bal-tec.com
	Bioptechs	www.bioptechs.com
	Bioquant	www.bioquant.com
	Canemco	www.canemco.com
	Fiberoptic-Heim	www.fiberheim.ch
	Improvision	www.improvision.com
	JEOL	www.jeol.com
	Kappa	www.kappa.de
	Leica	www.leica.com or www.leica-microsystems.com
	Menzel	www.menzel.de
	Nikon	www.nikon-instruments.com
	Olympus	www.olympus.com
	Panasonic	www.panasonic.cm
	Philips	www.philips.com
	Pieper	www.pieper-video.de
	Plano	www.plano-em.de
	Scientific Instruments Co	www.simicroscopes.com
	Soft Imaging System	www.soft-imaging.de
	Ted Pella	www.tedpella.com
	Universal Imaging Corp.	www.image1.com
	Zeiss	www.zeiss.com
Molecular **biology reagents**	Ambion	www.ambion.com
	Dharmacon	www.dharmacon.com
	Gibco/Invitrogen	www.gibcobrl.com www.invitrogen.com
	MWG Biotechnology	www.mwg-biotech.com
	New England Biolabs	www.neb.com
	Promega	www.promega.com
	QBIOgene	www.qbiogene.com
	Qiagen	www.qiagen.com
	Roche	www.roche-applied-science.com
	Trevigen Inc	www.trevigen.com
	Upstate Biotechnologies	www.upstatebiotech.com
	ViraQuest Inc	www.viraquest.com
Software, **electronic** **equipment**	Bilaney	www.bilaney.co.uk
	Deneba [ACD Systems]	www.deneba.com
	Bürklin	www.buerklin.com
	Dentaltrey	www.dentaltrey.com
	Enprotech	www.enprotech.com
	OpenLab	www.openlab.ch

Endothelial Cell Biology on the Web

Vascular webpages

- *//cbi.swmed.edu/ryburn/sato/protocols.htm* – A collection of vascular in vivo imaging protocols from Dr. Tom Sato's lab (Dallas)
- *//microcirc.org* – The Microcirculatory Society
- *//vlib.org/Science/Cell_Biology/angiogenesis.shtml* – Virtual Library of Cell Biology: Angiogenesis
- *www.angio.org* – The Angiogenesis Foundation
- *www.angiogenese.de* – Homepage of the German angiogenesis program
- *www.ibl.fr/angiogenese/Sommaire.htm* – Homepage of French angiogenesis program
- *www.kcl.ac.uk/depsta/biomedical/lvbf.htm* – London Vascular Biology Forum
- *www.med.unibs.it/~airc* – Homepage of the Italian angiogenesis program
- *www.microcirculation.org.uk* – The British Microcirculatory Society
- *www.navbo.org* – North American Vascular Biology Organization
- *www.pharma.ethz.ch/bmm/protocols/endo.html* – A collection of endothelial cell biology protocols adapted from the original Clonetics manual maintained by Dr. Dario Neri (Zürich)
- *www.spherogenex.de* – Description of spheroid-based endothelial cell culture techniques
- *www.svmb.org/* – Society for Vascular Medicine and Biology
- *www.targetvegf.com* – Everything related to VEGF by Genentech

General protocol and tools webpages

- *//bric.postech.ac.kr/protocols* – Very extensive collection of links to all fields of science
- *//dept.kent.edu/projects/cell/Fluoro.htm* – A guide to flurescence microscopy by Dr. Douglas Kline, Ohio
- *//dept.kent.edu/projects/cell/media.htm* – Media and reagent manual by Dr. Douglas Kline, Ohio

- *www.bioscience.org/knockout/alphabet.htm* – Gene knockout database
- *www.cellsignal.com/reference* – A guide to cell signaling maintained by Cell Signaling Technology
- *www.mshri.on.ca/nagy/cre.htm* – Floxed gene and Cre-transgenic database by Dr. Andras Nagy, Toronto
- *www.protocol-online.org* – Gateway for protocols from all disciplines of the life sciences
- *www.protocol-online.org/prot/Cell_Biology/Cell_Culture* – Collection of cell culture protocols
- *www.roche-applied-science.com/LabFAQs* – A well structured question and answer manual for work with DNA, RNA, proteins, media, and buffers maintained by Roche Applied Science
- *www.science.mcmaster.ca/biochem/faculty/andrews/lab/projects/methodsandprograms/labman* – An extensive collection of laboratory protocols from Dr. David Andrews laboratory
- *www.signaling-gateway.org* – The signaling gateway maintained by Nature

Subject Index

Printing: Mercedes-Druck, Berlin
Binding: Stein+Lehmann, Berlin